Theory of Bridge Aerodynamics

Einar N. Strømmen

Theory of Bridge Aerodynamics

Springer

Prof.Dr. Einar N. Strømmen
Norwegian University of Science and Technology
Department of Structural Engineering
7491 Trondheim
Norway
E-mail: einar.strommen@ntnu.no

ISBN-13 978-3-642-44813-3 Springer Berlin Heidelberg NewYork

Springer is a part of Springer Science+Business Media
springer.com

Softcover re-print of the Hardcover 2nd edition 2010

Typesetting: by the authors and TechBooks using a Springer LATEX macro package

Cover design: Erich Kirchner, Heidelberg

Printed on acid-free paper SPIN: 80013058 89/TechBooks 5 4 3 2 1 0

To
Mary, Hannah, Kristian and Sigrid

PREFACE TO 2ND EDITION

In this second edition a new chapter has been added covering the buffeting theory in a finite element format. The motivation for this has been that a finite element format is becoming more and more dominant in all areas of structural mechanics. It is streamlined for computer programming, and it facilitates the use of general purpose routines that are applicable in several types of structural engineering problems. In this book the finite element formulation of the problem of dynamic response calculations follows the general principle of virtual work, a general principle which may be found in many other text books. While the buffeting wind load itself has with no trouble been included in a finite element format, the main challenge has been to obtain a consistent formulation that includes all the relevant motion induced forces. This has been important, because, while many structures (e.g. long-span suspension bridges) may suffer greatly and become unstable at high wind velocities, the same structures may also benefit from these effects at the design wind velocity. It is well known that motion induced forces will change the stiffness and damping properties of the combined structure and flow system. If calculations are performed for a suitably close set of increasing mean wind velocities and the changing mechanical properties (stiffness and damping) are updated from one velocity to the next, then the response of the system may be followed up to wind velocities close to the stability limit, i.e. up to response values that are perceived as unduly large. Finite element calculations may be performed in time domain, in frequency domain or converted into a modal format. All these options have been included. Pursuing a time domain solution strategy requires the use of the so-called indicial functions. The theory behind such a formulation is also covered, and the determination of these functions from aerodynamic derivatives has been included in a separate appendix.

A comment regarding the use of aerodynamic derivatives obtained from aeroelastic wind tunnel experiments to predict structural response has been included.

It goes without saying that typing errors and calculation errors in examples that I so far have come across in the first edition have been corrected.

Trondheim, January 2010

Einar N. Strömmen
einar.strommen@ntnu.no

PREFACE TO 1ST EDITION

This text book is intended for studies in wind engineering, with focus on the stochastic theory of wind induced dynamic response calculations for slender bridges or other line–like civil engineering type of structures. It contains the background assumptions and hypothesis as well as the development of the computational theory that is necessary for the prediction of wind induced fluctuating displacements and cross sectional forces. The simple cases of static and quasi-static structural response calculations are for the sake of completeness also included.

The text is at an advanced level in the sense that it requires a fairly comprehensive knowledge of basic structural dynamics, particularly of solution procedures in a modal format. None of the theory related to the determination of eigen–values and the corresponding eigen–modes are included in this book, i.e. it is taken for granted that the reader is familiar with this part of the theory of structural dynamics. Otherwise, the reader will find the necessary subjects covered by e.g. Clough & Penzien [2] and Meirovitch [3]. It is also advantageous that the reader has some knowledge of the theory of statistical properties of stationary time series. However, while the theory of structural dynamics is covered in a good number of text books, the theory of time series is not, and therefore, the book contains most of the necessary treatment of stationary time series (chapter 2).

The book does not cover special subjects such as rain-wind induced cable vibrations. Nor does it cover all the various available theories for the description of vortex shedding, as only one particular approach has been chosen. The same applies to the presentation of time domain simulation procedures. Also, the book does not contain a large data base for this particular field of engineering. For such a data base the reader should turn to e.g. Engineering Science Data Unit (ESDU) [7] as well as the relevant standards in wind and structural engineering.

The writing of this book would not have been possible had I not had the fortune of working for nearly fifteen years together with Professor Erik Hjorth–Hansen on a considerable number of wind engineering projects.

The drawings have been prepared by Anne Gaarden. Thanks to her and all others who have contributed to the writing of this book.

Trondheim, August 2005

Einar N. Strömmen
einar.strommen@ntnu.no

CONTENTS

NOTATION

Matrices and vectors:

Matrices are in general bold upper case Latin or Greek letters, e.g. $\mathbf{Q}$ or $\mathbf{\Phi}$.

Vectors are in general bold lower case Latin or Greek letters, e.g. $\mathbf{q}$ or $\boldsymbol{\varphi}$.

$diag[\cdot]$ is a diagonal matrix whose content is written within the brackets.

$det(\cdot)$ is the determinant of the matrix within the brackets.

Statistics:

$E[\cdot]$ is the average value of the variable within the brackets.

$Pr[\cdot]$ is the probability of the event given within the bracket.

$P(x)$ is the cumulative probability function, $P(x) = Pr[X \leq x]$.

$p(x)$ is the probability density function of variable x.

$Var(\cdot)$ is the variance of the variable within the brackets.

$Cov(\cdot)$ is the covariance of the variable within the brackets.

$Coh(\cdot)$ is the coherence function of the content within the brackets.

$R(\cdot)$ is the auto- or cross-correlation function.

R_p is short for return period.

$\rho(\cdot)$ is the covariance (or correlation) coefficient of content within brackets.

$\boldsymbol{\rho}$ is a cross covariance or correlation matrix between a set of variables.

σ, σ^2 is the standard deviation, variance.

μ is a quantified small probability.

Imaginary quantities:

i is the imaginary unit (i.e. $i = \sqrt{-1}$).

$Re[\cdot]$ is the real part of the variable within the brackets.

$Im[\cdot]$ is the imaginary part of the variable within the brackets.

Superscripts and bars above symbols:

Super-script T indicates the transposed of a vector or a matrix.

Super-script * indicates the complex conjugate of a quantity.

Dots above symbols (e.g. $\dot{\mathbf{r}}$, $\ddot{\mathbf{r}}$) indicates time derivatives, i.e. d/dt , d^2/dt^2 .

A prime on a variable (e.g. C_L' or ϕ') indicates its derivative with respect to a relevant variable (except t), e.g. $C_L' = dC_L/d\alpha$ and $\phi' = d\phi/dx$. Two primes is then the second derivative (e.g. $\phi'' = d^2\phi/dx^2$) and so on.

Line (–) above a variable (e.g. $\bar{C}_D$) indicates its average value.

A tilde (~) above a symbol (e.g. $\tilde{M}_i$) indicates a modal quantity.

A hat (∧) above a symbol (e.g. $\hat{\mathbf{B}}$) indicates a normalised quantity.

The use of indexes and superscript:

Index x,y or z refers to the corresponding structural axis.

x_f,y_f or z_f refers to the corresponding flow axis.

u,v or w refer to flow components.

i and j are mode shape numbers.

m refers to y,z or θ directions, n refers to u,v or w flow components.

p and k are in general used as node numbers.

F represents a cross sectional force component.

D,L,M refer to drag, lift and moment.

tot,B,R indicate total, background or resonant.

ae is short for aerodynamic, i.e. it indicates a flow induced quantity.

cr is short for critical.

max,min are short for maximum and minimum.

int or *ext* are short for internal and external.

pv is short for peak value.

r is short for response.

s indicates quantities associated with vortex shedding.

Abbreviations:

CC and SC are short for cross-sectional neutral axis centre and shear centre.

FFT is short for Fast Fourier Transform. Sym. is short for symmetry.

$\int_{L_{\exp}}$ means integration over the wind exposed part of the structure.

$\int_{L}$ means integration over the entire length of the structure.

Latin letters

A	Area, cross sectional area
A_n or A_p	Wind spectrum coefficient (n or $p=u$, v or w)
A_{mn}	Aerodynamic admittance functions ($m = y$, z or θ, $n = u$ or w)
$A_1^* - A_6^*$	Aerodynamic derivatives associated with the motion in torsion
$\mathbf{A}, \mathbf{A}_m, \mathbf{A}_n$	Connectivity matrix (associated with element m or n
a	Constant or Fourier coefficient, amplitude
$\mathbf{a}_r, \mathbf{a}_R$	Fourier coefficient vector associated with response or load
$\mathbf{a}_\eta, \mathbf{a}_{\tilde{R}}$	Fourier coefficient vector associated with modal response or load
B	Cross sectional width
$\mathbf{B}_q$, $\hat{\mathbf{B}}_q$	Buffeting dynamic load coefficient matrix at cross sectional level
$\mathbf{B}_{Q_0}, \mathbf{B}_{Q_n}$	Buffeting dynamic load coefficient matrix at element level
b	Constant, coefficient, band-width parameter
$\mathbf{b}_q$, $\hat{\mathbf{b}}_q$	Mean wind load coefficient vector
C , $\mathbf{C}$	Damping coefficient or matrix containing damping coefficient
C_{ae} , $\mathbf{C}_{ae}$	Aerodynamic damping, aerodynamic damping matrix
$\bar{C}$	Force coefficients at mean angle of incidence
C'	Slope of load coefficient curves at mean angle of incidence
c	Constant, coefficient, Fourier amplitude
$\mathbf{c}_0$	Damping matrix at a cross sectional level
$\mathbf{c}, \mathbf{c}_{ae}$	Damping matrix at element level, aerodynamic damping matrix
Co, $\mathbf{Co}$	Co-spectral density, co-spectral density matrix
$\mathbf{Cov}_{rr}$	Matrix containing covariance of response quantities
$\mathbf{Cov}_{\hat{v}\hat{v}}$	Matrix containing covariance of reduced wind velocity components
D	Cross sectional depth
d	Constant or coefficient
$\mathbf{d}$, d_k	Element displacement vector, element end displacement (k=1, 2,…,12)

E	Modulus of elasticity
$\hat{E}$, $\hat{\mathbf{E}}$	Impedance, impedance matrix
e	Eccentricity, distance between shear centre and cetroid
$\mathbf{F}$, F	Element force vector, force at element level
f, f_i	Frequency [Hz], eigen–frequency associated with mode i
$f(\cdot)$	Function of variable within brackets
G	Modulus of elasticity in shear
G_F, $\mathbf{G}_F$	Influence function or matrix ($F = V_y$, V_z, V_x, M_x, M_y or M_x)
$g(\cdot)$	Function of variable within brackets
$H_1^* - H_6^*$	Aerodynamic derivatives associated with the across-wind motion
H, $\mathbf{H}$, $\mathbf{H}_r$	Frequency response function, frequency response matrix
$\tilde{H}_\eta$, $\tilde{\mathbf{H}}_\eta$	Modal frequency response functions, modal frequency response matrix
I_p	Polar moment of inertia
I_t, I_w	St Venant torsion and warping constants
I_u, I_v, I_w	Turbulence intensity of flow components u, v or w
I_y, I_z	Moment of inertia with respect to y or z axis
$\mathbf{I}$	Identity matrix
$\mathbf{I}_v$	Turbulence matrix ($\mathbf{I}_v = diag[I_u \quad I_w]$ or $\mathbf{I}_v = diag[I_u \quad I_v \quad I_w]$)
i	The imaginary unit (i.e. $i = \sqrt{-1}$) or index variable
J	Joint acceptance function
j	Index variable
K, $\mathbf{K}$	Stiffness, stiffness matrix
K_{ae}, $\mathbf{K}_{ae}$	Aerodynamic stiffness, aerodynamicstiffness matrix
k	Index variable, node or sample number
k_p	Peak factor
k_T	Terrain roughness coefficient
$\mathbf{k}$, $\mathbf{k}_{ae}$	Stiffness matrix at element level, aerodynamic stiffness matrix
L, $L_{\exp}$	Length, wind exposed length
${}^mL_n$	Integral length scales ($m = y$, z or θ, $n = u$,v or w)
M_m, M_u	Bending moment (m=x, y, z), ultimate bending moment strength
m	Index variable
m or $\mathbf{M}$	Mass or mass matrix
$\tilde{m}_i$	Modally equivalent and evenly distributed mass

$\mathbf{m}_0$	Mass matrix at a cross sectional level
$\mathbf{m}$	Mass matrix at element level
N	Number, number of nodes or number of elements in series
$\mathbf{N}$, $N_i(x)$	Shape function matrix, polynomial shape function ($i = 1,2,\ldots,12$)
$\hat{\mathbf{N}}$	Contains the first and second order derivatives of $N_i(x)$
n	Index variable
$P_1^* - P_6^*$	Aerodynamic derivatives associated with the along-wind motion
p	Index variable, node or sample number
Q, $\mathbf{Q}$	Wind load or wind load vector at system level
q, $\mathbf{q}$	Wind load or wind load vector at cross sectional level
q_{ae}, $\mathbf{q}_{ae}$	Aerodynamic (motion induced) load at cross sectional level
q_U, q_V	Velocity pressure, i.e. $q_U = \rho U^2/2$, $q_V = \rho V^2/2$
R, $\mathbf{R}$, $\mathbf{R}_n$	External load, external load vector at system level, at element level
$\tilde{R}$, $\tilde{\mathbf{R}}$	Modal load, Modal load vector
Re	Reynolds number
R_p	Return period.
r, $\mathbf{r}$	Cross sectional displacement or rotation, displacement vector
r_{el}, $\mathbf{r}_{el}$	Element cross sectional displacement, displacement vector
r_p	Polar radius
St	Strouhal number
S, $\mathbf{S}$	Auto or cross spectral density, cross-spectral density matrix
$\mathbf{S}_{rr}$, $\mathbf{S}_{RR}$	Cross spectral density matrix associated with response or load
$\mathbf{S}_{\eta\eta}$, $\mathbf{S}_{\tilde{R}\tilde{R}}$	Cross spectral density matrix associated with modal response or load
$S_x(\omega)$	Single side auto-spectral density of variable x
$S_x(\pm\omega)$	Double side auto-spectral density of variable x
S_{xy}	Cross-spectral density between components x and y
s	Relative time ($s = t - \tau$)
$\mathbf{s}$, s_m	Stress vector, stress component ($m = x,y,z$ or θ)
t, T	Time, total length of time series
T_n	Turbulence time scales ($n = u,v$ or w)
U	Instantaneous wind velocity in the main flow direction
u	Fluctuating along-wind horizontal velocity component
V, V_R	Mean wind velocity, resonance mean wind velocity

v	Fluctuating across wind horizontal velocity component
$\mathbf{v}, \mathbf{v}_0, \mathbf{v}_n$	Wind velocity vector containing fluctuating components
W_{ext}, W_{int}	External, internal work
w	Fluctuating across wind vertical velocity component
X, Y, x, y	Arbitrary variables, e.g. functions of t
X, Y, Z	Cartesian structural global axis
x, y, z	Cartesian structural element cross sectional main neutral axis (with origo in the shear centre, x in span-wise direction and z vertical)
x_f, y_f, z_f	Cartesian flow axis (x_f in main flow direction and z_f vertical)
x_r	Chosen span-wise position for response calculation
z_0	Terrain roughness length
$z_{\min}$	Minimum height for the use of a logarithmic wind profile

Greek letters

α	Coefficient, angle of incidence
β	Constant, coefficient
$\boldsymbol{\beta}$	Matrix containing mode shape derivatives
γ	Coefficient, safety coefficient
γ_θ	Shear strain associated with torsion
δ	Incremental displacement operator
$\boldsymbol{\varepsilon}$, ε_m	Strain vector, strain component ($m = x, y$ or z)
ε_0	Mean wind velocity band width parameter
ζ or $\boldsymbol{\zeta}$	Damping ratio or damping ratio matrix
η or $\boldsymbol{\eta}$	Generalised coordinate or vector containing N_{mod} η components
θ	Index indicating cross sectional rotation (about shear centre)
κ	Constant, statistic variable
$\boldsymbol{\kappa}_{ae}$	Matrix containing aerodynamic modal stiffness contributions
ν	Kinematic viscosity of air
λ	Non–dimensional coherence length scale of vortices
μ	A quantified small probability.
μ_n	Spectral moment
$\boldsymbol{\mu}_{ae}$	Matrix containing aerodynamic modal mass contributions

ρ	Coefficient or density (e.g. of air)
$\rho(\cdot)$	Covariance (or correlation) coefficient of content within brackets
ρ_{pp}	Covariance associated with components $p = u, v$ or w
$\boldsymbol{\rho}$	Cross covariance or correlation matrix between a set of variables
σ, σ^2	Standard deviation, variance
τ	Time shift (lag) or dummy time variable
$\mathbf{\Phi}$	$3 \cdot N_{\text{mod}}$ by N_{mod} matrix containing all mode shapes $\boldsymbol{\varphi}_i$
$\mathbf{\Phi}_r$	3 by N_{mod} matrix containing the content of $\mathbf{\Phi}$ at $x = x_r$
Φ_{mn}	Indicial function ($m = D, L$ or M, $n = \dot{y}, \dot{z}$ or θ)
$\boldsymbol{\varphi}$	3 by 1 mode shape vector containing components $\phi_y, \phi_z, \phi_\theta$
$\phi_{y_i}, \phi_{z_i}, \phi_{\theta_i}$	Mode shape components in y, z and θ directions associated with mode shape i (continuous functions of x or N by 1 vectors)
φ_{xy}	Phase spectrum between components x and y
ψ	Phase angle
$\psi(\cdot)$	Function of the variable within the brackets
$\boldsymbol{\Psi}, \boldsymbol{\Psi}_n, \boldsymbol{\Psi}_0$	Matrix containing element orientation properties
ω	Circular frequency (rad/s)
ω_i	Still air eigen-frequency associated with mode shape i
$\omega_i(V)$	Resonance frequency assoc. with mode i at mean wind velocity V

Symbols with both Latin and Greek letters:

$\Delta f, \Delta\omega$	Frequency segment
Δt	Time step
Δs	Separation ($s = x$, y or z)
Δx	Span–wise integration step

Chapter 1

INTRODUCTION

1.1 General considerations

This text book focuses exclusively on the prediction of wind induced static and dynamic response of slender line-like civil engineering structures. Throughout the main part of the book it is taken for granted that the structure is horizontal, i.e. a bridge, but the theory is generally applicable to any line–like type of structure, and thus, it is equally applicable to e.g. a vertical tower. It is a general assumption that structural behaviour is linear elastic and that any non linear part of the relationship between load and structural displacement may be disregarded. It is also taken for granted that the main flow direction throughout the entire span of the structure is perpendicular to the axis in the direction of its span. The wind velocity vector is split into three fluctuating orthogonal components, U in the main flow along–wind direction, and v and w in the across wind horizontal and vertical directions. For a relevant structural design situation it is assumed that U may be split into a mean value V that only varies with height above ground level and a fluctuating part u, i.e. $U = V + u$. V is the commonly known mean wind velocity, and u, v and w are the zero mean turbulence components, created by friction between the terrain and the flow of the main weather system. It is taken for granted that the instantaneous wind velocity pressure is given by Bernoulli's equation

$$q_U(t) = \frac{1}{2}\rho\left[U(t)\right]^2 \tag{1.1}$$

If an air flow is met by the obstacle of a more or less solid line–like body, the flow/structure interaction will give raise to forces acting on the body. Unless the body is extremely streamlined and the speed of the flow is very low and smooth, these forces will fluctuate. Firstly, the oncoming flow in which the body is submerged contains turbulence, i.e. it is itself fluctuating in time and space. Secondly, on the surface of the body additional flow turbulence and vortices are created due to friction, and if the body has sharp edges the flow will separate on these edges and the flow passing the body is unstable in the sense that a variable part of it will alternate from one side to the other, causing vortices to be shed in the wake of the body. And finally, if the body is flexiblethe fluctuating forces may cause the body to oscillate, and the alternating flow and the oscillating body may interact and generate further forces.

Thus, the nature of wind forces may stem from pressure fluctuations (turbulence) in the oncoming flow, vortices shed on the surface and into the wake of the body, and from the interaction between the flow and the oscillating body itself. The first of these effects is known as buffeting, the second as vortex shedding, and the third is usually labelled motion induced forces. In literature, the corresponding response calculations are usually treated separately. The reason for this is that for most civil engineering structures they occur at their strongest in fairly separate wind velocity regions, i.e. vortex shedding is at its strongest at fairly low wind velocities, buffeting occur at stronger wind velocities, while motion induced forces are primarily associated with the highest wind velocities. Surely, this is only for convenience as there are really no regions where they exclusively occur alone. The important question is to what extent they are adequately included in the mathematical description of the loading process.

In structural engineering the wind induced fluctuating forces and corresponding response quantities are usually assumed stationary, and thus, response calculations may be split into a time invariant and a fluctuating part (static and dynamic response). An illustration of what can be expected is shown in Fig. 1.1.

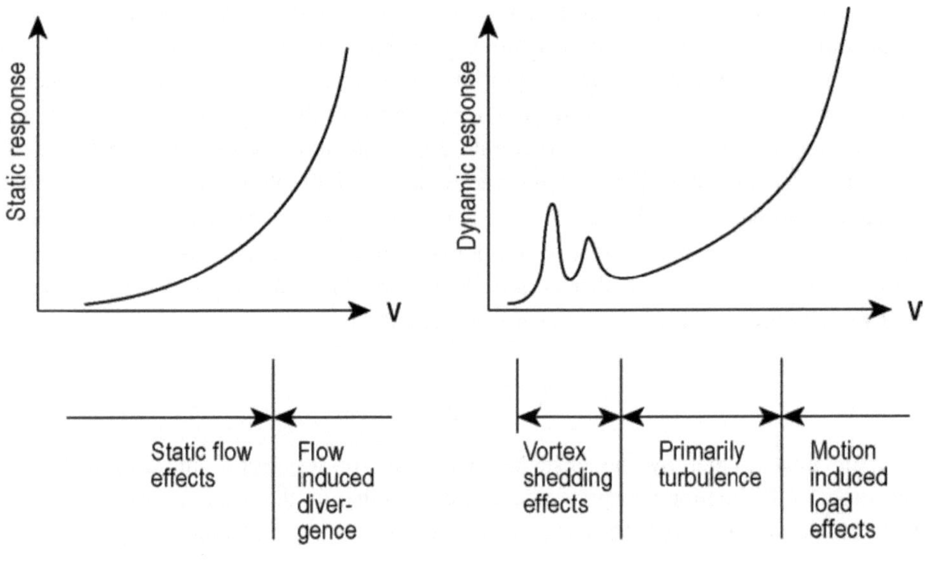

Fig. 1.1 *Typical response behaviour of slender civil engineering structures*

For a mathematical description of the process from a fluctuating wind field to a corresponding load that causes a fluctuating load effect (e.g. displacements or cross sectional stress resultants) a solution strategy in time domain is possible but demanding. The reason for this is that the wind field is a complex process that is randomly distributed in time and space. A far more convenient mathematical model may be established in frequency domain. This requires the establishment of a frequency domain description of the wind field as well as the structural properties, and it involves the establishment of frequency domain transfer functions, one from the wind field velocity pressure distribution to the corresponding load, and one from load to structural response. We shall see that this implies the perception of wind as a stochastic process, and a structural response calculation based on its modal frequency-response-properties. The important input parameters to this solution strategy are the statistical properties of the wind field in time and space, and the eigen-modes and corresponding eigen-frequencies of the structural system in question. The outcome is the statistical characteristics of the structural response.

Thus, apart from the geometry and mass properties of the structural system, it is necessary to know its eigen–modes and corresponding eigen– frequencies. These are the results of eigen–value calculations. The theory of such calculations may be found in many classical text books, see e.g. Timoshenko, Young & Weaver [1], Clough & Penzien [2] and Meirovitch [3]. It has been considered unnecessary to include any of such theory in this book, except for a simple example shown in chapter 4.1. I.e., it will be taken for granted that sufficient information regarding the eigen–value solution has already been provided. Most often, such information has been obtained from a finite element calculation of a discretised structural system, and thus, the eigen–modes are given as more or less ample vectors representing eigen–mode displacements along the span. In the following it is tacitly assumed that such an eigen–value analysis has been performed in vacuum or in still air.

It should be acknowledged that in the mathematical development of the basic theory in this book it is for convenience assumed that eigen–modes are continuous functions. This simplifies and helps on the comprehension of the various steps behind the theory. After the final expressions of response are developed, the vector-matrix operations involved in a purely numerical format of the solution strategy are presented wherever it is considered necessary.

In structural dynamics where a modal solution procedure is adopted it is also necessary to quantify modal eigen–damping properties. This is another subject that will not be treated in this book. It is taken for granted that the modal damping ratio is known from elsewhere (e.g. standards or handbooks).

1.2 Random variables and stochastic processes

A physical process is called a stochastic process if its numerical outcome at any time or position in space is random and can only be predicted with a certain probability. A data set of observations of a stochastic process can only be regarded as one particular set of realisations of the process, none of which can with certainty be repeated even if the conditions are seemingly the same. In fact, the observed numerical outcome of all physical processes is more or less random. The outcome of a process is only deterministic in so far as it represents a mathematical simulation whose input parameters has all been predetermined and remains unchanged.

The physical characteristics of a stochastic process are described by its statistical properties. If it is the cause of another process, this will also be a stochastic process. I.e. if a physical event may mathematically be described by certain laws of nature, a stochastic input will provide a stochastic output. Thus, statistics constitute a mathematical description that provides the necessary parameters for numerical predictions of the random variables that are the cause and effects of physical events. The instantaneous wind velocity pressure (see Eq. 1.1) at a particular time and position in space is such a stochastic process. This implies that an attempt to predict its value at a certain position and time can only be performed in a statistical sense. An observed set of records can not precisely be repeated, but it will follow a certain pattern that may only be mathematically represented by statistics.

Since wind in our built environment above ground level is omnipresent, it is necessary to distinguish between short and long term statistics, where the short term random outcome are time domain representatives for the conditions within a certain weather situation, e.g. the period of a low pressure passing, while the long term conditions are ensemble representatives extracted from a large set of individual short term conditions. For a meaningful use in structural engineering it is a requirement that the short term wind statistics are stationary and homogeneous. Thus, it represents a certain time–space–window that is short and small enough to render sufficiently constant statistical properties. The space window is usually no problem, as the weather conditions surrounding most civil engineering structures may be considered homogeneous enough, unless the terrain surrounding the structure has an unusually strong influence on the immediate wind environment that cannot be ignored in the calculations of wind load effects. The time window is often set at a period of $T = 10$ minutes.

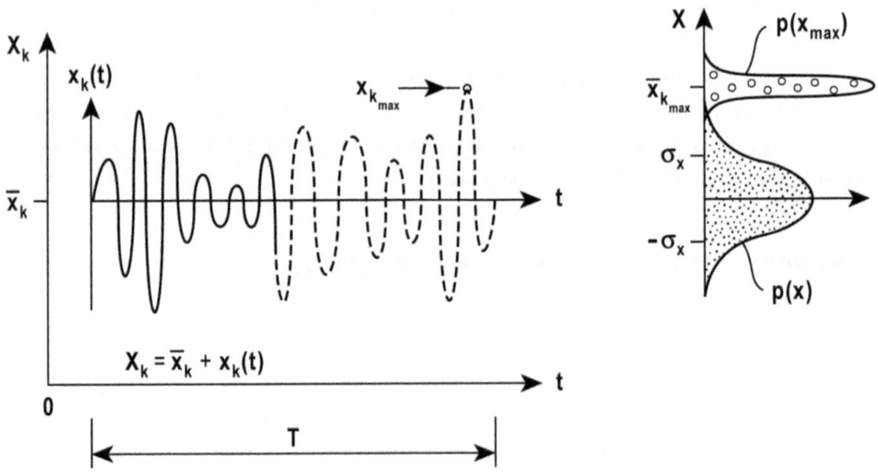

Fig. 1.2 *Short term stationary random process*

Such a typical stochastic process is illustrated in Fig. 1.2. It may for instance be a short term representation of the fluctuating along wind velocity, or the fluctuating structural displacement response at a certain point along its span. As can be seen, it is taken for granted that the process may be split into a constant mean and a stationary fluctuating part. There are two levels of randomness in this process. Firstly, it is random with respect to the instantaneous value within the short term period between 0 and T. I.e., regarding it as a set of successive individual events rather than a continuous function, the process observations are stored by two vectors, one containing time coordinates and another containing the instantaneous recorded values of the process. The stochastic properties of the process may then be revealed by performing statistical investigations to the sample vector of recorded values. For the fluctuating part, it is a general assumption herein that the sample vector of a stochastic process will render a Gaussian probability distribution as illustrated to the right in the figure. This type of investigation is in the following labelled time domain statistics.

The second level of randomness pertains to the simple fact that the sample set of observations shown in Fig. 1.2 is only one particular realisation of the process. I.e. there is an infinite number of other possible representatives of the process. Each of these may look similar and have nearly the same statistical properties, but they are random in the sense that they are never precisely equal to the one singled out in Fig. 1.2. From each of a particular set of different realisations we may for instance only be interested in the mean value and the maximum value. Collecting a large number of different realisations will render a sample set of these values, and thus, statistics may also be performed on the mean value and the maximum value of the process. This is in the following labelled ensemble statistics.

In wind engineering $X_k = \bar{x}_k + x_k(t)$ may be a representative of the wind velocity fluctuations in the main flow direction. The time invariant part $\bar{x}_k$ is then the commonly known mean wind velocity, given at a certain reference height (e.g. at 10 m) and increasing with increasing height above the ground, but at this height assumed constant within a certain area covered by the weather system. The fluctuating part $x_k(t)$ represents the turbulence component in the along wind direction. The mean wind velocity is a typical stochastic variable for which long term ensemble statistics are applicable, while the turbulence component is a stochastic variable whose statistical properties are primarily interesting only within a short term time domain window.

Likewise, the relevant structural response quantities, such as displacements and cross sectional stress resultants, may be regarded as stochastic processes. In the following, it is to be taken for granted that the calculation of structural response, dynamic or non-dynamic, are performed within a time window where the load effects are stationary [i.e. the static (mean) load effects are constant and the dynamic (fluctuating) load effects are Gaussian with a constant standard deviation].

1.3 Basic flow and structural axis definitions

The instantaneous wind velocity vector is described in a Cartesian coordinate system $\left[x_f, y_f, z_f\right]$, where x_f is in the direction of the main flow and z_f is in the vertical direction as shown in Fig. 1.3.a. Accordingly, the wind velocity vector is divided into three components.

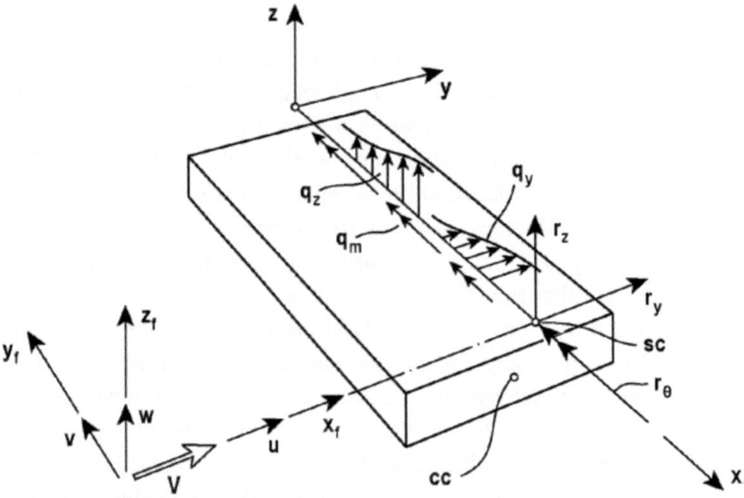

a) *Definition of flow and structural axes, displacements and loads*

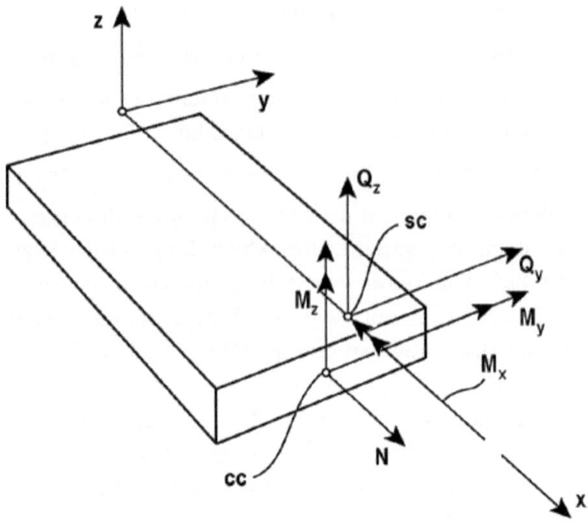

b) Definition of cross sectional forces (stress resultants)

Fig. 1.3 *Basic definitions of flow and structural axes*

As mentioned above, the relevant time window is of limited length such that the component in the main flow direction may be split into a time invariant mean value and a fluctuating part. Thus, the instantaneous wind velocity vector is defined by

$$\left.\begin{array}{l} U(x_f,y_f,z_f,t) = V(x_f,y_f,z_f) + u(x_f,y_f,z_f,t) \\ v(x_f,y_f,z_f,t) \\ w(x_f,y_f,z_f,t) \end{array}\right\} \tag{1.2}$$

where V is the mean value in the main flow direction, and u, v and w are the turbulence components whose time domain mean values are zero. Since the main flow direction is assumed perpendicular to the span of the structure, the velocity vector may be greatly simplified depending on structural orientation. Thus, Eq. 1.2 may be reduced to

$$\left.\begin{array}{c} U(y_f,t) = V + u(y_f,t) \\ w(y_f,t) \end{array}\right\} \tag{1.3}$$

for a line–like horizontal structure (e.g. a bridge), and into

$$\left.\begin{array}{c} U(z_f,t) = V(z_f) + u(z_f,t) \\ v(z_f,t) \end{array}\right\} \tag{1.4}$$

for a vertical structure (e.g. a tower). As shown in Fig. 1.3 the structure is described in a Cartesian coordinate system $[x,y,z]$, with origo at the shear centre of the cross section, x is in the span direction and with y and z parallel to the main neutral structural axis (i.e. the neutral axis with respect to cross sectional bending). Correspondingly, the wind load drag, lift and pitching moment components (per unit length along the span) are all referred to the shear centre and split into a mean and a fluctuating part, i.e.

$$\bar{\mathbf{q}} + \mathbf{q} = \begin{bmatrix} \bar{q}_y(x) \\ \bar{q}_z(x) \\ \bar{q}_\theta(x) \end{bmatrix} + \begin{bmatrix} q_y(x,t) \\ q_z(x,t) \\ q_\theta(x,t) \end{bmatrix} \tag{1.5}$$

Similarly, the response displacements

$$\bar{\mathbf{r}} + \mathbf{r} = \begin{bmatrix} \bar{r}_y(x) \\ \bar{r}_z(x) \\ \bar{r}_\theta(x) \end{bmatrix} + \begin{bmatrix} r_y(x,t) \\ r_z(x,t) \\ r_\theta(x,t) \end{bmatrix} \tag{1.6}$$

and cross sectional stress resultants

$$\begin{bmatrix} \bar{Q}_y(x) \\ \bar{Q}_z(x) \\ \bar{M}_x(x) \end{bmatrix} + \begin{bmatrix} Q_y(x,t) \\ Q_z(x,t) \\ M_x(x,t) \end{bmatrix} \tag{1.7}$$

are also referred to the shear centre, while bending moment and axial stress resultants

$$\begin{bmatrix} \bar{M}_y(x) \\ \bar{M}_z(x) \\ \bar{N}(x) \end{bmatrix} + \begin{bmatrix} M_y(x,t) \\ M_z(x,t) \\ N(x,t) \end{bmatrix} \tag{1.8}$$

are referred to the centroid of the cross section (where, as shown above, the centriod is defined as the origo of main neutral structural axis).

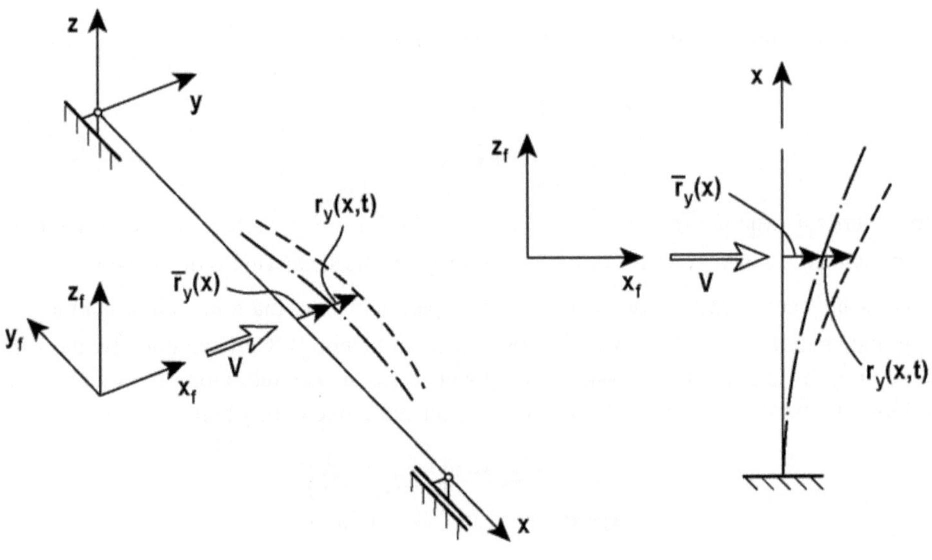

Fig. 1.4 *Structural axes and displacement components*

Thus, it is assumed that structural response in general can be predicted as the sum of a mean value and a fluctuating part, as illustrated in Fig. 1.4. It is assumed that within the time window considered the mean values are constant as well as the statistical properties of the fluctuating parts.

As shown in Fig. 1.3, all flow and structural quantities are treated as vectors within the coordinate system they belong, except the quantities associated with torsion. Cross sectional rotation and the corresponding torsion moment are considered positive with the windward front face up, which is a long standing practice in wind engineering. However, it should be noted that such an inconsistency can not be accepted in a finite element format, and thus, in chapter 9 a strictly vector description has been adopted, see Fig. 9.2.

1.4 *Structural design quantities*

Design calculations are intended to cover a certain unfavourable loading condition, e.g. an extreme storm situation, that is characteristic to the particular place where the structure is located, and whose probability of occurrence is suitably small. In this situation it is the comparison of structural strength or capacity to the extreme value of some critical stress or stress resultant that is of interest. The situation is illustrated in Fig. 1.5. Since structural behaviour is assumed linear elastic, these quantities may in general be obtained from the extreme values of the displacements

$$\bar{r}_k(x) + \left[r_k(x,t)\right]_{max} \qquad \text{where} \qquad k = y,z,\theta \tag{1.9}$$

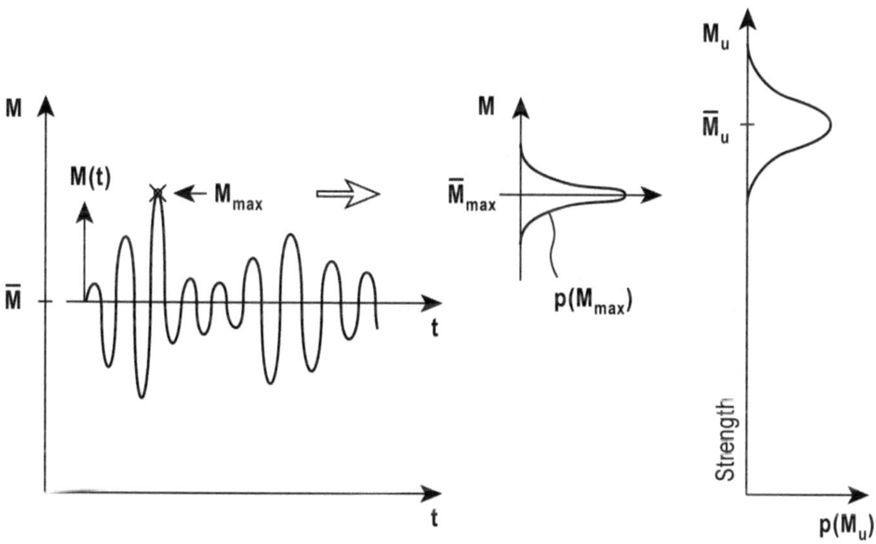

Fig. 1.5 *Bending moment design quantities*

However, the mean values in this situation are time invariants, and the response calculations have inevitably been based on predetermined values taken from standards or other design specifications. They have been established from authoritative sources to represent the characteristic values within a certain short term weather condition chosen for the special purpose of design safety considerations. Therefore, in a particular design situation time invariant quantities may be considered as deterministic quantities, and thus, the mean values of displacements or stress resultants may be obtained directly from simple linear static calculations. I.e., it is only the fluctuating part of the response quantities that requires treatment as stochastic processes. It may be shown (see chapter 2.4) that if a zero mean stochastic process is stationary and Gaussian, then its extreme value is proportional to its standard deviation σ_{r_k}, i.e.

$$\left[r_k(x,t)\right]_{\max} = k_p \cdot \sigma_{r_k} \tag{1.10}$$

where k_p is a time invariant peak factor between about 1.5 and 4.5, and thus

$$r_{k_{\max}} = \bar{r}_k(x) + \left[r_k(x,t)\right]_{\max} = \bar{r}_k(x) + k_p \cdot \sigma_{r_k} \tag{1.11}$$

Similarly, the extreme values of bending moment and shear force stress resultants may be expressed by

$$M_{k_{\max}} = \bar{M}_k + k_p \cdot \sigma_{M_k} \qquad \text{where} \qquad k = x,y,z \tag{1.12}$$

Therefore, the main focus is in the following on the calculation of the standard deviation to fluctuating components, σ_{r_k} and σ_{M_k}, whether they contain dynamic amplification or not. However, in many design situations it is necessary to consider the combined effects of stresses or stress resultants, and therefore, it is not only the standard deviation of processes that are of interest in structural design considerations but also the covariance between fluctuating components. For instance, let a fluctuating (dynamic) displacement response at arbitrary position x

$$\begin{bmatrix} r_y \\ r_z \\ r_\theta \end{bmatrix} = \begin{bmatrix} \bar{r}_y(x) \\ \bar{r}_z(x) \\ \bar{r}_\theta(x) \end{bmatrix} + \begin{bmatrix} r_y(x,t) \\ r_z(x,t) \\ r_\theta(x,t) \end{bmatrix} \tag{1.13}$$

be associated with corresponding cross sectional moment and shear force components

$$\begin{bmatrix} M_y \\ M_z \\ M_x \\ Q_y \\ Q_z \end{bmatrix} = \begin{bmatrix} \bar{M}_y(x) \\ \bar{M}_z(x) \\ \bar{M}_x(x) \\ \bar{Q}_y(x) \\ \bar{Q}_z(x) \end{bmatrix} + \begin{bmatrix} M_y(x,t) \\ M_z(x,t) \\ M_x(x,t) \\ Q_y(x,t) \\ Q_z(x,t) \end{bmatrix} \tag{1.14}$$

Then the normal stress and shear stress components at cross sectional position (y,z) are given by

$$\begin{bmatrix} s_x \\ s_{yz} \end{bmatrix} = \begin{bmatrix} \bar{s}_x \\ \bar{s}_{yz} \end{bmatrix} + \begin{bmatrix} s_x(t) \\ s_{yz}(t) \end{bmatrix} = \begin{bmatrix} \dfrac{\bar{M}_y}{I_y} z + \dfrac{\bar{M}_z}{I_z} y \\ \dfrac{\bar{Q}_y}{A_y} + \dfrac{\bar{Q}_z}{A_z} + \dfrac{\bar{M}_x}{2A_m t_0} \end{bmatrix} + \begin{bmatrix} \dfrac{M_y}{I_y} z + \dfrac{M_z}{I_z} y \\ \dfrac{Q_y}{A_y} + \dfrac{Q_z}{A_z} + \dfrac{M_x}{2A_m t_0} \end{bmatrix} \tag{1.15}$$

where I_y and I_z are moments of inertia associated with bending about y and z axis, A_y and A_z are the cross sectional shear areas (rendering averaged values of shear stresses) and, for simplicity assuming that we are dealing with a closed box type of cross section, A_m is the sector area inscribed by the cross section and t_0 is the material thickness at position (y,z). The variance of the normal stress is then given by

$$\begin{aligned} \sigma_{s_x}^2 &= E\left[s_x^2(t)\right] = E\left[\left(\frac{M_y}{I_y} z + \frac{M_z}{I_z} y\right)^2\right] \\ &= E\left[\left(\frac{M_y}{I_y} z\right)^2 + 2\left(\frac{M_y}{I_y} z\right)\cdot\left(\frac{M_z}{I_z} y\right) + \left(\frac{M_z}{I_z} y\right)^2\right] \\ &= \left(\frac{\sigma_{M_y}}{I_y} z\right)^2 + 2\frac{Cov_{M_y M_z}}{I_y I_z} yz + \left(\frac{\sigma_{M_z}}{I_z} y\right)^2 \end{aligned} \tag{1.16}$$

where $\sigma_{M_y}^2 = E\left[M_y^2\right]$, $\sigma_{M_z}^2 = E\left[M_z^2\right]$ and $Cov_{M_y M_z} = E\left[M_y M_z\right]$, which may be further developed into

$$\sigma_{s_x}^2 = \left(\frac{\sigma_{M_y}}{I_y} z\right)^2 + 2\left(\frac{\sigma_{M_y}}{I_y} z\right)\left(\frac{\sigma_{M_z}}{I_z} y\right)\rho_{M_y M_z} + \left(\frac{\sigma_{M_z}}{I_z} y\right)^2 \tag{1.17}$$

where $\rho_{M_yM_z} = Cov_{M_yM_z} \big/ \left(\sigma_{M_y}\sigma_{M_z}\right)$ is the covariance coefficient between M_y and M_z fluctuations. Similarly, the variance of the shear stress is given by

$$\sigma_{s_{yz}}^2 = E\left[s_{yz}^2(t)\right] = E\left[\left(\frac{Q_y}{A_y} + \frac{Q_z}{A_z} + \frac{M_x}{2A_m t_0}\right)^2\right]$$

$$= E\left[\left(\frac{Q_y}{A_y}\right)^2 + \left(\frac{Q_z}{A_z}\right)^2 + \left(\frac{M_x}{2A_m t_0}\right)^2 + 2\left(\frac{Q_y}{A_y}\frac{Q_z}{A_z}\right) + 2\left(\frac{Q_y}{A_y}\frac{M_x}{2A_m t_0}\right) + 2\left(\frac{Q_z}{A_z}\frac{M_x}{2A_m t_0}\right)\right]$$

$$= \left(\frac{\sigma_{Q_y}}{A_y}\right)^2 + \left(\frac{\sigma_{Q_z}}{A_z}\right)^2 + \left(\frac{\sigma_{M_x}}{2A_m t_0}\right)^2 + 2\frac{Cov_{Q_yQ_z}}{A_yA_z} + 2\frac{Cov_{Q_yM_x}}{A_y 2A_m t_0} + 2\frac{Cov_{Q_zM_x}}{A_z 2A_m t_0}$$

$$= \left(\frac{\sigma_{Q_y}}{A_y}\right)^2 + \left(\frac{\sigma_{Q_z}}{A_z}\right)^2 + \left(\frac{\sigma_{M_x}}{2A_m t_0}\right)^2$$

$$+ 2\frac{\sigma_{Q_y}\sigma_{Q_z}}{A_yA_z}\rho_{Q_yQ_z} + 2\frac{\sigma_{Q_y}\sigma_{M_x}}{A_y 2A_m t_0}\rho_{Q_yM_x} + 2\frac{\sigma_{Q_z}\sigma_{M_x}}{A_z 2A_m t_0}\rho_{Q_zM_x} \tag{1.18}$$

Where
$$\begin{bmatrix} \sigma_{Q_y}^2 \\ \sigma_{Q_z}^2 \\ \sigma_{M_x}^2 \end{bmatrix} = E\begin{bmatrix} Q_y^2 \\ Q_z^2 \\ M_x^2 \end{bmatrix} \qquad \begin{bmatrix} Cov_{Q_yQ_z} \\ Cov_{Q_yM_x} \\ Cov_{Q_zM_x} \end{bmatrix} = E\begin{bmatrix} Q_yQ_z \\ Q_yM_x \\ Q_zM_x \end{bmatrix}$$

and
$$\begin{bmatrix} \rho_{Q_yQ_z} \\ \rho_{Q_yM_x} \\ \rho_{Q_zM_x} \end{bmatrix} = \begin{bmatrix} Cov_{Q_yQ_z} \big/ \left(\sigma_{Q_y}\sigma_{Q_z}\right) \\ Cov_{Q_yM_x} \big/ \left(\sigma_{Q_y}\sigma_{M_x}\right) \\ Cov_{Q_zM_x} \big/ \left(\sigma_{Q_z}\sigma_{M_x}\right) \end{bmatrix}$$

Thus, the main focus is in the following on the calculation of the standard deviation and the covariance to fluctuating response components, whether they contain dynamic amplification or not. Simple linear static calculations are considered trivial. However, some mention of the calculation of time invariant mean values has been included for the sake of completeness.

Chapter 2

SOME BASIC STATISTICAL CONCEPTS IN WIND ENGINEERING

2.1 Parent probability distributions, mean value and variance

For a continuous random variable *X*, its probability density function $p(x)$ is defined by

$$\Pr[x \leq X \leq x+dx] = P(x+dx) - P(x) - \frac{dP(x)}{dx}dx = p(x)dx \tag{2.1}$$

where $P(x)$ is the cumulative probability function, from which it follows that

$$\Pr[X \leq x] = P(x) = \int_{-\infty}^{x} p(x)dx \tag{2.2}$$

and that $\lim_{x\to\infty} P(x) = 1$. Similarly, for two random variables *X* and *Y* the joint probability density function is defined by

$$p(x,y) = \frac{d^2P(x,y)}{dx\ dy} \tag{2.3}$$

where $P(x,y) = \Pr[X \leq x, Y \leq y]$. The mean value and variance of *X* are given by

$$\left.\begin{aligned} \bar{x} &= E[X] = \int_{-\infty}^{+\infty} x \cdot p(x)dx \\ Var(X) &= \sigma_x^2 = E\left[(X-\bar{x})^2\right] = \int_{-\infty}^{+\infty} (x-\bar{x})^2 \cdot p(x)dx \end{aligned}\right\} \tag{2.4}$$

Equivalent definitions apply to a discrete random variable X. It is in the following assumed that each realisation X_k of X has the same probability of occurrence, and thus, the mean value and variance of X may be estimated from a large data set of N individual realisations:

$$\left.\begin{aligned} \bar{x} &= \lim_{N\to\infty} \frac{1}{N}\sum_{k=1}^{N} X_k \\ Var(X) = \sigma_x^2 &= \lim_{N\to\infty} \frac{1}{N}\sum_{k=1}^{N} (X_k - \bar{x})^2 \end{aligned}\right\} \tag{2.5}$$

The square root of the variance, σ_x, is called the standard deviation. Recalling that $E[X] = \bar{x}$, the expression for the variance may be further developed into

$$\sigma_x^2 = E\left[(X-\bar{x})^2\right] = E\left[X^2 - 2\bar{x}X + \bar{x}^2\right] = E\left[X^2\right] - \bar{x}^2 \tag{2.6}$$

There are three probability density distributions that are of primary importance in wind engineering. These are the Gaussian (normal), Weibull and Rayleigh distributions, each defined by the following expressions:

$$\left.\begin{aligned} p(x) &= \frac{1}{\sqrt{2\pi}\sigma_x}\exp\left[-\frac{1}{2}\left(\frac{x-\bar{x}}{\sigma_x}\right)^2\right] \\ p(x) &= \beta\frac{x^{\beta-1}}{\gamma^{\beta}}\exp\left[-\left(\frac{x}{\gamma}\right)^{\beta}\right] \\ p(x) &= \frac{x}{\gamma^2}\exp\left[-\frac{1}{2}\left(\frac{x}{\gamma}\right)^2\right] \end{aligned}\right\} \tag{2.7}$$

They are graphically illustrated in Fig. 2.1. It is seen that a Rayleigh distribution is the Weibull distribution with β=2.

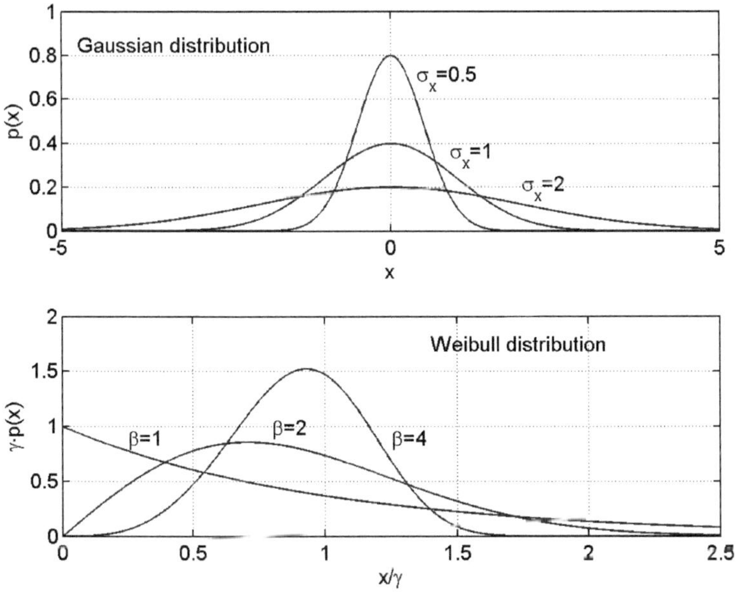

Fig. 2.1 *Gauss (with $\bar{x} = 0$) and Weibull distributions*

2.2 Time domain and ensemble statistics

As mentioned in Chapter 1 there are two types of statistics dealt with in wind engineering: *time domain statistics* and *ensemble statistics*. Illustrating time domain statistics, a typical realisation of the outcome of a stochastic process over a period T is illustrated in Fig. 2.2. This may for instance represent a short term recording of the wind velocity at some point in space, or it may equally well represent the displacement response somewhere along the span of the structure. Considering consecutive and for practical purposes equidistant points along the time series as individual random observations of the process, then time domain statistics may be performed on this realisation.

It will in the following be assumed that any time domain statistics are based on a continuous or discrete time variable X, which theoretically may attain values between $-\infty$ and $+\infty$ and are applicable over a limited time range between 0 and T, within which the process is stationary and homogeneous (i.e. have constant statistical properties) such that

$$X = \bar{x} + x(t) \tag{2.8}$$

Its mean value and variance are then given by

$$\left.\begin{aligned}\bar{x} &= \lim_{T\to\infty}\frac{1}{T}\int_0^T X dt\\ \sigma_x^2 &= \lim_{T\to\infty}\frac{1}{T}\int_0^T [x(t)]^2 dt\end{aligned}\right\}\tag{2.9}$$

It will in the following also be assumed that the individual observations of the fluctuating part $x(t)$ within the time window between 0 and T may with sufficient accuracy be fitted to a Gaussian probability distribution, as illustrated on the right hand side of Fig. 2.2.

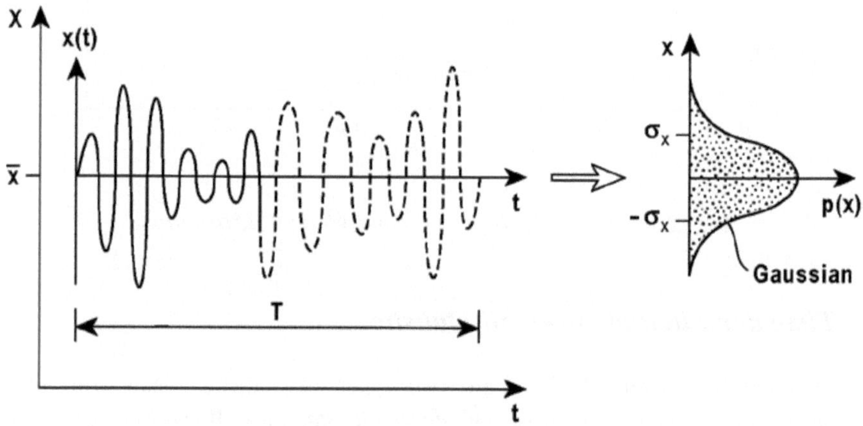

Fig. 2.2 *Time domain statistics*

Example 2.1

Given a zero mean variable $x(t) = a \cdot \cos(\omega_0 t)$, where $\omega_0 = 2\pi/T_0$. Its variance is then given by

$$\sigma_x^2 = \lim_{T\to\infty}\frac{1}{T}\int_0^T x^2(t)dt = \lim_{T\to\infty}\frac{1}{T}\int_0^T [a\cos(\omega_0 t)]^2 dt$$

Substituting $T = n \cdot T_0$, where n is an integer, then

$$\sigma_x^2 = \lim_{n\to\infty}\frac{1}{n\cdot T_0}\cdot\left[n\cdot\int_0^{T_0}\left[a\cos\left(\frac{2\pi}{T_0}t\right)\right]^2 dt\right] = \frac{a^2}{2}$$

Similarly, given a zero mean variable $x(t) = a \cdot \sin(\omega_0 t)$, where $\omega_0 = 2\pi/T_0$. Its variance is then given by

$$\sigma_x^2 = \lim_{n\to\infty} \frac{1}{n \cdot T_0} \cdot \left[n \cdot \int_0^{T_0} \left[a \sin\left(\frac{2\pi}{T_0} t \right) \right]^2 dt \right] = \frac{a^2}{2}$$

Given a second zero mean variable comprising two cosine functions with different amplitudes and frequencies, i.e.:

$$x(t) = a_1 \cdot \cos(\omega_1 t) + a_2 \cdot \cos(\omega_2 t)$$

where $\omega_1 = 2\pi/T_1$ and $\omega_2 = 2\pi/T_2$. It is easily seen that if $T_1/T_2 = 1$ then

$$x(t) = (a_1 + a_2) \cdot \cos(\omega_1 t)$$

and thus, the calculation of its variance is identical to the solution given above, i.e.:

$$\sigma_x^2 = \frac{(a_1 + a_2)^2}{2}$$

If $T_1/T_2 \neq 1$ then the variance of $x(t)$ is given by

$$\sigma_x^2 = \lim_{T\to\infty} \frac{1}{T} \int_0^T \left[a_1^2 \cos^2(\omega_1 t) + 2a_1 a_2 \cos(\omega_1 t) \cos(\omega_2 t) + a_2^2 \cos^2(\omega_2 t) \right] dt$$

Substituting $T = n_1 \cdot T_1$ into the integration of the first two terms and $T = n_2 \cdot T_2$ into the third, where n_1 and n_2 are integers, then

$$\sigma_x^2 = \lim_{n_1\to\infty} \frac{1}{n_1 T_1} \cdot n_1 \int_0^{T_1} a_1^2 \cos^2(\omega_1 t) dt + \lim_{n_1\to\infty} \frac{1}{n_1 T_1} \cdot n_1 \int_0^{T_1} 2a_1 a_2 \cos(\omega_1 t) \cos(\omega_2 t) dt$$
$$+ \lim_{n_2\to\infty} \frac{1}{n_2 T_2} \cdot n_2 \int_0^{T_2} a_2^2 \cos^2(\omega_2 t) dt$$

It is seen that the first and the third integrals are identical to the integral of a single cosine squared shown above, and thus, they are equal to $a_1^2/2$ and $a_2^2/2$, respectively. The second integral, containing the product of two cosine functions, may most effectively be solved by the substitution $\hat{t} = \omega_1 t = (2\pi/T_1) t$, in which case it is given by

$$\frac{2a_1 a_2}{T_1} \cdot \left[\frac{T_1}{2\pi} \int_0^{2\pi} \cos(\hat{t}) \cdot \cos\left(\frac{T_1}{T_2} \hat{t} \right) d\hat{t} \right] = \frac{a_1 a_2}{\pi} \left[\frac{\sin 2\pi \left(1 - \frac{T_1}{T_2} \right)}{2\left(1 - \frac{T_1}{T_2} \right)} + \frac{\sin 2\pi \left(1 + \frac{T_1}{T_2} \right)}{2\left(1 + \frac{T_1}{T_2} \right)} \right]$$

$$= \frac{a_1 a_2}{\pi} \frac{\sin\left(2\pi \frac{T_1}{T_2} \right)}{\frac{T_1}{T_2} - \frac{T_2}{T_1}} \quad \text{which is} \begin{cases} = 0 \text{ if } T_1/T_2 \text{ is an integer unequal to 1} \\ \neq 0 \text{ if } T_1/T_2 \text{ is not an integer} \end{cases}$$

Thus, the variance of $x(t) = a_1 \cdot \cos(\omega_1 t) + a_2 \cdot \cos(\omega_2 t)$ is then given by

$$\sigma_x^2 = \begin{cases} (a_1 + a_2)^2 / 2 \text{ if } \omega_2/\omega_1 = 1 \\ \frac{a_1^2}{2} + \frac{a_2^2}{2} \text{ if } \omega_2/\omega_1 \text{ is an integer } \neq 1 \\ \frac{a_1^2}{2} + \frac{a_1 a_2}{\pi} \cdot \frac{\sin(2\pi\,\omega_2/\omega_1)}{\omega_2/\omega_1 - \omega_1/\omega_2} + \frac{a_2^2}{2} \text{ if } \omega_2/\omega_1 \text{ is not an integer} \end{cases}$$

Similar results would have been obtained if the cosines had been replaced by sinus functions. Thus, if for instance $x(t) = a_1 \cdot \cos(\omega_1 t) + a_2 \cdot \cos(\omega_2 t)$ and ω_2/ω_1 is an integer $\neq 1$, then the variance of

$$\dot{x}(t) = \frac{dx}{dt} = -a_1\omega_1 \cdot \sin(\omega_1 t) - a_2\omega_2 \cdot \sin(\omega_2 t)$$

is given by

$$\sigma_{\dot{x}}^2 = \frac{(-a_1\omega_1)^2}{2} + \frac{(-a_2\omega_2)^2}{2} = \omega_1^2 \cdot \frac{a_1^2}{2} + \omega_2^2 \cdot \frac{a_2^2}{2}$$

Likewise, the variance of the nth derivative of $x(t)$, $f_n(t) = \frac{d^n x}{dt^n}$, is given by

$$\sigma_f^2 = \omega_1^{2n} \cdot \frac{a_1^2}{2} + \omega_2^{2n} \cdot \frac{a_2^2}{2}$$

Illustrating ensemble statistics, a situation where *N* different recordings of a stochastic process within a time window between 0 and *T* are shown in Fig. 2.3. These may for instance represent *N* simultaneous realisations of the along wind velocity in space, i.e. they represent the wind velocity variation taken simultaneously and at a certain distance (horizontal or vertical) between each of them. Extracting the recorded values at a given time from each of these realisations will render a set of data $X_k(t), k = 1,....,N$. On this data set ensemble statistics may be performed. This is the type of statistics that provides a stochastic description of the wind field distribution in space.

Another example of ensemble statistics is illustrated in Fig. 2.4.a, where the situation is illustrated that *N* different observations of a stochastic process have been recorded, each taken within a certain time window but in this case not necessarily at the same time. Each of these time series is assumed to be stationary and Gaussian within the short term period that has been considered. In wind engineering this may be an illustration of the situation when a number of time series have been recorded of the wind velocity at a certain point in space, each taken during different weather conditions. In that case one may only be interested in performing statistics on the mean values and discard the rest of the recordings.

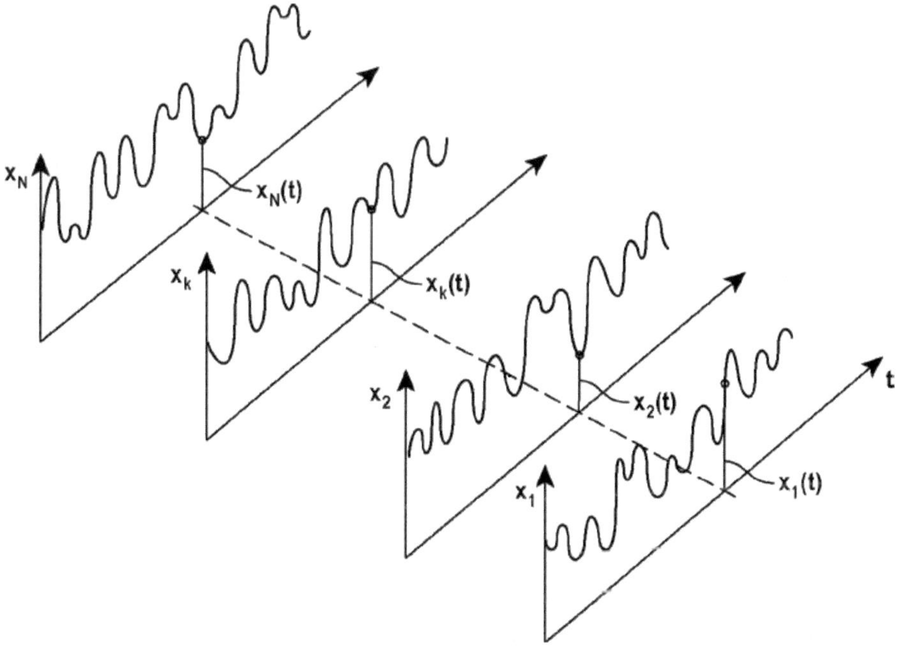

Fig. 2.3 *Ensemble statistics of simultaneous events*

The statistical properties of the data set of extracted mean values will then represent an example of long term ensemble statistics. Typically, the probability density distribution of a data set of mean values may attain a shape that may be fitted to a Weibull or a Rayleigh distribution as illustrated in Fig. 2.4.b.

Apart from fitting the data from a random variable to a suitable parent probability distribution and estimating its mean value and variance (see chapter 2.1 above), it is the properties of correlation and covariance that are of particular interest. These are both providing information about possible relationships in the time domain or ensemble data that have been extracted from the process. Correlation estimates are taken on the full value of the process variable, i.e. on $X(t) = \bar{x} + x(t)$, while covariance is estimated from zero mean variables $x_i(t)$.

Given two realisations $X_1(t) = \bar{x}_1 + x_1(t)$ and $X_2(t) = \bar{x}_2 + x_2(t)$, either two of the same process at different time or location, or of two entirely different processes. Then the correlation and covariance between these two process variables are defined by

$$R_{x_1x_2} = E\left[X_1(t) \cdot X_2(t)\right] = \lim_{T\to\infty} \frac{1}{T} \int_0^T X_1(t) \cdot X_2(t) dt \tag{2.10}$$

$$Cov_{x_1x_2} = E\left[x_1(t)\cdot x_2(t)\right] = \lim_{T\to\infty}\frac{1}{T}\int_0^T x_1(t)\cdot x_2(t)dt \tag{2.11}$$

Similarly, given two data sets of N individual and equally probable realisations that have been extracted from two random variables, X_1 and X_2, then the ensemble correlation and covariance are defined by:

$$R_{x_1x_2} = E\left[X_1\cdot X_2\right] = \lim_{N\to\infty}\frac{1}{N}\sum_{k=1}^{N} X_{1_k}\cdot X_{2_k} \tag{2.12}$$

$$\begin{aligned} Cov_{x_1x_2} &= E\left[(X_1-\bar{x}_1)\cdot(X_2-\bar{x}_2)\right] \\ &= \lim_{N\to\infty}\frac{1}{N}\sum_{k=1}^{N}\left(X_{1_k}-\bar{x}_1\right)\cdot\left(X_{2_k}-\bar{x}_2\right) \end{aligned} \tag{2.13}$$

However, correlation and covariance estimates may also be taken on the process variable itself. Thus, defining an arbitrary time lag τ, the time domain auto correlation and auto covariance functions are defined by

$$R_x(\tau) = E\left[X(t)\cdot X(t+\tau)\right] = \lim_{T\to\infty}\frac{1}{T}\int_0^T X(t)\cdot X(t+\tau)dt \tag{2.14}$$

$$Cov_x(\tau) = E\left[x(t)\cdot x(t+\tau)\right] = \lim_{T\to\infty}\frac{1}{T}\int_0^T x(t)\cdot x(t+\tau)dt \tag{2.15}$$

These are defined as functions because τ is perceived as a continuous variable. As long as τ is considerably smaller than T

$$E\left[X(t)\right] = E\left[X(t+\tau)\right] = \bar{x} \tag{2.16}$$

and thus, the relationship between R_x and Cov_x is the following

$$Cov_x(\tau) = E\left[\{X(t)-\bar{x}\}\cdot\{X(t+\tau)-\bar{x}\}\right] = R_x(\tau)-\bar{x}^2 \tag{2.17}$$

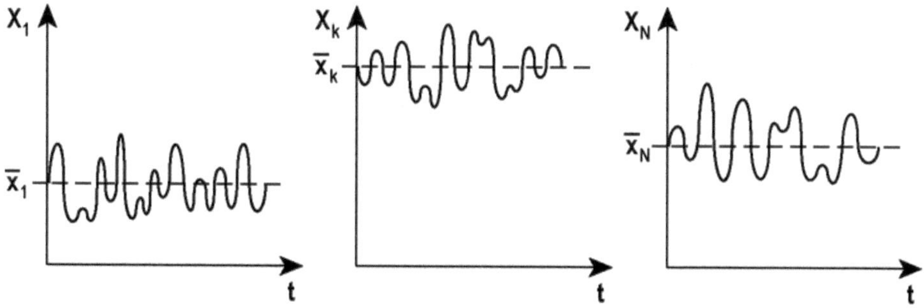

a) *N independent short term realisations*

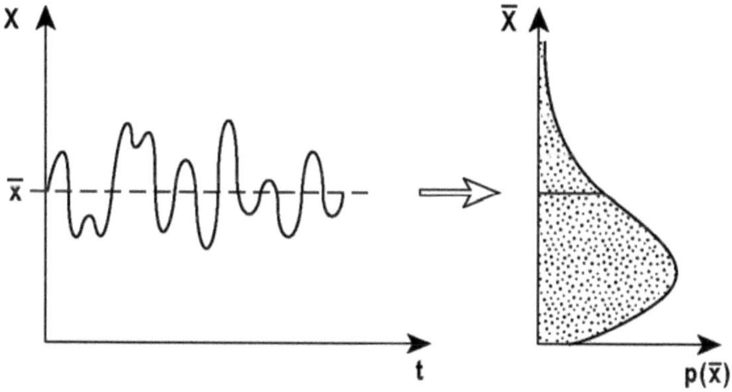

b) *The probability of mean values*

Fig. 2.4 *Ensemble statistics of mean value recordings*

There is no reason why τ may not attain negative as well as positive values, and since

$$E\left[x(t)\cdot x(t-\tau)\right]=E\left[x(t-\tau)\cdot x(t)\right]=E\left[x(t-\tau)\cdot x(t-\tau+\tau)\right] \quad (2.18)$$

then

$$Cov_x(\tau)=Cov_x(-\tau) \quad (2.19)$$

Thus, Cov_x is symmetric with respect to its variation with τ.

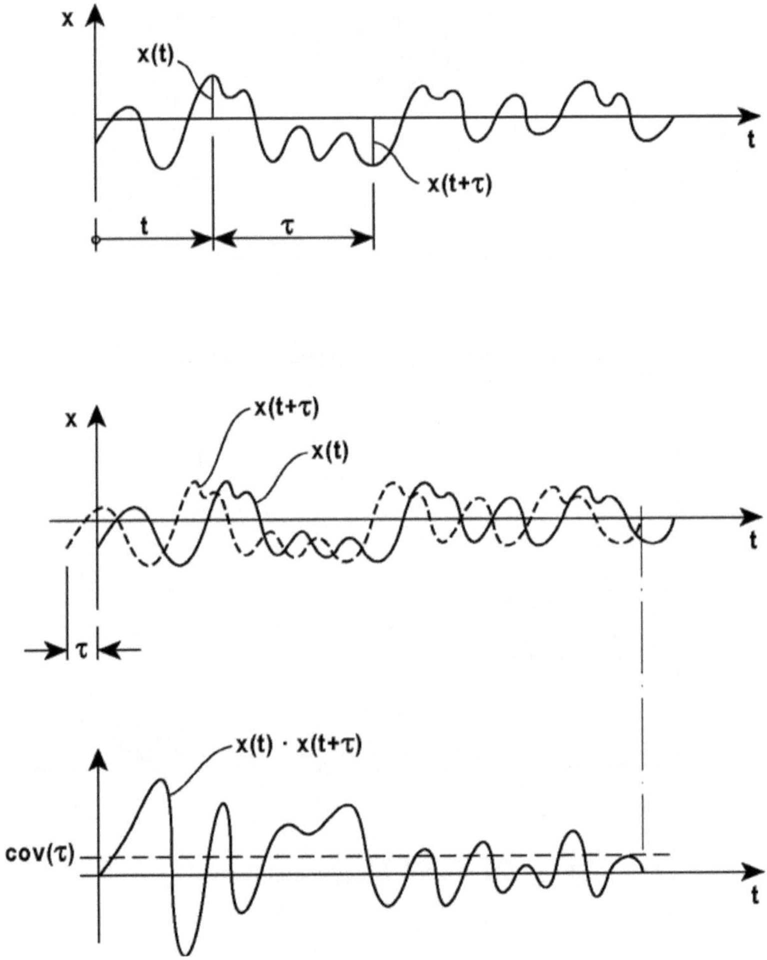

Fig. 2.5 *The auto covariance function*

As illustrated in Fig. 2.5 the auto covariance function is the mean value of the time series multiplied by itself at a time shift equal to τ. Theoretically τ may vary between 0 and *T*, but the practical significance of $Cov_x(\tau)$ seizes to exist long before τ is in the vicinity of *T*. The reason is that while it in theoretical developments is convenient to consider $x(t)$ as a continuous function, it will in practical calculations only occur as a discrete and finite vector of random values x_k, usually taken at regular intervals Δt. If *T* is large and Δt is small, then the number of elements in this vector is $N \approx T/\Delta t$, in which case the continuous integral in Eq. 2.15 may be replaced by its discrete counterpart

$$Cov_x(\tau = j \cdot \Delta t) = E\left[x(t) \cdot x(t+\tau)\right] = \frac{1}{N-j}\sum_{k=1}^{N-j} x_{k+j} \cdot x_k \qquad (2.20)$$

from which it is seen that j must be considerably smaller than N for a meaningful outcome of the auto covariance estimate. The same is true for the auto correlation function in Eq. 2.14.

Example 2.2

Given a variable: $x(t) = a_1 \cdot \sin(\omega_1 t)$, $\omega_1 = 2\pi/T_1$. Using the substitutions $T = nT_1$ (where n is an integer) and $\hat{t} = (2\pi/T_1)t$, then the auto covariance of x is given by

$$Cov_x(\tau) = \lim_{T\to\infty}\frac{1}{T}\int_0^T x(t)\cdot x(t+\tau)dt = \lim_{n\to\infty}\frac{1}{nT_1}\cdot\left[n\int_0^{T_1} a_1\sin(\omega_1 t)\cdot a_1\sin(\omega_1 t + \omega_1\tau)dt\right]$$

$$= \frac{a_1^2}{2\pi}\int_0^{2\pi}\sin\hat{t}\cdot\sin(\hat{t}+\omega_1\tau)d\hat{t} = \frac{a_1^2}{2\pi}\int_0^{2\pi}\left[\sin^2\hat{t}\cdot\cos(\omega_1\tau) + \sin\hat{t}\cdot\cos\hat{t}\cdot\sin(\omega_1\tau)\right]d\hat{t}$$

$$= \frac{a_1^2}{2\pi}\left[\cos(\omega_1\tau)\int_0^{2\pi}\sin^2\hat{t}d\hat{t} + \frac{\sin(\omega_1\tau)}{2}\int_0^{2\pi}\sin 2\hat{t}d\hat{t}\right]$$

The first of these integrals is equal to π, while the second is zero, and thus:

$$Cov_x(\tau) = \frac{a_1^2}{2}\cos(\omega_1\tau)$$

Since the variance of $x(t)$ is $\sigma_x^2 = a_1^2/2$ (see example 2.1), then the auto covariance coefficient is given by:

$$\rho_x(\tau) = \frac{Cov_x(\tau)}{\sigma_x^2} = \cos(\omega_1\tau)$$

As can be seen: $\rho_x(\tau = 0) = 1$.

Similar to the definitions above, cross correlation and cross covariance functions may be defined between observations that have been obtained from two short term realisations $X_1(t) = \bar{x}_1 + x_1(t)$ and $X_2(t) = \bar{x}_2 + x_2(t)$ of the same process or alternatively from realisations of two different processes:

$$R_{X_1X_2}(\tau) = E\left[X_1(t)\cdot X_2(t+\tau)\right] = \lim_{T\to\infty}\frac{1}{T}\int_0^T X_1(t)\cdot X_2(t+\tau)dt \qquad (2.21)$$

$$Cov_{x_1x_2}(\tau) = E\left[x_1(t)\cdot x_2(t+\tau)\right] = \lim_{T\to\infty}\frac{1}{T}\int_0^T x_1(t)\cdot x_2(t+\tau)dt \qquad (2.22)$$

A normalised version of the cross covariance between the fluctuating parts of the realisations is defined by the cross covariance coefficient

$$\rho_{x_1x_2}(\tau) = \frac{Cov_{x_1x_2}(\tau)}{\sigma_{x_1}\sigma_{x_2}} \qquad (2.23)$$

where σ_{x_1} and σ_{x_2} are the standard deviations of the two zero mean time variables.

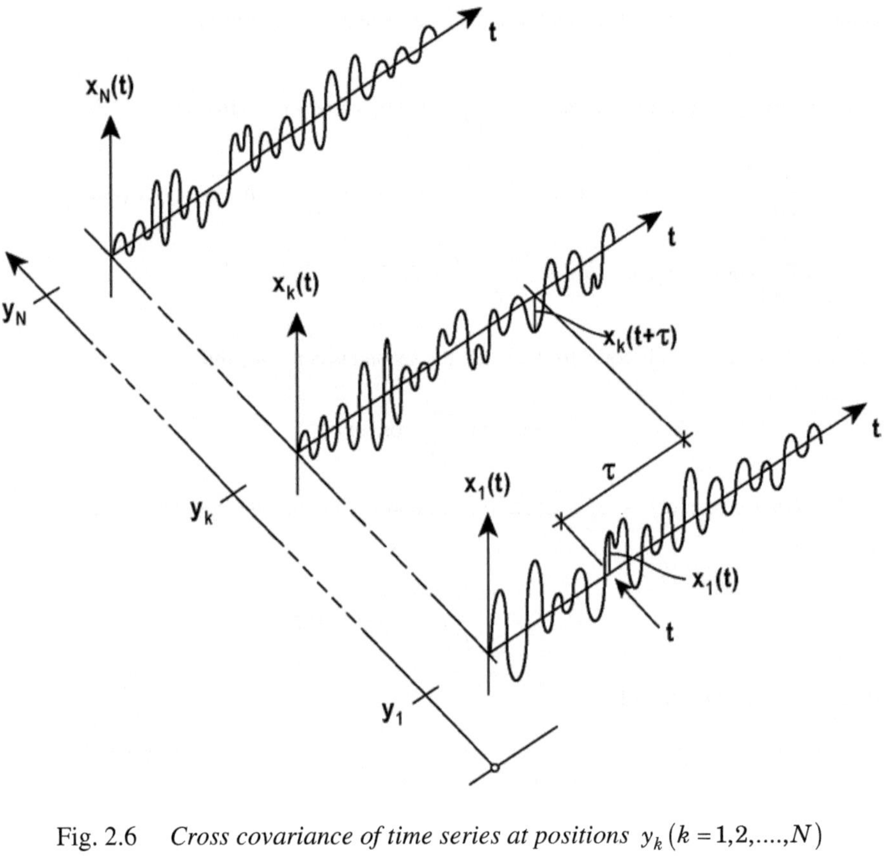

Fig. 2.6 *Cross covariance of time series at positions* $y_k\ (k = 1,2,....,N)$

If such cross covariance estimates are taken from a set of simultaneous realisations of a process distributed in space, e.g. as illustrated in Fig. 2.6 where the N realisations of the process is assumed to be taken at arbitrary positions y in the horizontal direction, then a cross covariance function between realisations at distance Δy may be defined:

$$Cov_{xx}(\Delta y,\tau) = E\left[x(y,t)\cdot x(y+\Delta y,t+\tau)\right] \\ = \lim_{T\to\infty}\frac{1}{T}\int_0^T x(y,t)\cdot x(y+\Delta y,t+\tau)dt \tag{2.24}$$

Obviously, $Cov_{xx}(\Delta y = 0, \tau = 0) = \sigma_x^2$. In wind engineering such covariance estimates will in general be a decaying function with increasing τ or spatial separation Δs, $s = x,y$ or z , as illustrated in Fig. 2.7. The covariance function may attain negative values at large values of Δs or τ .

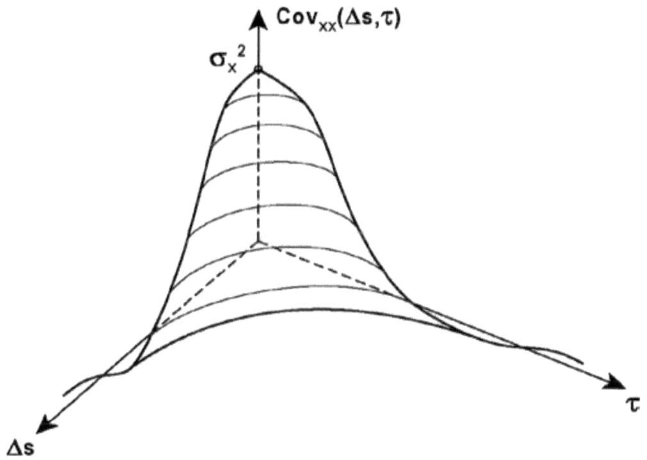

Fig. 2.7 *Typical spatial separation and time lag covariance function*

As previously indicated, the statistical properties defined above may also be applied to functions that are obtained from realisations of two different processes. Then, by simple arithmetic, the variance of the sum of two zero mean variables, $x_1(t)$ and $x_2(t)$, is given by

$$Var(x_1 + x_2) = E\left[(x_1 + x_2)\cdot(x_1 + x_2)\right] \\ = Var(x_1) + Var(x_2) + 2\cdot Cov(x_1 \cdot x_2) \tag{2.25}$$

Similarly, the variance of the sum of N different variables, $x_i(t)$, is given by

$$Var\left(\sum_{i=1}^{N} x_i\right) = E\left[(x_1 + x_2 + \ldots + x_i + \ldots + x_N)\cdot(x_1 + x_2 + \ldots + x_j + \ldots + x_N)\right]$$

$$\Rightarrow Var\left(\sum_{i=1}^{N} x_i\right) = \sum_{i=1}^{N}\sum_{j=1}^{N} Cov\left(x_i \cdot x_j\right) = \sum_{i=1}^{N}\sum_{j=1}^{N} \rho\left(x_i \cdot x_j\right) \cdot \sigma_i \sigma_j \tag{2.26}$$

If $x_i(t)$ are independent (i.e. uncorrelated) then the variance of the sum of the processes is the sum of the variances of the individual processes, i.e.

$$\text{if } Cov\left(x_i \cdot x_j\right) = \begin{cases} \sigma_{x_i}^2 \text{ when } i = j \\ 0 \text{ when } i \neq j \end{cases} \qquad \text{then} \qquad Var\left(\sum_{i=1}^{N} x_i\right) = \sum_{i=1}^{N} \sigma_{x_i}^2 \tag{2.27}$$

Example 2.3

Given an ensemble variable: $x = a \cdot \sin(\omega t + \theta)$, where the probability density distribution of θ

is: $p(\theta) = \begin{cases} \dfrac{1}{2\pi} \text{ for } 0 \le \theta \le 2\pi \\ 0 \text{ elsewhere} \end{cases}$

The ensemble covariance of x_k at a time lag τ is then given by

$$Cov_x(\tau) = E\left[x(t,\theta) \cdot x(t+\tau,\theta)\right] = \int_0^{2\pi} p(\theta) \cdot x(t,\theta) \cdot x(t+\tau,\theta) d\theta$$

$$= \int_0^{2\pi} \frac{1}{2\pi} \cdot a \sin(\omega t + \theta) \cdot a \sin(\omega t + \omega\tau + \theta) d\theta$$

$$= \frac{a^2}{2\pi} \int_0^{2\pi} \sin(\omega t + \theta) \cdot \left[\sin(\omega t + \theta) \cdot \cos(\omega\tau) + \cos(\omega t + \theta) \cdot \sin(\omega\tau)\right] d\theta$$

$$= \frac{a^2}{2\pi}\left[\cos(\omega\tau) \int_0^{2\pi} \sin^2(\omega t + \theta) d\theta + \frac{\sin(\omega\tau)}{2} \int_0^{2\pi} \sin 2(\omega t + \theta) d\theta\right]$$

which, after the substitution $\hat{\theta} = \omega t + \theta$, renders

$$Cov_x(\tau) = \frac{a^2}{2} \cos(\omega\tau) \int_{\omega t}^{\omega t + 2\pi} \sin^2 \hat{\theta} d\hat{\theta} + \frac{\sin(\omega\tau)}{2} \int_{\omega t}^{\omega t + 2\pi} \sin 2\hat{\theta} d\hat{\theta}$$

As shown in example 2.2, the first of these integrals is equal to π, while the second is zero, and thus:

$$Cov_x(\tau) = \frac{a^2}{2} \cos(\omega\tau)$$

There are still other types of time domain and ensemble statistics that are of great importance in wind engineering and that have not yet been mentioned. These comprise the properties of threshold crossing, the distributions of peaks and extreme values, and finally, the auto and cross spectral densities, which are frequency domain properties of the process, i.e. they are frequency domain counterparts to the concepts of variance and covariance. These are dealt with below.

2.3 Threshold crossing and peaks

In Fig. 2.8 is illustrated a time series realisation $x(t)$ of a Gaussian stationary and homogeneous process (for simplicity with zero mean value), taken over a period *T*. First we seek to develop an estimate of the average frequency $f_x(a)$ between the events that $x(t)$ is crossing the threshold a in its upward direction.

Let a single upward crossing take place in a time interval Δt that is small enough to justify the approximation

$$x(t+\Delta t) \cong x(t) + \dot{x}(t) \cdot \Delta t \tag{2.28}$$

The probability of an up crossing event during Δt is then given by

$$P\left[x(t) \le a \text{ and } x(t) + \dot{x}(t) \cdot \Delta t > a\right] = f_x(a) \cdot \Delta t \tag{2.29}$$

from which it follows that

$$f_x(a) = \lim_{\Delta t \to 0} \frac{1}{\Delta t} \int_0^{\infty} \left[\int_{a - \dot{x} \cdot \Delta t}^{a} p_{x\dot{x}}(x, \dot{x}) dx \right] d\dot{x} \tag{2.30}$$

where $p_{x\dot{x}}(x,\dot{x})$ is the probability density function for the joint events $x(t)$ and $\dot{x}(t)$. As $\Delta t \to 0$ the following approximation applies

$$\int_{a - \dot{x} \cdot \Delta t}^{a} p_{x\dot{x}}(x, \dot{x}) dx \cong \dot{x} \cdot \Delta t \cdot p_{x\dot{x}}(a, \dot{x}) \tag{2.31}$$

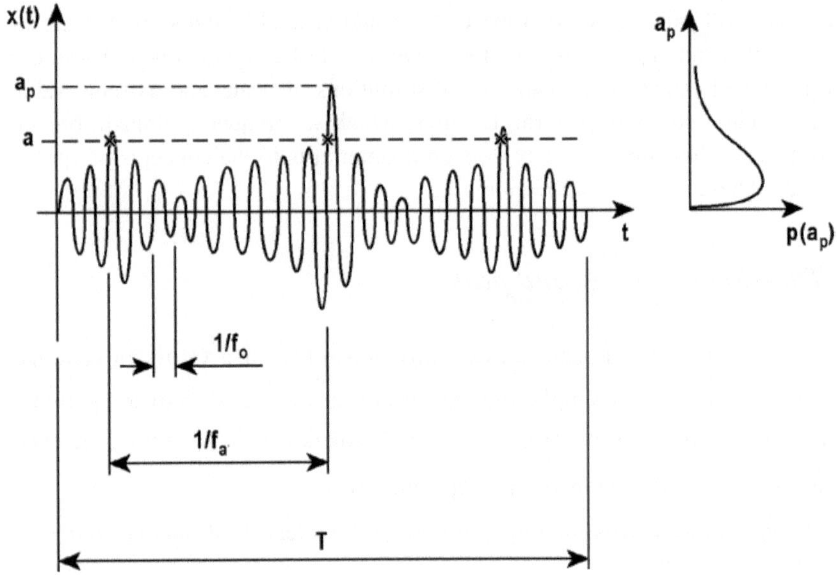

Fig. 2.8 *Threshold crossing and peaks*

For the type of processes covered herein it is a reasonable assumption that the joint events of $x(t)$ and $\dot{x}(t)$ are statistically independent, and thus, $p_{x\dot{x}}(x,\dot{x}) = p_x(x) \cdot p_{\dot{x}}(\dot{x})$. The average up crossing event that $x(t) = a$ is then given by

$$f_x(a) = \int_0^\infty \dot{x} \cdot p_{x\dot{x}}(a,\dot{x}) d\dot{x} = p_x(a) \cdot \int_0^\infty \dot{x} \cdot p_{\dot{x}}(\dot{x}) d\dot{x} \tag{2.32}$$

For each threshold up-crossing there is a corresponding down-crossing event, i.e $f_x(+a) = f_x(-a)$, although there may be several consecutive positive or negative peaks in the process. Assuming that both x and $\dot{x}$ are Gaussian, then

$$f_x(a) = \frac{1}{\sqrt{2\pi}\sigma_x} \exp\left[-\frac{1}{2}\left(\frac{a}{\sigma_x}\right)^2\right] \int_0^\infty \dot{x} \cdot \frac{1}{\sqrt{2\pi}\sigma_{\dot{x}}} \exp\left[-\frac{1}{2}\left(\frac{\dot{x}}{\sigma_{\dot{x}}}\right)^2\right] d\dot{x}$$

$$\Rightarrow f_x(a) = \frac{1}{2\pi} \cdot \frac{\sigma_{\dot{x}}}{\sigma_x} \cdot \exp\left[-\frac{1}{2}\left(\frac{a}{\sigma_x}\right)^2\right] = f_x(0) \cdot \exp\left[-\frac{1}{2}\left(\frac{a}{\sigma_x}\right)^2\right] \tag{2.33}$$

where : $$f_x(0) = \frac{1}{2\pi} \cdot \frac{\sigma_{\dot{x}}}{\sigma_x} \tag{2.34}$$

is the average zero up–crossing frequency of the process (see Eq. 2.95). If $x(t)$ is also narrow banded, such that a zero up crossing and a peak x_p (larger than zero) are simultaneous events (as shown for the process in Fig. 2.8), then the expected number of peaks $x_p > a_p$ is $f_x(a_p) \cdot T$, while the total number of peaks is $f_x(0) \cdot T$. Thus

$$\Pr\left[x_p \le a_p\right] = P(a_p) = 1 - \frac{f_x(a_p)}{f_x(0)} \tag{2.35}$$

from which it follows that the probability density distribution to a_p is given by

$$p(a_p) = \frac{d}{da_p} P(a_p) = \frac{d}{da_p}\left[1 - \frac{f_x(a_p)}{f_x(0)}\right] = -\frac{1}{f_x(0)} \cdot \frac{df_x(a_p)}{da_p}$$

$$\Rightarrow p(a_p) = \frac{a_p}{\sigma_x^2} \exp\left[-\frac{1}{2}\left(\frac{a_p}{\sigma_x}\right)^2\right] \tag{2.36}$$

Thus, the probability density $p(a_p)$ of peaks to a narrow banded Gaussian process is a Rayleigh distribution (see Eq. 2.7). The distribution is illustrated on the right hand side of Fig. 2.8 (see also Fig. 2.1).

2.4 *Extreme values*

Fig. 2.9.a shows a collection of N short term time series, each a short term realisation of the fluctuating part $x(t)$ of a stochastic variable $X(t) = \bar{x} + x(t)$. It is assumed that they are all stationary and ergodic, and for the validity of the development below it is a necessary requirement that they are fairly broad banded. From this ensemble of realisations it may be of particular interest to develop the statistical properties of extreme values, as illustrated in Fig. 2.9.b. Referring to Eq. 2.33 and Fig. 2.8, an extreme peak value $a_p = x_{max}$ within each short term realisation occur when

$$\left[f_x(a_p)\right]^{-1} \to T \tag{2.37}$$

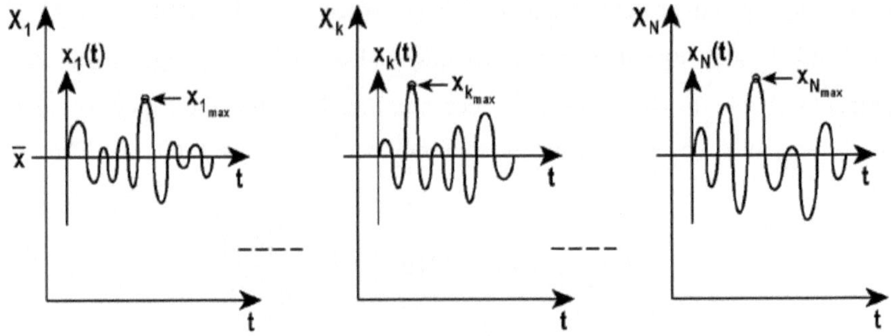

a) *N short term independent realisations*

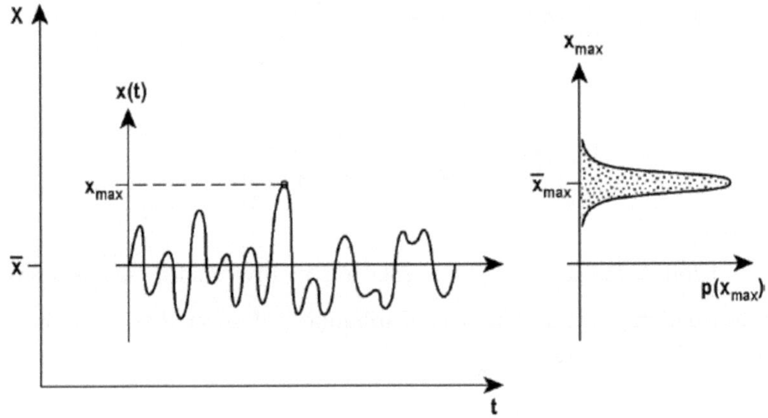

b) *The distribution of extremes*

Fig. 2.9 *Distribution of extreme values*

Let therefore

$$\kappa = f_x\left(x_{\max}\right) \cdot T \tag{2.38}$$

be an ensemble variable signifying the event that $x\left(0 \le t \le T\right)$ exceeds a given value $x_{\max}$. The probability that κ occurs only once within each realisation is an event that coincides with the occurrence of $x_{\max}$, i.e. they are simultaneous events. They are rare events at the tail of the peak distribution given in Eq. 2.36, and for the statistics of such events it is a reasonable assumption that they will also comply to an exponential distribution, i.e. that

$$P_\kappa(1,T) = P_{x_{max}}(x_{max}|T) = \exp(-\kappa) \tag{2.39}$$

Introducing Eqs. 2.33 (with $a = x_{max}$) into 2.38 and solving for x_{max}, then the following is obtained

$$\begin{aligned} x_{max} &= \sigma_x \cdot \left\{2 \cdot \ln\left[f_x(0)T\right] - 2 \cdot \ln \kappa\right\}^{\frac{1}{2}} \\ &\approx \sigma_x \cdot \sqrt{2 \cdot \ln\left[f_x(0) \cdot T\right]}\left\{1 - \frac{\ln \kappa}{2 \cdot \ln\left[f_x(0) \cdot T\right]}\right\} \end{aligned} \tag{2.40}$$

where the approximation $(1-x)^n \approx 1 - n \cdot x$ has been applied, assuming that $\ln\left[f_x(0) \cdot T\right]$ is large as compared to $\ln \kappa$. Thus, observing that $x_{max} = 0$ corresponds to $\kappa = \infty$, while $x_{max} = \infty$ corresponds to $\kappa = 0$, the mean value of x_{max} may be estimated from

$$\begin{aligned} \bar{x}_{max} &= \int_0^\infty x_{max} \cdot \left(\frac{dP_{x_{max}}}{dx_{max}}\right) dx_{max} = \int_0^\infty x_{max} \cdot \left(\frac{dP_{x_{max}}}{d\kappa}\right) \cdot \left(\frac{d\kappa}{dx_{max}}\right) dx_{max} \\ &= \int_\infty^0 x_{max} \cdot \left[-\exp(-\kappa)\right] d\kappa = \int_0^\infty x_{max} \cdot \exp(-\kappa) d\kappa \\ \Rightarrow \bar{x}_{max} &= \sigma_x \cdot \sqrt{2 \cdot \ln\left[f_x(0) \cdot T\right]} \cdot \left[\int_0^\infty \exp(-\kappa) d\kappa - \frac{\int_0^\infty \ln \kappa \cdot \exp(-\kappa) d\kappa}{2 \cdot \ln\left[f_x(0) \cdot T\right]}\right] \end{aligned} \tag{2.41}$$

Thus, the mean value of x_{max} is given by

$$\bar{x}_{max} = \sigma_x \cdot \left\{\sqrt{2 \cdot \ln\left[f_x(0) \cdot T\right]} + \frac{\gamma}{\sqrt{2 \cdot \ln\left[f_x(0) \cdot T\right]}}\right\} \tag{2.42}$$

where $\gamma = -\int_0^\infty \ln \kappa \cdot \exp(-\kappa) \approx 0.5772$ is the Euler constant. Similarly, it may be shown that the variance of x_{max} is given by

$$\sigma_{x_{max}}^2 = \frac{\pi^2}{12 \cdot \ln\left[f_x(0) \cdot T\right]} \cdot \sigma_x^2 \tag{2.43}$$

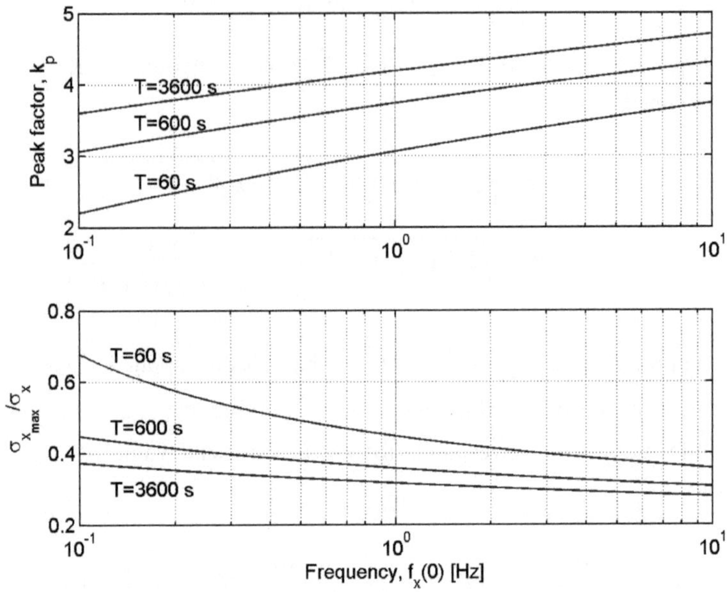

Fig. 2.10 *Plots of* k_p *and* $\sigma_{x_{\max}}/\sigma_x$

Given a stochastic variable $X(t) = \bar{x} + x(t)$, the expected value of its largest peak during a realisation with length *T* may then be estimated from

$$X_{\max} = \bar{x} + k_p \cdot \sigma_x \tag{2.44}$$

where the peak factor k_p is given by

$$k_p = \sqrt{2 \cdot \ln\left[f_x(0) \cdot T\right]} + \frac{\gamma}{\sqrt{2 \cdot \ln\left[f_x(0) \cdot T\right]}} \tag{2.45}$$

For fairly broad banded processes this peak factor will render values between 2 and 5. Plots of k_p and $\sigma_{x_{\max}}/\sigma_x$ are shown in Fig. 2.10. It should be acknowledged that when $x(t)$ becomes ultra narrow banded then $k_p \to \sqrt{2}$, because for a single harmonic component

$$x(t) = c_x \cdot \cos(\omega_x t), \; 0 < t < T \tag{2.46}$$

the variance

$$\begin{aligned}\sigma_x^2 &= \lim_{T\to\infty}\frac{1}{T}\int_0^T\left[c_x\cdot\cos(\omega_x t)\right]^2 dt\\ &= \lim_{n\to\infty}\frac{1}{n\cdot T_x}\cdot n\cdot\int_0^{T_x}\left[c_x\cdot\cos\left(\frac{2\pi}{T_x}t\right)\right]^2 dt = \frac{c_x^2}{2}\end{aligned} \tag{2.47}$$

and thus, for such a process $x_{\max} = c_x = \sigma_x\cdot\sqrt{2}$. Therefore, Eq. 2.45 is only applicable for fairly broad banded processes.

2.5 *Auto spectral density*

The auto spectral density contains the frequency domain properties of the process, i.e. it is the frequency domain counterpart to the concept of variance. The various steps in the development of an auto spectral density function are illustrated in Fig. 2.11.

Given a zero mean time variable $x(t)$ with length T and performing a Fourier transformation of $x(t)$ implies that it may be approximated by a sum of harmonic components $X_k(\omega_k,t)$, i.e.

$$x(t) = \lim_{N\to\infty}\sum_{k=1}^{N}X_k(\omega_k,t) \qquad \text{where} \qquad \begin{cases}\omega_k = k\cdot\Delta\omega\\ \Delta\omega = 2\pi/T\end{cases} \tag{2.48}$$

The harmonic components in Eq. 2.48 are given by

$$X_k(\omega_k,t) = c_k\cdot\cos(\omega_k t + \varphi_k) \tag{2.49}$$

where the amplitudes $c_k = \sqrt{a_k^2 + b_k^2}$ and phase angles $\varphi_k = \arctan(b_k/a_k)$, and where the constants a_k and b_k are given by

$$\begin{bmatrix}a_k\\ b_k\end{bmatrix} = \frac{2}{T}\int_0^T x(t)\begin{bmatrix}\cos\omega_k t\\ \sin\omega_k t\end{bmatrix}dt \tag{2.50}$$

As shown in Fig. 2.11 the auto-spectral density of $x(t)$ is intended to represent its variance density distribution in the frequency domain. Hence, the definition of the single-sided auto-spectral density S_x associated with the frequency ω_k is

$$S_x(\omega_k) = \frac{E\left[X_k^2\right]}{\Delta\omega} = \frac{\sigma_{X_k}^2}{\Delta\omega} \tag{2.51}$$

which, when T becomes large, is given by

$$S_x(\omega_k) = \lim_{T\to\infty} \frac{1}{\Delta\omega}\cdot\frac{1}{T}\int_0^T \left[c_k \cos(\omega_k t + \varphi_k)\right]^2 dt \tag{2.52}$$

Introducing the period of the harmonic component, $T_k = 2\pi/\omega_k$, and replacing T with $n\cdot T_k$, $n\to\infty$, then the following is obtained

$$S_x(\omega_k) = \lim_{n\to\infty} \frac{1}{\Delta\omega}\cdot\frac{1}{n\cdot T_k}\cdot n\cdot \int_0^{T_k} \left[c_k \cos\left(\frac{2\pi}{T_k}\cdot t + \varphi_k\right)\right]^2 dt = \frac{c_k^2}{2\Delta\omega} \tag{2.53}$$

In Fig. 2.11, the arrival at $S_x(\omega_k)$ is shown via the amplitude spectrum (or the Fourier amplitude diagram) to ease the understanding of the concept of spectral density representations. It is seen from this illustration that it is not possible to retrieve the parent time domain variable from the spectral density function alone, because it does not contain the necessary phase information (unless a corresponding phase spectrum is also established). From its very definition the spectrum contains information about the variance distribution in frequency domain, and from Eqs. 2.51 and 2.53 it is seen that

$$\sigma_x^2 = \lim_{N\to\infty}\sum_{k=1}^{N}\sigma_{X_k}^2 = \lim_{N\to\infty}\sum_{k=1}^{N}\frac{c_k^2}{2} = \lim_{N\to\infty}\sum_{k=1}^{N} S_x(\omega_k)\cdot\Delta\omega \tag{2.54}$$

In a continuous format, i.e. in the limit of both N and T approaching infinity, the single-sided auto-spectral density is defined by

$$S_x(\omega) = \lim_{T\to\infty}\lim_{N\to\infty}\frac{E\left[X^2(\omega,t)\right]}{\Delta\omega} \tag{2.55}$$

where $X(\omega,t)$ is an arbitrary Fourier component of $x(t)$. In the limit $\Delta\omega \to d\omega$, and thus, the variance of the process may be calculated from

$$\sigma_x^2 = \int_0^\infty S_x(\omega)d\omega \tag{2.56}$$

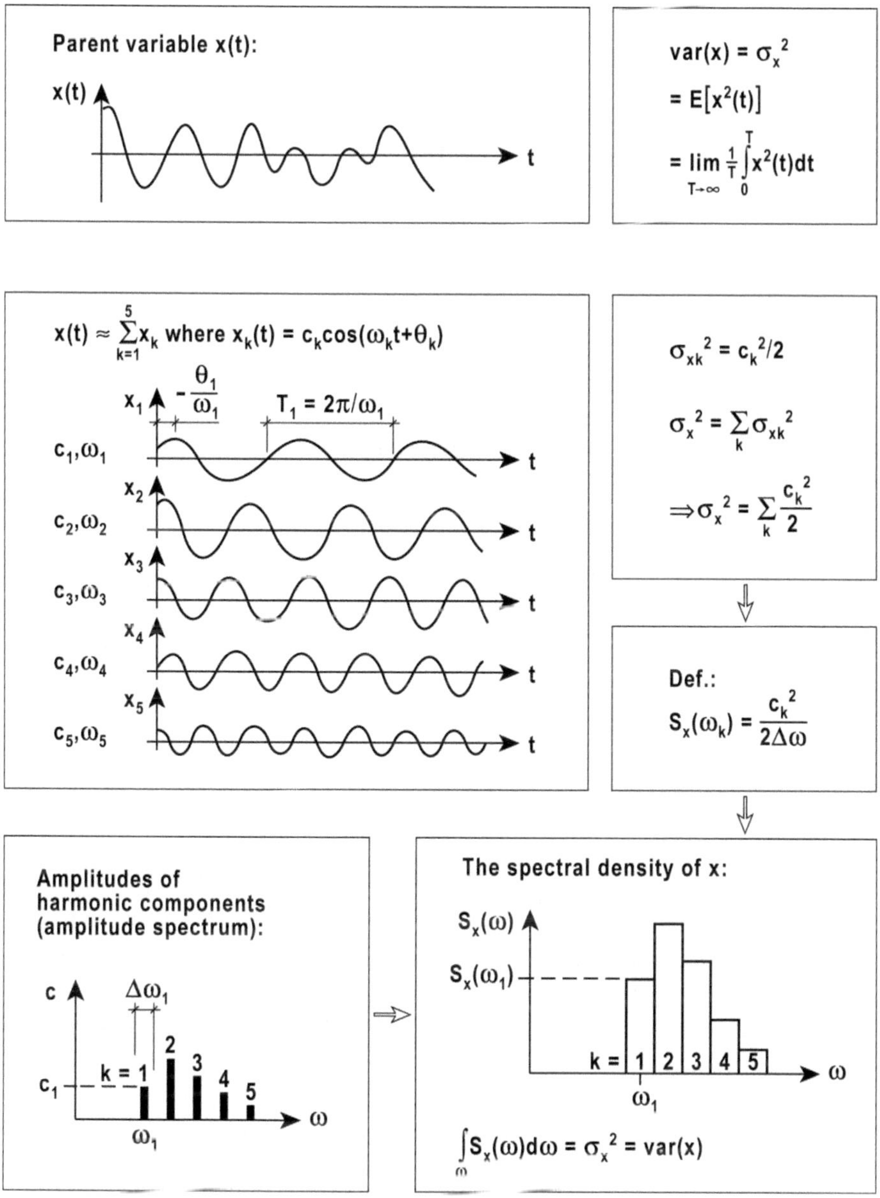

Fig. 2.11 *The definition of auto spectral density from a Fourier decomposition*

The development above may more conveniently be expressed in a complex format. Adopting a frequency axis spanning the entire range of both positive and (imaginary) negative values, introducing the Euler formulae

$$\begin{bmatrix} e^{i\omega t} \\ e^{-i\omega t} \end{bmatrix} = \begin{bmatrix} 1 & i \\ 1 & -i \end{bmatrix} \begin{bmatrix} \cos\omega t \\ \sin\omega t \end{bmatrix} \tag{2.57}$$

(where $i = \sqrt{-1}$) and defining the complex Fourier amplitude

$$d_k = \frac{1}{2}(a_k - i \cdot b_k) \tag{2.58}$$

then:

$$x(t) = \sum_{-\infty}^{\infty} X_k(\omega_k, t) = \sum_{-\infty}^{\infty} d_k(\omega_k) \cdot e^{i \cdot \omega_k t} \tag{2.59}$$

Taking the variance of the complex Fourier components in Eq. 2.59 and dividing by $\Delta\omega$,

$$\frac{E\left[X_k^* \cdot X_k\right]}{\Delta\omega} = \frac{1}{T} \int_0^T \frac{\left(d_k^* e^{-i\omega_k t}\right)\left(d_k e^{i\omega_k t}\right)}{\Delta\omega} dt = \frac{d_k^* d_k}{\Delta\omega} \tag{2.60}$$

which may be further developed into

$$\Rightarrow \frac{E\left[X_k^* \cdot X_k\right]}{\Delta\omega} = \frac{1}{4} \frac{(a_k + i \cdot b_k) \cdot (a_k - i \cdot b_k)}{\Delta\omega} = \frac{c_k^2}{4\Delta\omega} \tag{2.61}$$

It is seen (see Eq. 2.53) that this is half the auto spectral value associated with ω_k. Thus, a symmetric double-sided auto spectrum associated with $-\omega_k$ as well as $+\omega_k$ may be defined with a value that is half the corresponding value of the single sided auto-spectrum. Extending the frequency axis from minus infinity to plus infinity and using the complex Fourier components X_k given in Eq. 2.59 above, this double sided auto spectrum is then defined by

$$S_x(\pm\omega_k) = \frac{E\left[X_k^* \cdot X_k\right]}{\Delta\omega} = \frac{d_k^* d_k}{\Delta\omega} = \frac{c_k^2}{4\Delta\omega} \tag{2.62}$$

which, in the limit of T and $N \to \infty$, becomes the continuous function $S_x(\pm\omega)$, and from which the variance of the process may be obtained by integration over the entire positive as well as negative (imaginary) frequency range

$$\sigma_x^2 = \int_{-\infty}^{+\infty} S_x(\pm\omega)d\omega \tag{2.63}$$

Thus, the connection between double- and single-sided spectra is simply that $S_x(\omega) = 2 \cdot S_x(\pm\omega)$. Assuming that the process is stationary and of infinite length, such that the position of the time axis for integration purposes is arbitrary, then it is in the literature of mathematics usually considered convenient to introduce a non-normalized amplitude (which may be encountered in connection with the theory of generalised Fourier series and identified as a Fourier constant)

$$a_k(\omega_k) = \int_0^T x(t) \cdot e^{-i \cdot \omega_k t} dt = T \cdot d_k \tag{2.64}$$

in which case the double-sided auto-spectral density associated with $\pm\omega_k$ is defined by

$$S_x(\pm\omega_k) = \frac{d_k^* \cdot d_k}{\Delta\omega} = \frac{\left(a_k^*/T\right) \cdot \left(a_k/T\right)}{2\pi/T} = \frac{1}{2\pi T} \cdot a_k^* a_k \tag{2.65}$$

In the limit of T and $N \to \infty$ this may be written on the following continuous form

$$S_x(\pm\omega) = \lim_{T\to\infty} \lim_{N\to\infty} \frac{1}{2\pi T} \cdot a^*(\omega) \cdot a(\omega) \tag{2.66}$$

and accordingly, the single sided version is given by

$$S_x(\omega) = \lim_{T\to\infty} \frac{1}{\pi T} \cdot a^*(\omega) \cdot a(\omega) \tag{2.67}$$

where it is taken for granted that N is sufficiently large. The auto-spectral density $S_x(\omega)$ is defined by use of circular frequency ω as shown above. It may be replaced by a corresponding definition $S_x(f)$ using frequency f (with unit $Hz = sek^{-1}$) Since $S_x(\omega) \cdot \Delta\omega$ and $S_x(f) \cdot \Delta f$ both represent the variance of the process at ω and f, they must give the same contribution to the total variance of the process, and thus

$$S_x(f) \cdot \Delta f = S_x(\omega) \cdot \Delta\omega = S_x(\omega) \cdot (2\pi \cdot \Delta f)$$

$$\Rightarrow S_x(f) = 2\pi \cdot S_x(\omega) = \lim_{T\to\infty} \lim_{N\to\infty} \frac{2}{T} \cdot a^*(f) \cdot a(f) \tag{2.68}$$

2.6 Cross-spectral density

The cross spectral density contains the frequency domain and coherence properties between processes, i.e. it is the frequency domain counterpart to the concept of covariance. Given two stationary time variable functions $x(t)$ and $y(t)$, both with length T and zero mean value (i.e. $E\left[x(t)\right]=E\left[y(t)\right]=0$), and performing a Fourier transformation (adopting a double-sided complex format) implies that $x(t)$ and $y(t)$ may be represented by sums of harmonic components $X_k(\omega_k,t)$ and $Y_k(\omega_k,t)$, i.e.

$$\begin{bmatrix} x(t) \\ y(t) \end{bmatrix} = \lim_{N\to\infty} \sum_{-N}^{N} \begin{bmatrix} X_k(\omega_k,t) \\ Y_k(\omega_k,t) \end{bmatrix} \tag{2.69}$$

where:

$$\begin{bmatrix} X_k(\omega_k,t) \\ Y_k(\omega_k,t) \end{bmatrix} = \frac{1}{T} \begin{bmatrix} a_{X_k}(\omega_k) \\ a_{Y_k}(\omega_k) \end{bmatrix} \cdot e^{i\cdot\omega_k t} \quad \text{and} \quad \begin{bmatrix} a_{X_k}(\omega_k) \\ a_{Y_k}(\omega_k) \end{bmatrix} = \lim_{T\to\infty} \int_{-T/2}^{T/2} \begin{bmatrix} x(t) \\ y(t) \end{bmatrix} \cdot e^{-i\cdot\omega_k t} dt$$

and where $\omega_k = k\cdot\Delta\omega$ and $\Delta\omega = 2\pi/T$. The definition of the double-sided cross-spectral density S_{xy} associated with the frequency ω_k is then

$$S_{xy}(\pm\omega_k) = \frac{E\left[X_k^* \cdot Y_k\right]}{\Delta\omega} = \frac{1}{2\pi T} a_{X_k}^* \cdot a_{Y_k} \tag{2.70}$$

Since the Fourier components are orthogonal

$$E\left[X_i(\omega_i,t)\cdot Y_j(\omega_j,t)\right] = \begin{cases} S_{xy}(\omega_k,t)\cdot\Delta\omega \text{ when } i=j=k \\ 0 \text{ when } i\neq j \end{cases} \tag{2.71}$$

it follows from Eqs. 2.69 and 2.70 that an estimate of the covariance between $x(t)$ and $y(t)$ are given by

$$Cov_{xy} = E\left[x(t)\cdot y(t)\right] = \lim_{N\to\infty} E\left[\left(\sum_{-N}^{N} X_i\right)\cdot\left(\sum_{-N}^{N} Y_j\right)\right] = \lim_{N\to\infty} \sum_{-N}^{N} \left(E\left[X_k \cdot Y_k\right]\right)$$

$$\Rightarrow Cov_{xy} = \lim_{N\to\infty} \sum_{-N}^{N} S_{xy}(\pm\omega_k)\cdot\Delta\omega \tag{2.72}$$

In a continuous format, i.e. in the limit of both N and T approaching infinity, the double-sided cross-spectral density is defined by

$$\begin{aligned} S_{xy}(\pm\omega) &= \lim_{T\to\infty}\lim_{N\to\infty} \frac{E\left[X^*(\omega,t)\cdot Y(\omega,t)\right]}{\Delta\omega} \\ &= \lim_{T\to\infty}\lim_{N\to\infty} \frac{1}{2\pi T} a_X^*(\omega)\cdot a_Y(\omega) \end{aligned} \tag{2.73}$$

The single sided version is then simply

$$S_{xy}(\omega) = 2\cdot S_{xy}(\pm\omega) = \lim_{T\to\infty}\lim_{N\to\infty} \frac{1}{\pi T} a_X^*(\omega)\cdot a_Y(\omega) \tag{2.74}$$

while the corresponding single-sided version using frequency f (Hz), is defined by

$$S_{xy}(f) = 2\pi\cdot S_{xy}(\omega) = \lim_{T\to\infty}\lim_{N\to\infty} \frac{2}{T}\cdot a_x^*(f)\cdot a_y(f) \tag{2.75}$$

Thus, the covariance between the two processes may be calculated from

$$Cov_{xy} = \int_{-\infty}^{+\infty} S_{xy}(\pm\omega)d\omega = \int_0^{\infty} S_{xy}(\omega)d\omega = \int_0^{\infty} S_{xy}(f)df \tag{2.76}$$

The cross-spectrum will in general be a complex quantity. With respect to the frequency argument, its real part is an even function labelled the co–spectral density $Co_{xy}(\omega)$, while its imaginary part is an odd function labelled the quad–spectrum $Qu_{xy}(\omega)$, i.e.

$$S_{xy}(\omega) = Co_{xy}(\omega) - i\cdot Qu_{xy}(\omega) \tag{2.78}$$

as illustrated in Fig. 2.12. Alternatively, $S_{xy}(\omega)$ may be expressed by its modulus and phase, i.e.

$$S_{xy}(\omega) = \left|S_{xy}(\omega)\right|\cdot e^{i\cdot\varphi_{xy}(\omega)} \tag{2.79}$$

where the phase spectrum $\varphi_{xy}(\omega) = \arctan\left[Qu_{xy}(\omega)/Co_{xy}(\omega)\right]$.

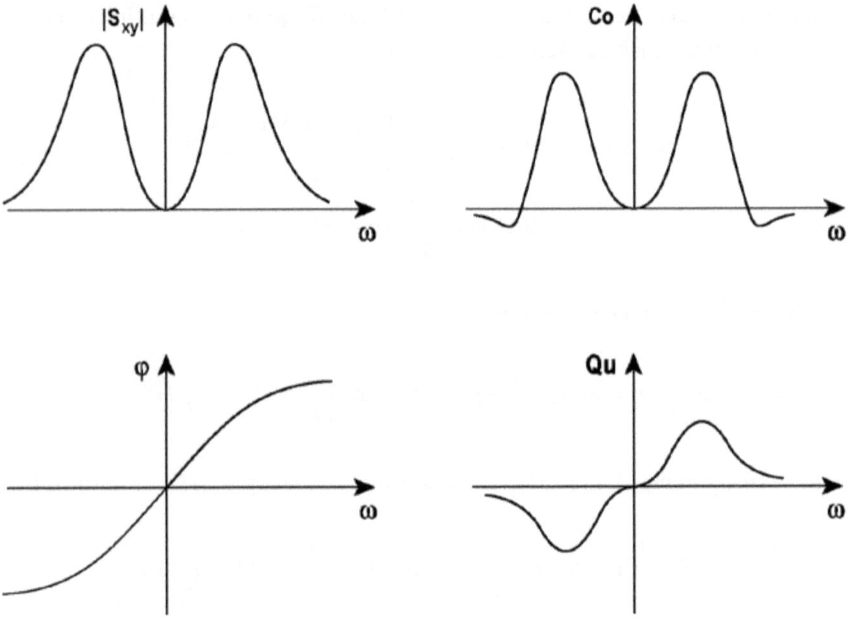

Fig. 2.12 *Cross spectrum decomposition into co-, quad-and phase spectra*

2.7 The connection between spectra and covariance

Auto-spectra $S_x(\omega)$ may also be calculated from the auto covariance function $Cov_x(\tau)$, see Eq. 2.15. Assuming that $x(t)$ is a stationary and zero mean stochastic variable, the following applies:

$$S_x(\omega) = \lim_{T\to\infty} \frac{E\left[X_k^* X_k\right]}{\Delta\omega} = \lim_{T\to\infty} \frac{E\left[\left(\frac{1}{T}\int_0^T x(t)e^{i\omega t}dt\right)\cdot\left(\frac{1}{T}\int_0^T x(t)e^{-i\omega t}dt\right)\right]}{2\pi/T}$$

$$= \lim_{T\to\infty} \frac{1}{2\pi T}\int_0^T\int_0^T E\left[x(t_1)\cdot x(t_2)\right]\cdot e^{-i\omega(t_2-t_1)}dt_1 dt_2$$

$$\Rightarrow S_x(\omega) = \lim_{T\to\infty} \frac{1}{2\pi T}\int_0^T\int_0^T Cov_x(t_2-t_1)\cdot e^{-i\omega(t_2-t_1)}dt_1 dt_2 \tag{2.80}$$

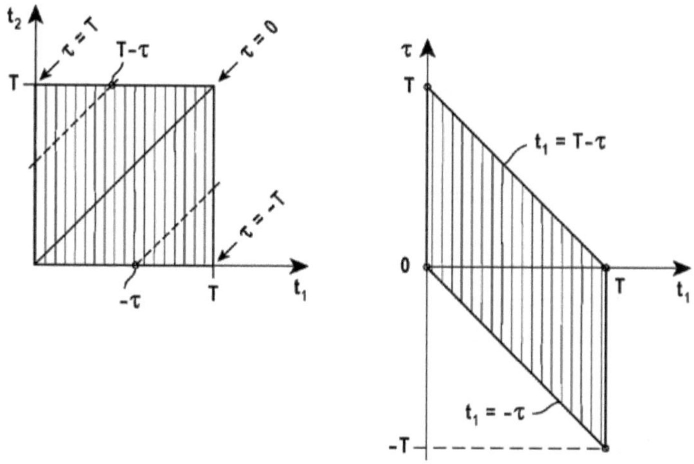

Fig. 2.13 *Substitution of variables and corresponding integration limits*

Replacing t_2 with $t_1 + \tau$, and changing the integration limits accordingly, implies (as illustrated in Fig. 2.13) that

$$\int_0^T \int_0^T dt_1 dt_2 = \int_{-T}^{0} \int_{-\tau}^{T} dt_1 d\tau + \int_0^T \int_0^{T-\tau} dt_1 d\tau \tag{2.81}$$

and thus

$$\begin{aligned} S_x(\omega) &= \lim_{T\to\infty} \frac{1}{2\pi T}\left[\int_{-T}^{0}\int_{-\tau}^{T} Cov_x(\tau)\cdot e^{-i\omega\tau} dt_1 d\tau + \int_0^T \int_0^{T-\tau} Cov_x(\tau)\cdot e^{-i\omega\tau} dt_1 d\tau\right] \\ &= \lim_{T\to\infty} \frac{1}{2\pi}\left[\int_{-T}^{0}\left(1+\frac{\tau}{T}\right)\cdot Cov_x(\tau)\cdot e^{-i\omega\tau} d\tau + \int_0^T \left(1-\frac{\tau}{T}\right) Cov_x(\tau)\cdot e^{-i\omega\tau} d\tau\right] \end{aligned}$$

$$\Rightarrow S_x(\omega) = \lim_{T\to\infty} \frac{1}{2\pi} \int_{-T}^{T} \left(1-\frac{|\tau|}{T}\right) Cov_x(\tau)\cdot e^{-i\omega\tau} d\tau \tag{2.82}$$

Provided the integral under the auto covariance function is finite, it is then seen that in the limit of $T \to \infty$, the following is obtained

$$S_x(\omega) = \frac{1}{2\pi} \int_{-\infty}^{+\infty} Cov_x(\tau)\cdot e^{-i\omega\tau} d\tau \tag{2.83}$$

This shows that the auto spectral density is the Fourier transform of the auto covariance function. Vice versa, it follows that the auto covariance function, which is the Fourier constant to the spectral density, is given by

$$Cov_x(\tau) = \int_{-\infty}^{+\infty} S_x(\omega) \cdot e^{i\omega\tau} d\omega \tag{2.84}$$

Similarly, the cross covariance function together with the cross spectral density will also constitute a pair of Fourier transforms:

$$S_{xy}(\omega) = \frac{1}{2\pi} \int_{-\infty}^{+\infty} Cov_{xy}(\tau) \cdot e^{-i\omega\tau} d\tau \quad \text{and} \quad Cov_{xy}(\tau) = \int_{-\infty}^{+\infty} S_{xy}(\omega) \cdot e^{i\omega\tau} d\omega \tag{2.85}$$

2.8 *Coherence function and normalized co-spectrum*

The coherence function is defined by

$$Coh_{xy}(\omega) = \frac{\left|S_{xy}(\omega)\right|^2}{S_x(\omega) \cdot S_y(\omega)} \tag{2.86}$$

If $x(t)$ and $y(t)$ are realisations of the same process, then $S_x(\omega) = S_y(\omega)$ and the cross-spectrum $S_{xy}(\omega) = S_{xx}(\omega)$ is given by

$$S_{xx}(\omega) = S_x(\omega) \cdot \sqrt{Coh_{xx}(\omega)} \cdot e^{i\varphi_{xx}(\omega)} \tag{2.87}$$

$\sqrt{Coh_{xx}(\omega)}$ is called the root–coherence function and φ_{xx} is the phase spectrum (see Eq. 2.79) . In the practical use of cross-spectra all imaginary parts will cancel out, and thus it is only the co-spectrum that is of interest. Therefore, a normalised co-spectrum is defined

$$\hat{C}o_{xy}(\omega) = \frac{\mathrm{Re}\left[S_{xy}(\omega)\right]}{\sqrt{S_x(\omega) \cdot S_y(\omega)}} \tag{2.88}$$

Again, if $x(t)$ and $y(t)$ are realisations of the same stationary and ergodic process, then $S_x(\omega) = S_y(\omega)$ and the real part of the cross-spectrum is given by

$$\operatorname{Re}\left[S_{xy}(\omega)\right] = S_x(\omega) \cdot \hat{C}o_{xy}(\omega) \tag{2.89}$$

2.9 *The spectral density of derivatives of processes*

It may in some cases be of interest to calculate the spectral density of the time derivatives [e.g. $\dot{x}(t)$ and $\ddot{x}(t)$] of processes. In structural engineering this is particularly relevant if $x(t)$ is a response displacement of such a character that it is necessary to evaluate as to whether or not it is acceptable with respect to human perception, in which case the design criteria most often will contain acceleration requirements. Since (see Eq. 2.59)

$$\dot{x}(t) = \sum_{-\infty}^{\infty} \dot{X}_k(\omega_k, t) = \sum_{-\infty}^{\infty} i\omega_k \cdot d_k(\omega_k) \cdot e^{i\,\omega_k t} \tag{2.90}$$

$$\ddot{x}(t) = \sum_{-\infty}^{\infty} \ddot{X}_k(\omega_k, t) = \sum_{-\infty}^{\infty} (i\omega_k)^2 \cdot d_k(\omega_k) \cdot e^{i \cdot \omega_k t} \tag{2.91}$$

and the double sided spectral density in general is given by the complex Fourier amplitude multiplied by its conjugated counterpart (see Eq. 2.62), then

$$\left.\begin{aligned} S_{\dot{x}}(\pm\omega_k) &= \frac{\left[i\omega_k d_k(\omega_k)\right]^* \cdot \left[i\omega_k d_k(\omega_k)\right]}{\Delta\omega} = \omega_k^2 \frac{d_k^* d_k}{\Delta\omega} = \omega_k^2 S_x(\pm\omega_k) \\ S_{\ddot{x}}(\pm\omega_k) &= \frac{\left[(i\omega_k)^2 d_k(\omega_k)\right]^* \cdot \left[(i\omega_k)^2 d_k(\omega_k)\right]}{\Delta\omega} = \omega_k^4 \frac{d_k^* d_k}{\Delta\omega} = \omega_k^4 S_x(\pm\omega_k) \end{aligned}\right\} \tag{2.92}$$

Similarly, cross spectral densities between a fluctuating displacement and its corresponding velocity and acceleration are given by

$$\left.\begin{aligned} S_{x\dot{x}}(\pm\omega_k) &= \frac{\left[d_k(\omega_k)\right]^* \cdot \left[i\omega_k d_k(\omega_k)\right]}{\Delta\omega} = i\omega_k \frac{d_k^* d_k}{\Delta\omega} = i\omega_k S_x(\pm\omega_k) \\ S_{x\ddot{x}}(\pm\omega_k) &= \frac{\left[d_k(\omega_k)\right]^* \cdot \left[(i\omega_k)^2 d_k(\omega_k)\right]}{\Delta\omega} = -\omega_k^2 \frac{d_k^* d_k}{\Delta\omega} = -\omega_k^2 S_x(\pm\omega_k) \\ S_{\dot{x}\ddot{x}}(\pm\omega_k) &= \frac{\left[i\omega_k d_k(\omega_k)\right]^* \cdot \left[(i\omega_k)^2 d_k(\omega_k)\right]}{\Delta\omega} = i\omega_k^3 \frac{d_k^* d_k}{\Delta\omega} = i\omega_k^3 S_x(\pm\omega_k) \end{aligned}\right\} \tag{2.93}$$

In the limit of T and $N \to \infty$ this may be written on the following continuous form

$$\begin{bmatrix} S_x(\pm\omega) & S_{x\dot{x}}(\pm\omega) & S_{x\ddot{x}}(\pm\omega) \\ & S_{\dot{x}}(\pm\omega) & S_{\dot{x}\ddot{x}}(\pm\omega) \\ sym. & & S_{\ddot{x}}(\pm\omega) \end{bmatrix} = \begin{bmatrix} 1 & i\omega & -\omega^2 \\ & \omega^2 & i\omega^3 \\ sym. & & \omega^4 \end{bmatrix} \cdot S_x(\pm\omega) \qquad (2.94)$$

Because $S_x(\pm\omega)$ is symmetric it is seen that for a stationary stochastic process

$$\begin{bmatrix} Cov_{x\dot{x}} \\ Cov_{\dot{x}\ddot{x}} \end{bmatrix} = \int_{-\infty}^{\infty} \begin{bmatrix} i\omega \\ i\omega^3 \end{bmatrix} \cdot S_x(\pm\omega) d\omega = \begin{bmatrix} 0 \\ 0 \end{bmatrix} \qquad (2.95)$$

Thus, the spectral density of time derivatives of processes may be obtained directly from the spectral density of the process itself. Since the single sided spectrum is simply twice the double sided, Eq. 2.94 will also hold if $S_x(\pm\omega)$, $S_{\dot{x}}(\pm\omega)$ and $S_{\ddot{x}}(\pm\omega)$ are replaced by $S_x(\omega)$, $S_{\dot{x}}(\omega)$ and $S_{\ddot{x}}(\omega)$.

From $S_x(\omega)$ and $S_{\dot{x}}(\omega)$ the average zero crossing frequency $f_x(0)$ of the process $x(t)$ may be found. Referring to Eq. 2.34, 2.56 and introducing $S_{\dot{x}}(\omega) = \omega^2 S_x(\omega)$, the following applies:

$$f_x(0) = \frac{1}{2\pi} \cdot \frac{\sigma_{\dot{x}}}{\sigma_x} = \frac{1}{2\pi} \cdot \left[\frac{\int_0^{\infty} \omega^2 S_x(\omega) d\omega}{\int_0^{\infty} S_x(\omega) d\omega} \right]^{1/2} = \frac{1}{2\pi} \cdot \sqrt{\frac{\mu_2}{\mu_0}} \qquad (2.96)$$

where for convenience the so–called n^{th} spectral moment $\mu_n = \int_0^{\infty} \omega^n \cdot S_x(\omega) d\omega$ has been introduced.

2.10 Spatial averaging in structural response calculations

A typical situation in structural engineering is illustrated in Fig. 2.14. A cantilevered tower–like beam is subject to a fluctuating short term (stationary) and distributed wind load. The problem at hand is to predict a load effect, e.g. the bending moment at the base. It is for simplicity assumed that the beam is so stiff that it is not necessary to include any dynamic amplification. It is taken for granted that the wind load may be split into a mean and a fluctuating part, i.e.

$$q_{y_{tot}} = \bar{q}_y(x) + q_y(x,t) \tag{2.97}$$

where $\bar{q}_y(x)$ is a deterministic quantity and $q_y(x,t)$ is a stochastic variable. Correspondingly, the load effect is split into a mean and a fluctuating part

$$M_{tot} = \bar{M} + M(t) \tag{2.98}$$

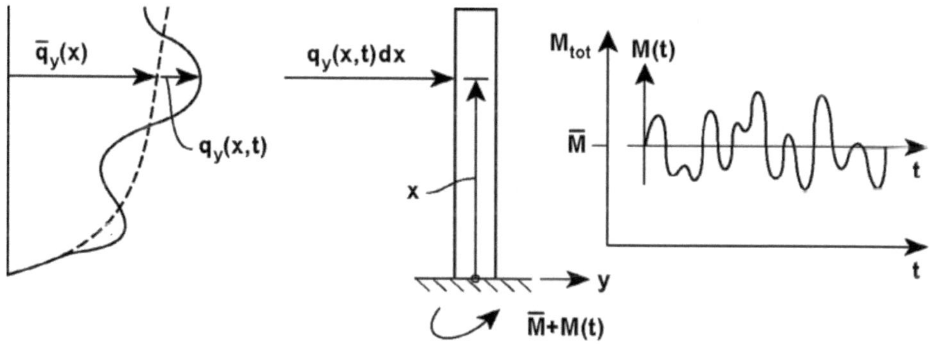

Fig. 2.14 *Cantilevered tower type of beam subject to fluctuating wind*

Since $\bar{M}$ may be obtained from $\bar{q}_y(x)$ alone it is then also a deterministic quantity. Thus, the prediction of $\bar{M}$ only involves the calculation of a simple static load effect, and it will not be pursued herein. The instantaneous value of $M(t)$ involves the same simple static calculation, but $M(t)$ is a stochastic variable, and it is only its statistical properties (i.e. its variance and auto spectral density) that can be predicted. From Fig. 2.14 it is readily seen that

$$M(t) = \int_0^L G_M(x) \cdot q_y(x,t) \cdot dx \tag{2.99}$$

where L is the total (or flow exposed) length of the beam and $G_M(x)$ is the influence function for the bending moment at the base [in this case $G_M(x) = x$]. The variance of $M(t)$ is then given by

$$\sigma_M^2 = \lim_{T\to\infty}\frac{1}{T}\int_0^T\left[M(t)\right]^2 dt = \lim_{T\to\infty}\frac{1}{T}\int_0^T\left[\int_0^L G_M(x)\cdot q_y(x,t)\cdot dx\right]^2 dt \tag{2.100}$$

It is desirable to perform statistics only on $q_y(x,t)$, as this is the only time domain variable on the right hand side of the equation. This may be obtained by splitting the squared integral into a product of two identical integrals, only made distinguishable by letting them contain different space variables, one labelled x_1 and the other x_2, Thus, the following is obtained

$$\begin{aligned}\sigma_M^2 &= \lim_{T\to\infty}\frac{1}{T}\int_0^T\left[\int_0^L G_M(x_1)\cdot q_y(x_1,t)\cdot dx_1\right]\cdot\left[\int_0^L G_M(x_2)\cdot q_y(x_2,t)\cdot dx_2\right]dt\\ &= \int_0^L\int_0^L G_M(x_1)\cdot G_M(x_2)\cdot\left[\lim_{T\to\infty}\frac{1}{T}\int_0^T q_y(x_1,t)\cdot q_y(x_2,t)dt\right]dx_1dx_2\end{aligned} \tag{2.101}$$

Recalling that the cross covariance function of $q_y(x,t)$ is given by

$$\begin{aligned}Cov_{q_yq_y}(\Delta x,\tau=0) &= \lim_{T\to\infty}\frac{1}{T}\int_0^T q_y(x,t)\cdot q_y(x+\Delta x,t)dt\\ &= \lim_{T\to\infty}\frac{1}{T}\int_0^T q_y(x_1,t)\cdot q_y(x_2,t)dt = Cov_{q_y}(\Delta x)\end{aligned} \tag{2.102}$$

where the separation $\Delta x = |x_2 - x_1|$, and introducing the covariance coefficient $\rho_{q_y}(\Delta x) = Cov_{q_y}(\Delta x)/\sigma_{q_y}^2$, it is seen that Eq. 2.101 simplifies into

$$\sigma_M^2 = \sigma_{q_y}^2\cdot\int_0^L\int_0^L G_M(x_1)\cdot G_M(x_2)\cdot\rho_{q_y}(\Delta x)dx_1dx_2 \tag{2.103}$$

The square root of the double integral

$$J_M = \left[\int_0^L\int_0^L G_M(x_1)\cdot G_M(x_2)\cdot\rho_{q_y}(\Delta x)dx_1dx_2\right]^{1/2} \tag{2.104}$$

is in wind engineering often called the joint acceptance function, because it contains the necessary statistical (i.e. variance) averaging in space. Thus,

$$\sigma_M = \sigma_{q_y} \cdot J_M \tag{2.105}$$

Similarly, the auto spectral density of $M(t)$ may be obtained by taking the Fourier transform on either side of Eq. 2.99

$$a_M(\omega) = \int_0^L G_M(x) \cdot a_{q_y}(x,\omega) \cdot dx \tag{2.106}$$

and applying Eq. 2.67

$$\begin{aligned} S_M(\omega) &= \lim_{T\to\infty} \frac{1}{\pi T} \cdot a_M^*(\omega) \cdot a_M(\omega) \\ &= \lim_{T\to\infty} \frac{1}{\pi T} \cdot \left(\int_0^L G_M(x) \cdot a_{q_y}^*(x,\omega) \cdot dx \right) \cdot \left(\int_0^L G_M(x) \cdot a_{q_y}(x,\omega) \cdot dx \right) \\ &= \int_0^L \int_0^L G_M(x_1) \cdot G_M(x_2) \cdot \left[\lim_{T\to\infty} \frac{1}{\pi T} \cdot a_{q_y}^*(x_1,\omega) \cdot a_{q_y}(x_2,\omega) \right] dx_1 dx_2 \\ \Rightarrow S_M(\omega) &= \int_0^L \int_0^L G_M(x_1) \cdot G_M(x_2) \cdot S_{q_y}(\Delta x,\omega) dx_1 dx_2 \end{aligned} \tag{2.107}$$

where $S_{q_y}(\Delta x,\omega)$ is the cross spectral density of the fluctuating part $q_y(x,t)$ of the distributed load, and $\Delta x = |x_2 - x_1|$ is spatial separation. Integrating over the entire frequency domain will then render the variance of $M(t)$:

$$\begin{aligned} \sigma_M^2 &= \int_0^\infty S_M(\omega) df = \int_0^\infty \left[\int_0^L \int_0^L G_M(x_1) \cdot G_M(x_2) \cdot S_{q_y}(\Delta x,\omega) dx_1 dx_2 \right] d\omega \\ &= \sigma_{q_y}^2 \cdot \int_0^\infty \left[\int_0^L \int_0^L G_M(x_1) \cdot G_M(x_2) \cdot \hat{S}_{q_y}(\Delta x,\omega) dx_1 dx_2 \right] d\omega \end{aligned} \tag{2.108}$$

where σ_{q_y} is defined above and $\hat{S}_{q_y}(\Delta x,\omega)$ is the normalised (but not non–dimensional) version of $S_{q_y}(\Delta x,\omega)$, i.e.

$$\hat{S}_{q_y}(\Delta x,\omega) = S_{q_y}(\Delta x,\omega) / \sigma_{q_y}^2 \tag{2.109}$$

Introducing $\hat{x} = x/L$ and correspondingly $\Delta\hat{x} = |\hat{x}_1 - \hat{x}_2|$, then a normalised frequency domain version of the joint acceptance function may be defined by

$$\hat{J}_M(\omega) = \left[\int_0^1 \int_0^1 G_M(\hat{x}_1) \cdot G_M(\hat{x}_2) \cdot \hat{S}_{q_y}(\Delta\hat{x}, \omega) d\hat{x}_1 d\hat{x}_2 \right]^{1/2} \tag{2.110}$$

in which case the following is obtained:

$$\sigma_M = \sigma_{q_y} \cdot L \cdot \left[\int_0^\infty \hat{J}_M^2(\omega) d\omega \right]^{1/2} \tag{2.111}$$

Under ideal conditions Eqs. 2.111 and 2.105 should render identical results. Obviously, Eq. 2.105 is the simpler choice, as 2.111 contains frequency domain integration as well as spatial averaging.

The necessity of a transition from the product of two line integrals into a volume integral in Eq. 2.101 (and similarly in Eq. 2.107), is better understood if the integral is replaced by a summation, as illustrated in Fig. 2.15.

I.e., the load is split into N concentrated loads

$$Q_k(t) = q_y(x_k, t) \cdot \Delta x \tag{2.112}$$

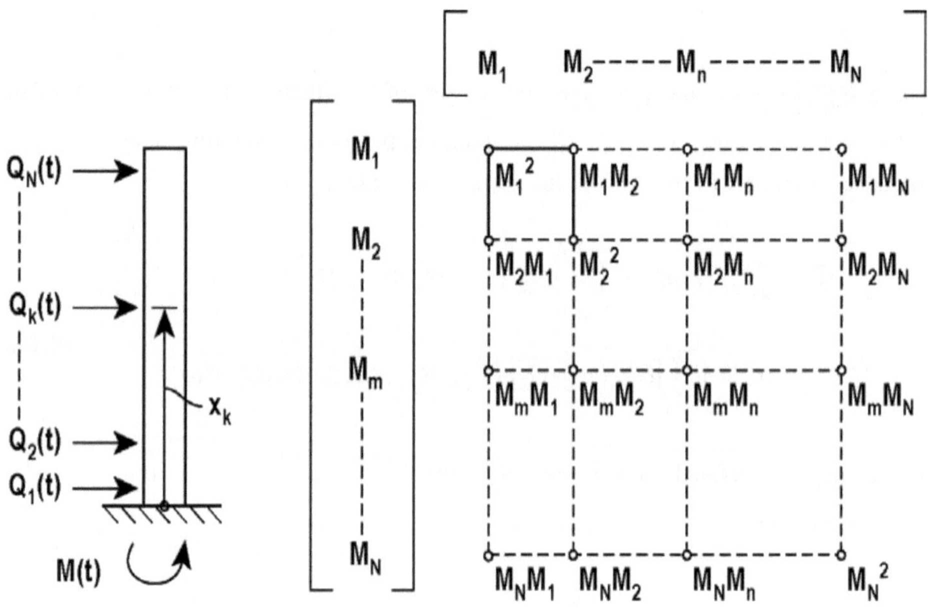

Fig. 2.15 *Calculation of base moment in cantilevered tower type of beam subject to fluctuating wind*

(see Fig. 2.14 and Eq. 2.99, and assuming for simplicity that Δx is a constant), each rendering a base moment contribution

$$M_k(t) = G_M(x_k) \cdot Q_k(t) \tag{2.113}$$

such that the total bending moment at the base is

$$M(t) = \sum_{k=1}^{N} M_k(t) \tag{2.114}$$

Its variance is then given by (see Eq. 2.100)

$$\sigma_M^2 = \lim_{T\to\infty} \frac{1}{T} \int_0^T [M(t)]^2 dt = \lim_{T\to\infty} \frac{1}{T} \int_0^T \left[\sum_{k=1}^{N} M_k(t) \right]^2 dt$$

$$= \lim_{T\to\infty} \frac{1}{T} \int_0^T [M_1 + \ldots + M_k + \ldots + M_N] \cdot [M_1 + \ldots + M_k + \ldots + M_N] dt$$

$$\Rightarrow \sigma_M^2 = \sum_{n=1}^{N} \sum_{m=1}^{N} \lim_{T\to\infty} \frac{1}{T} \int_0^T [M_n(t) \cdot M_m(t)] dt \tag{2.115}$$

As can be seen, the transition from a single summation to a double summation is necessary to capture all the cross products. Introducing Eqs. 2.112 and 2.113, the following is obtained:

$$\sigma_M^2 = \sum_{n=1}^{N} \sum_{m=1}^{N} \lim_{T\to\infty} \frac{1}{T} \int_0^T \left[G_M(x_n) q_y(x_n,t) \frac{L}{N} \cdot G_M(x_m) q_y(x_m,t) \frac{L}{N} \right] dt$$

$$= \left\{ \sum_{n=1}^{N} \sum_{m=1}^{N} G_M(x_n) \cdot G_M(x_m) \cdot \lim_{T\to\infty} \frac{1}{T} \int_0^T [q_y(x_n,t) \cdot q_y(x_m,t)] dt \right\} \cdot \left(\frac{L}{N} \right)^2$$

$$\Rightarrow \sigma_M^2 = \left(\frac{\sigma_{q_y} \cdot L}{N} \right)^2 \cdot \left\{ \sum_{n=1}^{N} \sum_{m=1}^{N} G_M(x_n) \cdot G_M(x_m) \cdot \rho_{q_y}(\Delta x) \right\} \tag{2.116}$$

where $\rho_{q_y}(\Delta x)$ is the covariance coefficient to the distributed load, and where $\Delta x = |x_m - x_n|$. The expression in Eq. 2.116 is equivalent to that which was obtained in Eq. 2.103.

Example 2.4

Considering the cantilevered beam shown in Fig. 2.15, then the reduced variance of the base moment fluctuations is given by:

$$\left(\frac{\sigma_M}{\sigma_{q_y} L^2}\right)^2 = \frac{1}{N^2} \sum_{n=1}^{N} \sum_{m=1}^{N} \hat{G}_M(\hat{x}_n) \cdot \hat{G}_M(\hat{x}_m) \cdot \rho_{q_y}(\Delta \hat{x})$$

where $\hat{G}_M(\hat{x}) = G_M/L = x/L = \hat{x}$, $\Delta x = |x_m - x_n|$ and $\Delta \hat{x} = \Delta x/L$. Assuming that the covariance coefficient $\rho_{q_y}(\Delta x) = \exp\left(-\Delta x/{}^zL_u\right)$ and setting for simplicity ${}^zL_u = L$, then $\rho_{q_y}(\Delta x) = \exp(-\Delta \hat{x})$. Choosing a reduced integration increment $\Delta x/L = 0.2$ and corresponding position vector $\hat{x} = [0.1 \quad 0.3 \quad 0.5 \quad 0.7 \quad 0.9]^T$ then the influence function multiplications $\hat{G}_M(\hat{x}_n) \cdot \hat{G}_M(\hat{x}_m)$ are given by

$\hat{G}_M(\hat{x}_n) \cdot \hat{G}_M(\hat{x}_m)$:

$\hat{G}_M(\hat{x}_m)$ \ $\hat{G}_M(\hat{x}_n)$	0.1	0.3	0.5	0.7	0.9
0.1	0.01	0.03	0.05	0.07	0.09
0.3	0.03	0.09	0.15	0.21	0.27
0.5	0.05	0.15	0.25	0.35	0.45
0.7	0.07	0.21	0.35	0.49	0.63
0.9	0.09	0.27	0.45	0.63	0.81

The covariance coefficient $\rho_{q_y}(\Delta \hat{x})$ is given by:

$\rho_{q_y}(\Delta \hat{x})$:

$\hat{x}_m$ \ $\hat{x}_n$	0.1	0.3	0.5	0.7	0.9
0.1	1	0.82	0.67	0.55	0.45
0.3	0.82	1	0.82	0.67	0.55
0.5	0.67	0.82	1	0.82	0.67
0.7	0.55	0.67	0.82	1	0.82
0.9	0.45	0.55	0.67	0.82	1

The inner product $\hat{G}_M(\hat{x}_n)\cdot\hat{G}_M(\hat{x}_m)\cdot\rho_{q_y}(\Delta\hat{x})$ is then:

m \ n	1	2	3	4	5
	$\hat{G}_M(\hat{x}_n)\cdot\hat{G}_M(\hat{x}_m)\cdot\rho_{q_y}(\Delta\hat{x})$				
1	0.01	0.025	0.034	0.039	0.041
2	0.025	0.09	0.123	0.141	0.149
3	0.034	0.123	0.25	0.287	0.302
4	0.039	0.141	0.287	0.49	0.517
5	0.041	0.149	0.302	0.517	0.81

As can be seen, the inner product $\hat{G}_M(\hat{x}_n)\cdot\hat{G}_M(\hat{x}_m)\cdot\rho_{q_y}(\Delta\hat{x})$ is symmetric about the diagonal $m=n$ and increasing with increasing distance from the base of the beam. The reduced variance of the base moment fluctuations is given by:

$$\left(\frac{\sigma_M}{\sigma_{q_y}L^2}\right)^2=\frac{1}{5^2}\left[0.01+0.09+0.25+0.49+0.81+2\cdot(0.025+0.034+0.039+0.041\right.$$

$$\left.+0.123+0.141+0.149+0.287+0.302+0.517)\right]\approx\underline{\underline{0.2}}$$

Thus, the standard deviation of the base moment is given by

$$\Rightarrow\sigma_M\approx0.45\cdot\sigma_{q_y}\cdot L^2$$

(See further calculations in Chapter 7.3.)

Chapter 3

STOCHASTIC DESCRIPTION OF TURBULENT WIND

The description of the wind field given below is only intended to provide the theoretical basis that is necessary for the ensuing calculations of structural response. More comprehensive descriptions have been presented by Simiu & Scanlan [4] and by Dyrbye & Hansen [5], where guidelines with respect to the choice of typical input parameters to the stochastic description of the wind field may be found. Such information has also been given by Solari & Piccardo [6]. The most comprehensive source of wind engineering data is provided by Engineering Science Data Unit [7]. Basic theory of turbulence may be found in many text books, see e.g. Batchelor [8] and Tennekes & Lumley [9]. As shown in Fig. 1.3.a the wind velocity vector at a certain point is described by its components (see Eqs. 1.2 – 1.4) in the Cartesian coordinate system $(x,y,z)_f$ with x_f in the main flow direction and z_f in the vertical direction. It is taken for granted that the wind field met by the structure is stationary and homogeneous within the time and space that is considered. A statistical description of the wind field comprises three levels: the long term variation of the mean wind velocity, the short term single point time domain variation of the turbulence components, and finally, the short term spatial distribution of the turbulence components.

3.1 Mean wind velocity

The statistical properties of the mean wind velocity $V(z_f)$ are required in order to establish a basis for the calculation of structural design load effects during the weather conditions that have been deemed representative for the purpose of obtaining sufficient safety against structural failure. A design check with respect to ultimate structural strength will only require information regarding the wind field properties under a characteristic extreme weather condition, but the properties under several representative weather conditions are required if vortex shedding may occur. If a fatigue design check is relevant even further information is required with respect to the wind climate on the construction site. Thus, mean wind statistics must be based on data covering numerous meteorological observations over several years, as it is the values of $V(z_f)$ under a large variation of weather conditions that are of interest (or ideally under any possible weather condition at the site in question). Such statistics are usually performed on the mean wind velocity at $z_f = 10$ m and averaged over a period of $T = 10$ min. A typical instantaneous wind velocity profile in the main flow direction is illustrated on the left hand side of Fig. 3.1, together with the mean velocity and turbulence variation with z_f.

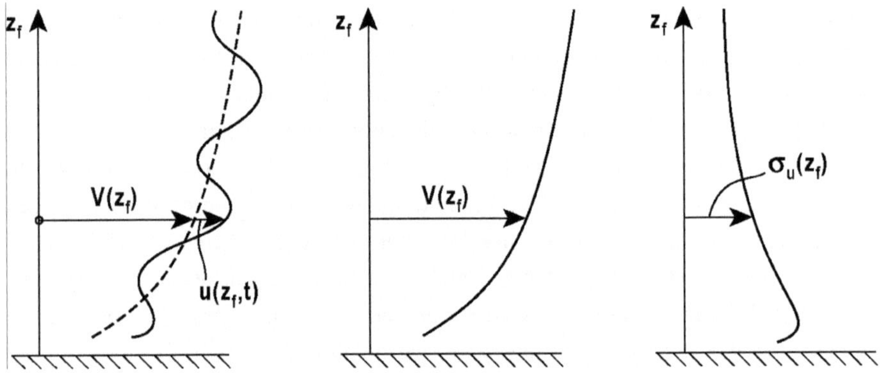

Fig. 3.1 *The wind velocity and turbulence profiles*

A theoretical approach renders a natural logarithmic profile for the height variation of the mean wind velocity (first shown by Millikan [30])

$$\frac{V_{10}(z_f)}{V_{10}(10)} = \begin{cases} k_T \cdot \ln\left(\dfrac{z_f}{z_0}\right) & \text{when } z_f > z_{min} \\ k_T \cdot \ln\left(\dfrac{z_{min}}{z_0}\right) & \text{when } z_f \le z_{min} \end{cases} \tag{3.1}$$

where the index 10 has been added to V, indicating an averaging period of 10 minutes, while k_T, z_0 and z_{min} are parameters characteristic to the terrain in question.

The height z_{min} has been introduced because such a velocity profile has a limited validity close to the ground, where turbulence and directional effects prevail.

z_0 is usually called the roughness length. It coincides with the height at which the velocity variation according to Eq. 3.1 is zero. Typical values of k_T and z_0 varies from about 0.15 and 0.01 for open sea and countryside without obstacles to about 0.25 and 1.0 for built up urban areas. Corresponding values of $z_{\min}$ varies between 2 and about 15 m. (Other profiles, e.g. the power law profile, may be found in the literature.)

Any statistical properties related to the mean wind velocity is in the following associated with $V_{10}\left(z_{ref}\right)$, where z_{ref} is a chosen reference height. In general, $z_{ref} = 10$ m as mentioned above, but for a bridge whose main girder is located at a certain height above the sea or terrain, z_{ref} will often be chosen at this height. To simplify notations $V_{10}\left(z_{ref}\right)$ is set equal to V_r or V_a for the remaining part of this chapter. The indexes r and a indicate whether the relevant statistical calculations have been performed on the parent population or on a reduced population of annual maxima. Data from a large population of parent observations may usually be fitted to a Weibull distribution, i.e. the cumulative and corresponding density distributions are given by

$$\left.\begin{aligned} P_r\left(V_r \le V, \varphi\right) &= 1 - \alpha(\varphi) \cdot \exp\left\{-\left[\frac{V}{\beta(\varphi)}\right]^{\gamma(\varphi)}\right\} \\ p_r\left(V, \varphi\right) &= \frac{dP_r}{dV} = \frac{\alpha(\varphi) \cdot \gamma(\varphi)}{\beta(\varphi)} \cdot \left[\frac{V}{\beta(\varphi)}\right]^{\gamma(\varphi)-1} \cdot \exp\left\{-\left[\frac{V}{\beta(\varphi)}\right]^{\gamma(\varphi)}\right\} \end{aligned}\right\} \tag{3.2}$$

where φ is the main flow direction and $\alpha(\varphi)$ and $\beta(\varphi)$ are parameters to be fitted to the relevant data. If the directionality effect is omitted, i.e. for omni-directional wind, the data may usually be fitted to a Rayleigh density distribution

$$p_r(V) = \frac{V}{V_m^2} \cdot \exp\left[-\frac{1}{2}\left(\frac{V}{V_m}\right)^2\right] \tag{3.3}$$

where V_m is the velocity at the apex of the distribution, as illustrated in Fig. 3.2. Thus, the probability μ of exceeding a certain limiting value V_s (see Fig. 3.2) is given by

$$\mu = P_r(V_r > V_s) = 1 - P_r(V_r \le V_s) = \exp\left[-\frac{1}{2}\left(\frac{V_s}{V_m}\right)^2\right]$$
$$\Rightarrow V_s = V_m\sqrt{-2\ln\mu} \tag{3.4}$$

Taken from the entire population of observations, independent of direction, V_s is then the velocity that has a probability μ of being exceeded.

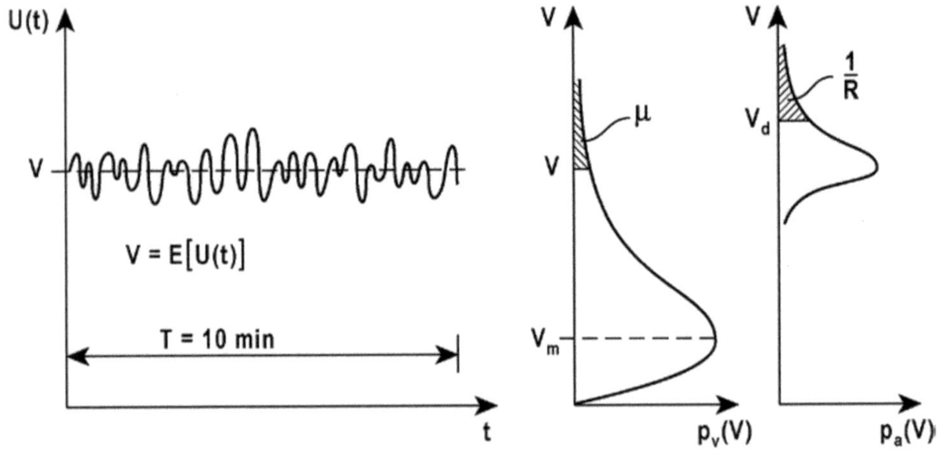

Fig. 3.2 *The probability density distribution of the mean wind velocity*

For a suitably small value of μ, V_s may be interpreted as what can be expected to be representative under severe weather conditions on the site. However, this is usually not considered the appropriate procedure for singling out a characteristic mean wind velocity for design checks against ultimate structural failure. For the purpose of structural design, the mean wind velocity that corresponds to an extreme weather condition with a certain small probability of occurrence is rather based on a limited data set of annual maxima, V_a, as illustrated on the right hand side of Fig. 3.2. This data is usually dealt with in the form of the mean wind velocity pressure $q_V = \rho V^2/2$ and fitted to a Fischer – Tippet Type I distribution

$$P_a\left(q_{V_a} \le q_V\right) = \exp\left[-\exp\left(-\frac{q_V - \alpha}{\beta}\right)\right] \tag{3.5}$$

where $\alpha = \bar{q}_V - \gamma \cdot \beta$ and $\beta = \left(\sqrt{6}/\pi\right) \cdot \sigma_q$, and where $\bar{q}_V$ is the mean value of the velocity pressure recordings, σ_q is the corresponding standard deviation and $\gamma \approx 0.5772$ is the Euler constant. α and β are parameters that are characteristic to the distribution of the recorded data. If the return period R_p is defined as the average number of years between rare q_{V_a} events, then a small probability μ of exceeding a certain limiting design value q_{V_d}

$$\mu = P_a\left(q_{V_a} > q_{V_d}\right) = 1 - P_a\left(q_{V_a} \le q_{V_d}\right) \approx \frac{1}{R_p} \tag{3.6}$$

and thus, the q_{V_d} that corresponds to such a return period is given by

$$\left.\begin{aligned} & 1 - \frac{1}{R_p} = P_a\left(q_{V_a} \le q_{V_d}\right) = \exp\left[-\exp\left(-\frac{q_{V_d} - \alpha}{\beta}\right)\right] \\ & \Rightarrow q_{V_d}\left(R_p\right) = \alpha - \beta \cdot \ln\left[-\ln\left(1 - 1/R_p\right)\right] \end{aligned}\right\} \tag{3.7}$$

It is the mean wind velocity V_d that corresponds to such a value of q_{V_d} that is used as a representative basis for the design of structures. R_p is in general subject to standardisation, e.g. $R_p = 50$ years, in which case $q_{V_d}(50) \approx \alpha + 3.9 \cdot \beta$. The ratio $\beta/\alpha \approx 0.2$ is frequently encountered in the literature. Since $q_{V_d} = \rho V_d^2/2$, then a change from $R_p = 50$ to another return period is given by

$$V_d\left(R_p\right)/V_d(50) \approx \sqrt{\left\{1 - (\beta/\alpha) \cdot \ln\left[-\ln\left(1 - 1/R_p\right)\right]\right\}/(1 + 3.9 \cdot \beta/\alpha)} \tag{3.8}$$

While the above considerations are concerned with the statistical properties of annual maxima, it should be mentioned that within any short term (10 min.) stationary weather window at high wind velocities it is possible to estimate an extreme value of the velocity fluctuations. For instance, at any chosen characteristic design value $V_d\left(R_p\right)$, the corresponding extreme value may be obtained by a simple linearization and the broad band type of process considerations shown in chapter 2.4. Since the instantaneous velocity pressure

$$q_u(t) = \frac{1}{2}\rho\left[U(t)\right]^2 = \frac{1}{2}\rho\left[V + u(t)\right]^2 - \frac{1}{2}\rho V^2\left\{1 + 2u(t)/V + \left[u(t)/V\right]^2\right\} \tag{3.9}$$

at low turbulence and high values of V can be approximated by

$$q_u(t) \approx \frac{1}{2}\rho V^2 \left[1 + 2\frac{u(t)}{V}\right] \tag{3.10}$$

it is seen that the mean value of q_u is $\bar{q}_u = q_V = \rho V^2/2$ while the fluctuating part is $\rho V u(t)$. The standard deviation of the velocity pressure is then $\sigma_{q_u} = \rho V \sigma_u$, where σ_u is the standard deviation of the along wind turbulence component. Thus, an extreme value of q_u may be obtained by

$$q_{u_{\max}} = \frac{1}{2}\rho V_{\max}^2 = \bar{q}_u + k_p \sigma_{q_u} \tag{3.11}$$

where k_p is a peak factor (see chapter 2.4, Eq. 2.45). The following is then obtained:

$$\frac{1}{2}\rho V_{\max}^2 = \frac{1}{2}\rho V^2 + k_p \rho V \sigma_u \qquad \Rightarrow V_{\max} = V\sqrt{1 + 2k_p\,\sigma_u/V} \tag{3.12}$$

3.2 *Single Point Statistics of Wind Turbulence*

While we in the previous chapter were dealing with long term statistics of ten minutes mean values, i.e. performing statistics on a data base covering many years of observations of V, we shall now return to short term statistics on the fluctuating flow components $u(t)$, $v(t)$ and $w(t)$. It is single point recordings of these variables within a stationary period of T=10 min that provide the source for determination of their time and frequency domain statistical properties. The sampling frequency within this period is in the following assumed to be large, rendering a sufficiently large data base for the extraction of reliable results. As shown in chapter 1.3, at a certain point $(x,y,z)_f$, e.g. at $z_f = 10m$ or at a reference point relevant to the structure in question, it is assumed that $U(t) = V + u(t)$, and that the turbulence components $u(t), v(t)$ and $w(t)$ are stationary and have zero mean values.

For the along wind u component the situation is illustrated in Fig. 3.3. It is taken for granted that statistics performed on time series recordings of each of the turbulence components will render three individual zero mean Gaussian probability density distributions with variances

$$\begin{bmatrix} \sigma_u^2 \\ \sigma_v^2 \\ \sigma_w^2 \end{bmatrix} = \frac{1}{T}\int_0^T \begin{bmatrix} u^2(t) \\ v^2(t) \\ w^2(t) \end{bmatrix} dt \tag{3.13}$$

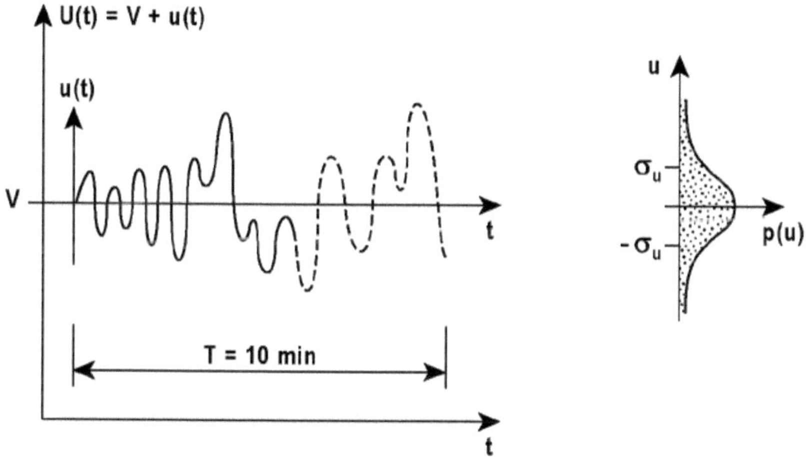

Fig. 3.3 *The probability distribution of the along–wind turbulence component*

The corresponding turbulence intensities are defined by

$$I_n(z_f) = \frac{\sigma_n(z_f)}{V(z_f)} \qquad \text{where} \qquad n = u, v, w \tag{3.14}$$

A typical variation of the turbulence intensity for the along wind u component is given by

$$I_u(z_f) \approx \begin{cases} 1/\ln(z_f/z_0) \text{ when } z_f > z_{\min} \\ 1/\ln(z_{\min}/z_0) \text{ when } z_f \le z_{\min} \end{cases} \tag{3.15}$$

where z_0 and $z_{\min}$ are defined in Eq. 3.1. Under isotropic conditions (e.g. high above the ground) $I_u \approx I_v \approx I_w$. In homogeneous terrain up to a height of about 200 m and not unduly close to the ground

$$\begin{bmatrix} I_v \\ I_w \end{bmatrix} \approx \begin{bmatrix} 3/4 \\ 1/2 \end{bmatrix} \cdot I_u \tag{3.16}$$

The auto covariance functions and corresponding auto covariance coefficients (see chapter 2.2) are defined by

$$\begin{bmatrix} Cov_u(\tau) \\ Cov_v(\tau) \\ Cov_w(\tau) \end{bmatrix} = \begin{bmatrix} E[u(t)\cdot u(t+\tau)] \\ E[v(t)\cdot v(t+\tau)] \\ E[w(t)\cdot w(t+\tau)] \end{bmatrix} = \frac{1}{T}\int_0^T \begin{bmatrix} u(t)\cdot u(t+\tau) \\ v(t)\cdot v(t+\tau) \\ w(t)\cdot w(t+\tau) \end{bmatrix} dt \tag{3.17}$$

$$\rho_n(\tau) = \frac{Cov_n(\tau)}{\sigma_n^2} \qquad \text{where} \qquad n = u,v,w \tag{3.18}$$

where τ is an arbitrary time lag that theoretically can take any value within $\pm T$. At $\tau = 0$ Eq. 3.17 becomes identical to 3.13, and thus

$$\rho_n(\tau = 0) = 1 \quad \text{where} \quad n = u,v,w \tag{3.19}$$

At increasing values of τ the auto covariance of the turbulence components diminish, and at large values of τ they asymptotically approach zero, i.e.

$$\lim_{\tau \to \infty} \rho_n(\tau) = 0 \quad \text{where} \quad n = u,v,w \tag{3.20}$$

As shown in Eq. 2.19,

$$Cov_n(\tau) = Cov_n(-\tau) \quad \text{where} \quad n = u,v,w \tag{3.21}$$

implying that also $\rho_n(\tau)$ is symmetric. A principal variation of the covariance coefficient for the along wind turbulence component is shown in Fig. 3.4. The time scale

$$T_n = \int_0^\infty \rho_n(\tau)\,d\tau \qquad \text{where} \qquad n = u,v,w \tag{3.22}$$

may be interpreted as the average duration of a u, v or w wind gust. Although the covariance coefficient in many practical cases may become negative at large values of τ it is a usual approximation to adopt

$$\rho_n(\tau) = \exp(-\tau / T_n) \qquad \text{where} \quad n = u, v, w \tag{3.23}$$

In homogeneous terrain, at heights below 100 m, T_u is usually in the range between 5 and 20 s, while T_v and T_w are in the ranges 2 – 5 and 0 – 2 s.

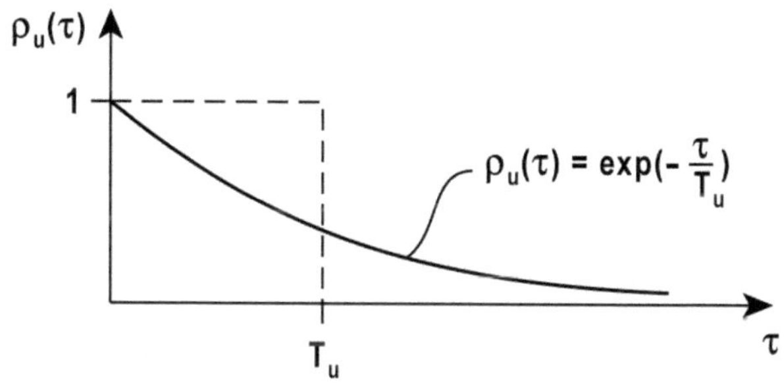

Fig. 3.4 *Auto covariance coefficient for the along–wind turbulence component*

Adopting Taylor's hypothesis that turbulence convection in the main flow direction takes place with the mean wind velocity (i.e. that flow disturbances travel with the average velocity V), then the average length scales of u, v and w in the x_f direction are given by

$$^{x_f}L_n = V \cdot T_n = V \cdot \int_0^\infty \rho_n(\tau) d\tau \qquad \text{where} \quad n = u, v, w \tag{3.24}$$

These turbulence length scales may be interpreted as the average eddy size of the u, v and w components in the direction of the main flow.

While auto covariance functions (or coefficients) represent the time domain properties of the turbulence components, it is the spectral densities that describe their frequency domain properties. In the literature many different expressions have been suggested to fit data from a variety of full scale recordings. The following non dimensional expression proposed by Kaimal et. al. [10] is often encountered in the literature:

$$\frac{f \cdot S_n\{f\}}{\sigma_n^2} = \frac{A_n \cdot \hat{f}_n}{\left(1 + 1.5 \cdot A_n \cdot \hat{f}_n\right)^{5/3}} \quad \text{where} \quad n = u, v, w \tag{3.25}$$

and where $\hat{f}_n = f \cdot {}^{xf}L_n / V$, and ${}^{xf}L_n$ is the integral length scale of the relevant turbulence component, as defined in Eq. 3.24 above.

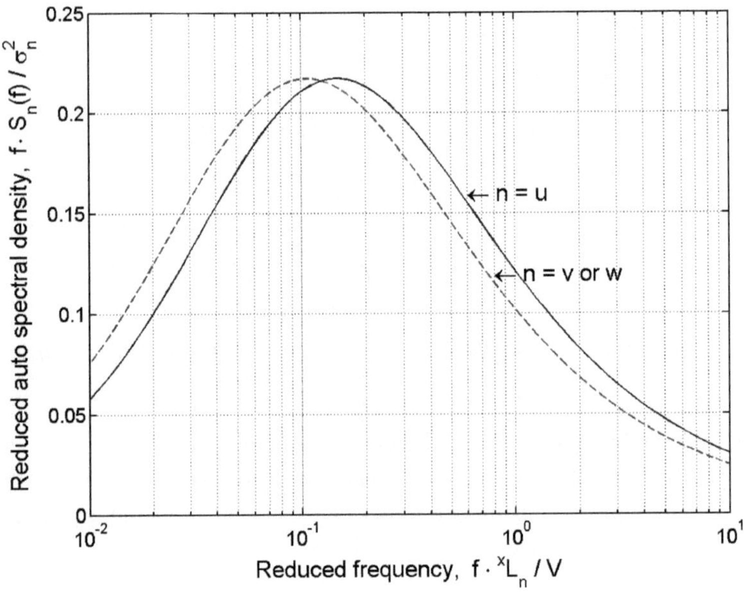

Fig. 3.5 *Kaimal auto spectra of turbulence components*

Unless full scale recordings indicate otherwise, the following values of the parameter A_n may be adopted: $A_u = 6.8$, $A_v = A_w = 9.4$. With these parameters, Eq. 3.25 has been plotted in Fig. 3.5. Alternatively, the von Kármán [11] spectra

$$\frac{f \cdot S_u\{f\}}{\sigma_u^2} = \frac{4 \cdot \hat{f}_u}{\left(1 + 70.8 \cdot \hat{f}_u^2\right)^{5/6}} \tag{3.26}$$

$$\frac{f \cdot S_n\{f\}}{\sigma_n^2} = \frac{4\hat{f}_n \cdot \left(1 + 755.2 \cdot \hat{f}_n^2\right)}{\left(1 + 283.2 \cdot \hat{f}_n^2\right)^{11/6}} \quad , \quad n = v,w \tag{3.27}$$

have the advantage that they contain only the length scales ${}^{xf}L_n$ that require fitting to the relevant data.

3.3 *The spatial properties of wind turbulence*

The spatial properties of wind turbulence are obtained from simultaneous two point recordings of the u, v and w components. It is taken for granted that the flow is homogeneous in space as well as stationary in time. Defining two vectors

$$\mathbf{u}_a = \begin{bmatrix} u(s,t) \\ v(s,t) \\ w(s,t) \end{bmatrix} \quad \text{and} \quad \mathbf{u}_b = \begin{bmatrix} u(s+\Delta s,t+\tau) \\ v(s+\Delta s,t+\tau) \\ w(s+\Delta s,t+\tau) \end{bmatrix} \tag{3.28}$$

where $s = x_f, y_f$ or z_f, τ is a time lag that theoretically can take any value within $\pm T$ and Δs is an arbitrary separation (between the two recordings) in the x_f, y_f or z_f directions. Thus, the following three by three covariance matrix may be defined

$$\mathbf{Cov}(\Delta s,\tau) = \begin{bmatrix} Cov_{uu} & Cov_{uv} & Cov_{uw} \\ Cov_{vu} & Cov_{vv} & Cov_{vw} \\ Cov_{wu} & Cov_{wv} & Cov_{ww} \end{bmatrix} = E\left[\mathbf{u}_a \cdot \mathbf{u}_b^T\right] = \frac{1}{T}\int_0^T \left(\mathbf{u}_a \cdot \mathbf{u}_b^T\right) dt \tag{3.29}$$

where all the relevant covariance functions

$$Cov_{mn}(\Delta s,\tau) \qquad \begin{cases} m,n = u,v,w \\ \Delta s = \Delta x_f, \Delta y_f, \Delta z_f \end{cases} \tag{3.30}$$

may contain separation in an arbitrary direction $s = x_f, y_f$ or z_f.

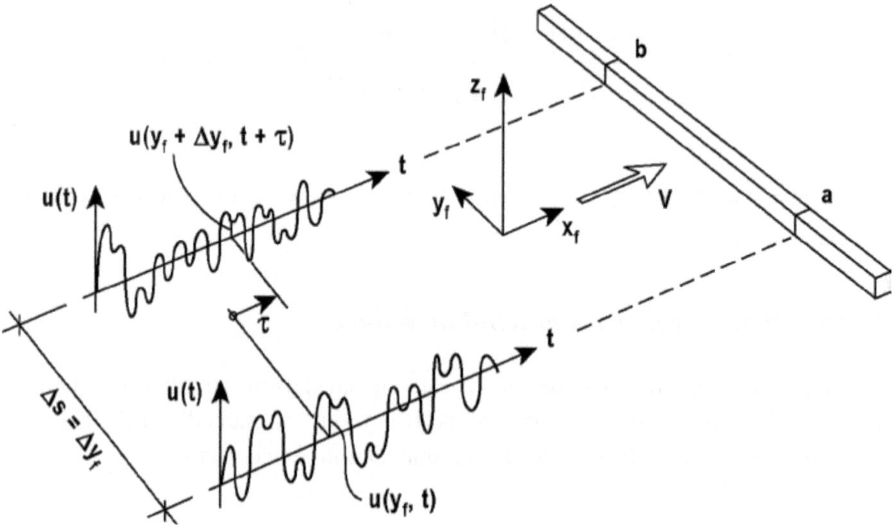

Fig. 3.6 *Cross covariance of along–wind u component*

The corresponding covariance coefficients are defined by

$$\rho_{mn}(\Delta s,\tau) = \frac{Cov_{mn}(\Delta s,\tau)}{\sigma_m \cdot \sigma_n} \qquad \begin{cases} m,n = u,v,w \\ \Delta s = \Delta x_f, \Delta y_f, \Delta z_f \end{cases} \tag{3.31}$$

(If the process is not ergodic, then σ_m should be taken at position s, while σ_n should be taken at $s+\Delta s$.) The situation is illustrated in Fig. 3.6 with $m=u$ and $s=y_f$, which is most relevant for a horizontal structure where time series of the turbulence components have been recorded at various positions *a* and *b* along the span of the structure. As can be seen from Eq. 3.29 (and 3.30), there are 27 possible covariance functions. However, it is a usual assumption that cross covariance between two different turbulence components may be neglected, at least beyond a certain distance above the ground. All off-diagonal terms in Eq. 3.29 will then become zero, and the number of possible covariance estimates is reduced to nine:

$$\begin{bmatrix} Cov_{uu}(\Delta s,\tau) \\ Cov_{vv}(\Delta s,\tau) \\ Cov_{ww}(\Delta s,\tau) \end{bmatrix} = E\begin{bmatrix} u(s,t)\cdot u(s+\Delta s,t+\tau) \\ v(s,t)\cdot v(s+\Delta s,t+\tau) \\ w(s,t)\cdot w(s+\Delta s,t+\tau) \end{bmatrix} = \frac{1}{T}\int_0^T \begin{bmatrix} u(s,t)\cdot u(s+\Delta s,t+\tau) \\ v(s,t)\cdot v(s+\Delta s,t+\tau) \\ w(s,t)\cdot w(s+\Delta s,t+\tau) \end{bmatrix} dt \tag{3.32}$$

where $s = x_f, y_f$ or z_f. The corresponding covariance coefficients are defined by

$$\rho_{nn}(\Delta s, \tau) = \frac{Cov_{nn}(\Delta s, \tau)}{\sigma_n^2} \qquad \begin{cases} n = u, v, w \\ \Delta s = \Delta x_f, \Delta y_f, \Delta z_f \end{cases} \tag{3.33}$$

The covariance properties in the wind field are in general decaying with increasing separation Δs and time lag τ. A typical decreasing curve at $\tau = 0$ is illustrated in Fig. 3.7.

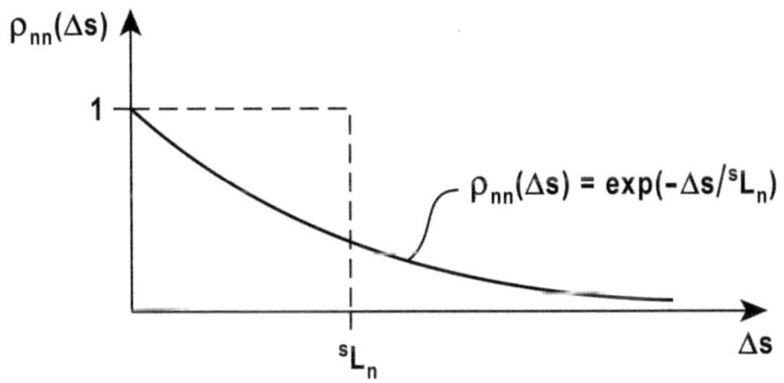

Fig. 3.7 *Spatial cross covariance properties of the wind field*

The situation at $\tau = 0$ is particularly interesting because

$$^{s}L_n = \int_0^\infty \rho_{nn}(\Delta s, \tau = 0) d(\Delta s) \qquad \begin{cases} n = u, v, w \\ \Delta s = \Delta x_f, \Delta y_f, \Delta z_f \end{cases} \tag{3.34}$$

is a characteristic length scale that may be interpreted as the average eddy size of component n in the direction of s. For instance, the length scales $^{xf}L_u$, $^{xf}L_v$ and $^{xf}L_w$ are quantities representing the average eddy size of the u, v and w components in the direction of the main flow. They have previously been presented in Eq. 3.24, and since they obviously can be extracted directly from two point data and Eq. 3.34, the use of Taylor's hypothesis behind Eq. 3.24 is obsolete. The remaining six length scales $^{s}L_n$ with $n = u, v, w$ and $s = y_f, z_f$ are the corresponding quantities that represent the spatial properties in a plane perpendicular to the main flow direction. Typical decay curves for the u component are shown in Fig. 3.8, illustrating the spatial interpretation of the integral length scales.

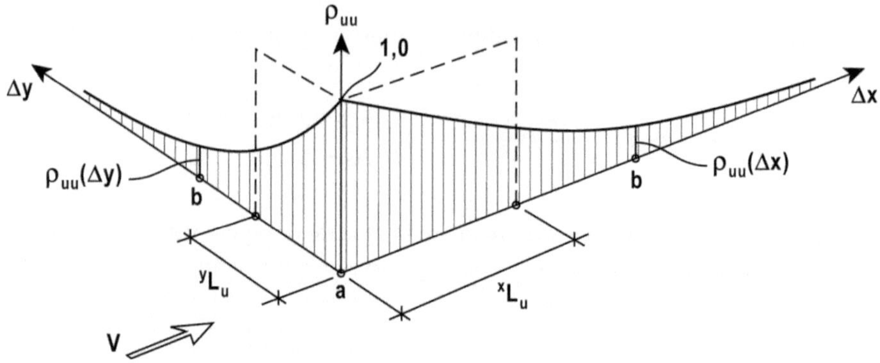

Fig. 3.8 *Spatial illustration of the integral length scales*

The spatial properties of turbulence are strongly dependent of the fetch, i.e. the up–wind terrain. In general, the determination of spatial properties of the turbulence components should be based on full scale recordings on the site in question. However, for a first approximation and under homogeneous conditions not unduly close to the ground, the following may be adopted

$$\rho_{nn}\left(\Delta s,\tau=0\right)\approx \exp\left(-\Delta s/{}^{s}L_{n}\right) \qquad \begin{cases} n=u,v,w \\ s=x_f,y_f,z_f \end{cases} \tag{3.35}$$

$$\begin{bmatrix} {}^{y_f}L_u \\ {}^{z_f}L_u \\ {}^{x_f}L_v \\ {}^{y_f}L_v \\ {}^{z_f}L_v \\ {}^{x_f}L_w \\ {}^{y_f}L_w \\ {}^{z_f}L_w \end{bmatrix} \approx \begin{bmatrix} 1/3 \\ 1/4 \\ 1/4 \\ 1/4 \\ 1/12 \\ 1/12 \\ 1/16 \\ 1/16 \end{bmatrix} \cdot {}^{x_f}L_u \qquad \text{where:} \quad \begin{cases} \dfrac{{}^{x_f}L_u\left(z_f\right)}{{}^{x_f}L_u\left(z_{f0}\right)} \approx \left(\dfrac{z_f}{z_{f0}}\right)^{0.3} \\ z_f \geq z_{f0} = 10m \\ {}^{x_f}L_u\left(z_{f0}\right) = 100m \end{cases} \tag{3.36}$$

While cross covariance functions (or coefficients) represent the time and space domain properties of the turbulence components, it is the auto and cross spectral densities that describe the frequency-space domain properties. In text books on mathematics, the double sided cross spectra are usually defined with ω as the frequency variable, in which case (see chapter 2.6 – 2.8)

$$S_{nn}(\Delta s,\omega)=\frac{1}{2\pi}\int_{-\infty}^{+\infty}Cov_{nn}(\Delta s,\tau)\cdot e^{-i\omega\tau}d\tau \qquad \begin{cases} n=u,v,w \\ \Delta s=\Delta x_f,\Delta y_f,\Delta z_f \end{cases} \tag{3.37}$$

but in wind engineering the frequency f (in Hz) is usually preferred, and then the double sided cross spectra are defined by (see Eqs. 2.68 and 2.75)

$$S_{nn}(\Delta s,f)=\int_{-\infty}^{+\infty}Cov_{nn}(\Delta s,\tau)\cdot e^{-2\pi f\tau}d\tau \qquad \begin{cases} n=u,v,w \\ \Delta s=\Delta x_f,\Delta y_f,\Delta z_f \end{cases} \tag{3.38}$$

The cross spectra are usually defined by the single point spectra, $S_n(f)$, the coherence function, $Coh_{nn}(\Delta s,f)$ and the phase spectra, $\varphi_{nn}(\Delta s,f)$, as shown in Eq. 2.87, i.e.

$$S_{nn}(\Delta s,f)=S_n(f)\cdot\sqrt{Coh_{nn}(\Delta s,f)}\cdot\exp[i\varphi] \qquad \begin{cases} n=u,v,w \\ \Delta s=\Delta x_f,\Delta y_f,\Delta z_f \end{cases} \tag{3.39}$$

Since the wind field is usually assumed homogeneous and perpendicular to the span of the (line-like) structure, phase spectra may be neglected. It should however be acknowledged that in structural response calculations spatial averaging takes place along the span of the structure (see chapters 6.4 and 6.5), and then all imaginary parts cancel out and only a double set of real parts remain. Taking it for granted that the single point spectrum $S_n(f)$ is known, it is then rather the normalised co-spectrum

$$\hat{C}o_{nn}(\Delta s,f)=\frac{\mathrm{Re}\left[S_{nn}(\Delta s,f)\right]}{S_n(f)} \qquad \begin{cases} n=u,v,w \\ \Delta s=\Delta x_f,\Delta y_f,\Delta z_f \end{cases} \tag{3.40}$$

that is necessary to give special attention to in wind engineering. Some general expressions occur in the literature. For a first approximation and under homogeneous conditions, the following may be adopted

$$\hat{C}o_{nn}(\Delta s,f)=\exp\left(-c_{ns}\cdot\frac{f\cdot\Delta s}{V(z_f)}\right) \qquad \begin{cases} n=u,v,w \\ s=x_f,y_f,z_f \\ \Delta s=\Delta x_f,\Delta y_f,\Delta z_f \end{cases} \tag{3.41}$$

where:

$$c_{ns}=\begin{cases} c_{uy_f}=c_{uz_f}\approx 9 \\ c_{vy_f}=c_{vz_f}=c_{wy_f}\approx 6 \\ c_{wz_f}\approx 3 \end{cases} \tag{3.42}$$

Caution should be exercised as the variation in c_{ns} values is considerable (see Solari & Piccardo [6]). The simple expression in Eq. 3.41 has the obvious disadvantage that the normalised co–spectrum becomes unity at all Δs when $f = 0$, whereas a typical normalised co–spectrum will decay at any value of f as illustrated in Fig.3.9. It also has the disadvantage that it is positive in the entire range of Δs. (It may be shown that this is in conflict with the definition of zero mean turbulence components.) Under the assumption of isotropic conditions, Krenk [12] has derived the following expression applicable for the along–wind u component

$$\hat{C}o_{uu}(\Delta s, f) = \left(1 - \frac{\kappa \cdot \Delta s}{2}\right) \cdot \exp(-\kappa \cdot \Delta s) \tag{3.43}$$

where
$$\kappa = \left[\left(\frac{2\pi f}{V}\right)^2 + \left(\frac{1}{1.34 \cdot {}^xL_u}\right)^2\right]^{1/2}$$

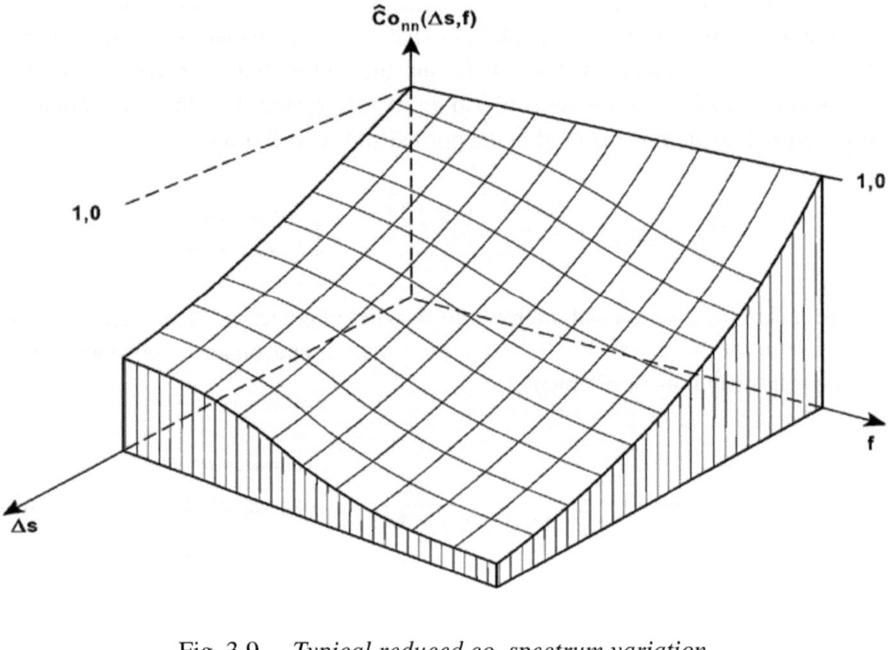

Fig. 3.9 *Typical reduced co–spectrum variation*

Chapter 4

BASIC THEORY OF STOCHASTIC DYNAMIC RESPONSE CALCULATIONS

The general finite element theory of wind induced dynamic response calculations for civil engineering line-like structures is presented in Chapter 9. In the chapter below the theory is presented in a continuous version tailored for the response and cross sectional force calculations in Chapter 6 and 7 where it is taken for granted that the necessary mode shapes are known and that they may be assembled in separate y , z and θ direction components. The reason for this choice is that it represents an effective solution focusing directly on the important wind load and displacement degrees of freedom.

4.1 Modal Analysis and Dynamic Equilibrium Equations

The relevant cross sectional displacements and forces are illustrated in Fig. 4.1. (See also Fig. 1.3.) It is assumed that displacements and loads (all referred to the

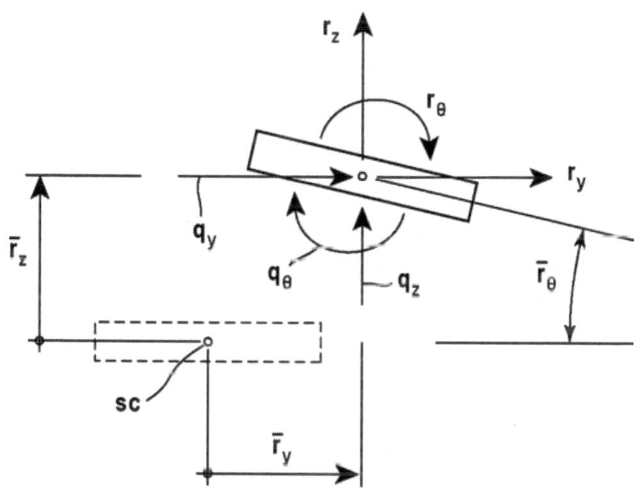

Fig. 4.1 *Cross sectional displacements and loads*

shear centre of the cross section) may be split into the sum of a time invariant mean part and a fluctuating part

$$\left.\begin{array}{l}\mathbf{r}_{tot}(x,t)=\bar{\mathbf{r}}(x)+\mathbf{r}(x,t)\\ \mathbf{q}_{tot}(x,t)=\bar{\mathbf{q}}(x)+\mathbf{q}(x,t)\end{array}\right\} \tag{4.1}$$

each containing three components (horizontal, vertical and torsion), i.e.:

$$\bar{\mathbf{r}}(x)=\begin{bmatrix}\bar{r}_y & \bar{r}_z & \bar{r}_\theta\end{bmatrix}^T \quad \mathbf{r}(x,t)=\begin{bmatrix}r_y & r_z & r_\theta\end{bmatrix}^T \tag{4.2}$$

$$\bar{\mathbf{q}}(x)=\begin{bmatrix}\bar{q}_y & \bar{q}_z & \bar{q}_\theta\end{bmatrix}^T \qquad \mathbf{q}(x,t)=\begin{bmatrix}q_y & q_z & q_\theta\end{bmatrix}^T \tag{4.3}$$

In the following the mean values of the response are considered trivial. The entire focus is on the calculation of the variances of the fluctuating displacement components. The solution will be based on a modal frequency domain approach. Thus, it is assumed that a sufficiently accurate eigen–value solution is available, and that it contains the necessary number of eigen-frequencies and corresponding eigen-modes. That they are orthogonal goes without saying. Scaling of mode shapes is optional, but consistency is required such that the relative difference between cross sectional displacement and rotation components is maintained. It is taken for granted that the eigen-value solution has been obtained in vacuum or in still air conditions. Such a solution has usually been obtained from some finite element formulation (see Chapter 9), and for line-like beam type of elements the eigen-modes will then occur as vectors usually containing six components in each element node, three displacements and three rotations. In the development of the theory below the number of eigen–value components is reduced, focusing on the degrees of freedom associated with r_y, r_z and r_θ. Thus, the mode shape components associated with an arbitrary mode is the displacements ϕ_y, ϕ_z and the cross sectional rotation ϕ_θ that has been extracted from a finite element type of solution. It should be noted that the mode components are formally treated as continuous functions, and therefore the two other rotation components may be retrieved from the first derivatives of ϕ_y and ϕ_z. It is then only the x–axis displacement (i.e. the component in the direction of the main span) that is entirely discarded, but this is not considered important since it is not directly associated with any flow induced load.

Example 4.1

Given a simply supported beam with a single symmetric channel type of cross section as shown in Fig. 4.2. The system contains three displacement components: $y(x,t)$, $z(x,t)$ and $\theta(x,t)$, all referred to the shear centre, which in this case does not coincide with the centroid. Disregarding any external loading and damping contributions, the differential equilibrium conditions are given by (see Timoshenco, Young & Weaver [1], chapter 5.21):

$$EI_z \frac{\partial^4 y}{\partial x^4} + m_y \frac{\partial^2}{\partial t^2}(y - e \cdot \theta) = 0$$

$$EI_y \frac{\partial^4 z}{\partial x^4} + m_z \frac{\partial^2 z}{\partial t^2} = 0$$

$$GI_t \frac{\partial^2 \theta}{\partial x^2} - EI_w \frac{\partial^4 \theta}{\partial x^4} + m_y \frac{\partial^2 y}{\partial t^2} \cdot e - m_\theta \frac{\partial^2 \theta}{\partial t^2} = 0$$

where EI_z and EI_y are cross sectional stiffness with respect to bending in the y and z directions, GI_t and EI_w are the corresponding torsion stiffness associated with St.Venant's torsion and warping, m_y and m_z are translational mass (per unit length), m_θ is rotational mass (with respect to the shear centre) and e is the vertical distance from the shear centre to the centroid. Obviously $m_y = m_z$ (for simplicity they are both set equal to m) and $m_\theta = \rho I_p + me^2$

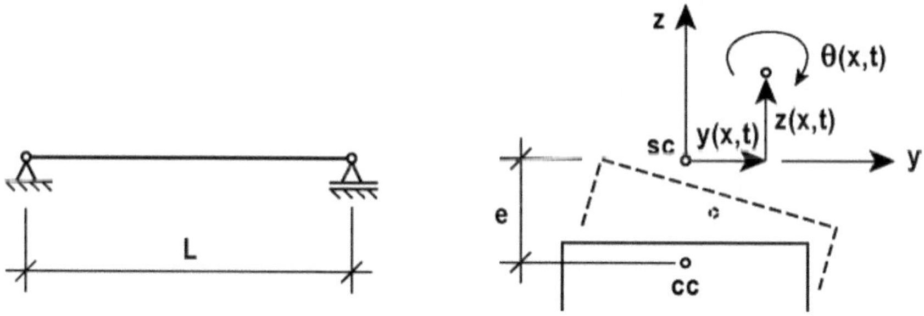

Fig. 4.2 *Simply supported beam with channel type of cross section*

These equations are satisfied over the entire span for the following displacement functions

$$\begin{bmatrix} y(x,t) \\ z(x,t) \\ \theta(x,t) \end{bmatrix} = \mathbf{a} \cdot f(x,t) \quad \text{where} \quad \begin{cases} \mathbf{a} = \begin{bmatrix} a_y & a_z & a_\theta \end{bmatrix}^T \\ f(x,t) = \sin \dfrac{n\pi x}{L} \cdot \exp(i\omega t) \end{cases}$$

and $n = 1, 2, \ldots\ldots, N$. Introducing this into the differential equations above, the following eigen-value type of problem is obtained: $(\mathbf{K} - \omega^2 \mathbf{M}) \cdot \mathbf{a} = \mathbf{0}$, where:

$$\mathbf{K} = \begin{bmatrix} \left(\frac{n\pi}{L}\right)^4 EI_z & 0 & 0 \\ 0 & \left(\frac{n\pi}{L}\right)^4 EI_y & 0 \\ 0 & 0 & \left(\frac{n\pi}{L}\right)^2 \left(GI_t + \frac{n^2\pi^2 EI_w}{L^2}\right) \end{bmatrix} \quad \text{and} \quad \mathbf{M} = \begin{bmatrix} m & 0 & -m \cdot e \\ 0 & m & 0 \\ -m \cdot e & 0 & m_\theta \end{bmatrix}$$

There are two independent eigen–value solutions to this problem. First, there is one that only involves $z(x,t)$ displacements, defined by

$$\left[\left(\frac{n\pi}{L}\right)^4 EI_y - \omega^2 m\right]\cdot a_z = 0$$

which will render n eigen–values and corresponding eigen–vectors

$$\omega_{1_n} = (n\pi)^2\sqrt{\frac{EI_y}{mL^4}} \qquad \text{and} \qquad \boldsymbol{\varphi}_{1_n}(x) = \sin\frac{n\pi x}{L}\begin{bmatrix}0\\1\\0\end{bmatrix}$$

The second solution involves a combined motion of $y(x,t)$ and $\theta(x,t)$ displacements. It is defined by

$$\begin{bmatrix}\left(\frac{n\pi}{L}\right)^4 EI_z - \omega^2 m & \omega^2 me \\ \omega^2 me & \left(\frac{n\pi}{L}\right)^2\left(GI_t + \frac{n^2\pi^2 EI_w}{L^2}\right) - \omega^2 m_\theta\end{bmatrix}\begin{bmatrix}a_y\\a_\theta\end{bmatrix} = \mathbf{0}$$

and it will render two different eigen–values and corresponding eigen–vectors:

$$\omega_{2_n} = \left\{\frac{K_\theta}{m_\theta - e^2 m}\left[\frac{1+\hat{\omega}}{2} + \sqrt{\left(\frac{1-\hat{\omega}}{2}\right)^2 + e^2\hat{K}}\right]\right\}^{1/2} \qquad \boldsymbol{\varphi}_{2_n}(x) = \sin\frac{n\pi x}{L}\begin{bmatrix}1\\0\\\hat{a}_{\theta_2}\end{bmatrix}$$

$$\omega_{3_n} = \left\{\frac{K_\theta}{m_\theta - e^2 m}\left[\frac{1+\hat{\omega}}{2} - \sqrt{\left(\frac{1-\hat{\omega}}{2}\right)^2 + e^2\hat{K}}\right]\right\}^{1/2} \qquad \boldsymbol{\varphi}_{3_n}(x) = \sin\frac{n\pi x}{L}\begin{bmatrix}1\\0\\\hat{a}_{\theta_3}\end{bmatrix}$$

where: $K_z = \left(\frac{n\pi}{L}\right)^4 EI_z$, $K_\theta = \left(\frac{n\pi}{L}\right)^2\left[GI_t + \left(\frac{n\pi}{L}\right)^2 EI_w\right]$, $\hat{K} = \frac{K_z}{K_\theta}$, $\hat{\omega} = \frac{K_z}{m}\frac{m_\theta}{K_\theta}$

$$\hat{a}_{\theta_2} = \frac{1}{e}\left[\frac{\hat{\omega}-1}{2} + \sqrt{\left(\frac{\hat{\omega}-1}{2} - e^2\hat{K}\right)}\right] \quad \hat{a}_{\theta_3} = \frac{1}{e}\left[\frac{\hat{\omega}-1}{2} - \sqrt{\left(\frac{\hat{\omega}-1}{2} - e^2\hat{K}\right)}\right]$$

It may be of some interest to develop the modal mass associated with these mode shapes. The cross sectional mass matrix is given by $\mathbf{M}_0 = diag\left[m \quad m \quad m_\theta\right]$, and thus

$$\tilde{M}_{1_n} = \int_0^L \boldsymbol{\varphi}_{1_n}^T \mathbf{M}_0 \boldsymbol{\varphi}_{1_n} dx = m\int_0^L \sin^2\frac{n\pi x}{L}dx = mL/2$$

$$\tilde{M}_{2_n} = \int_0^L \boldsymbol{\varphi}_{2_n}^T \mathbf{M}_0 \boldsymbol{\varphi}_{2_n} dx = \left(m + \hat{a}_{\theta_2}^2 m_\theta\right)\int_0^L \sin^2\frac{n\pi x}{L}dx = \left(m + \hat{a}_{\theta_2}^2 m_\theta\right)L/2$$

$$\tilde{M}_{3_n} = \int_0^L \boldsymbol{\varphi}_{3_n}^T \mathbf{M}_0 \boldsymbol{\varphi}_{3_n} dx = \tilde{M}_{2_n}$$

In this case mode shapes have been normalised with the displacement component, and therefore the rotation component in the mode shape vector has the unit m^{-1} while the modal mass has the unit kg.

In a general structural eigen–value problem

$$\left(\mathbf{K} - \omega^2 \mathbf{M}\right) \cdot \mathbf{\Phi} = \mathbf{0} \tag{4.4}$$

the modes are usually defined M-orthonormal, i.e. such that

$$\mathbf{\Phi}^T \cdot \mathbf{M} \cdot \mathbf{\Phi} = \mathbf{I} \tag{4.5}$$

where $\mathbf{I}$ is the (diagonal) identity matrix. It should be acknowledged that prior to any scaling of the modes their components have units meters or radians, and that after any scaling has taken place relative units must be maintained (as shown in the example above). It should also be noted that from a finite element solution the eigen-vectors will emerge in accordance with the chosen degrees of freedom in the system. Below, these original mode shape vectors have been rearranged into separate ϕ_y, ϕ_z and ϕ_θ components, each associated with the r_y, r_z and r_θ displacement components as illustrated in Fig. 4.3. The reason for this choice is that in Chapters 6 and 7 it will facilitate an effective solution strategy focusing directly on the important q_y, q_z and q_θ load components and corresponding displacement degrees of freedom.

In the mathematical development of a frequency domain response calculation theory that follows below, the cross sectional displacement and load components are as mentioned above formally taken as continuous function. The motivation behind this choice is mainly convenience, but it is also for practical reasons as spatial load integration will most often require mode shape vectors in a considerably finer element mesh than what is considered sufficient for the eigen–value solution from which they have been obtained. After the theory has been developed the return to discrete vectors will be shown wherever this is necessary for a convenient numerical solution. The basic assumption behind a modal approach is that the structural displacements $\mathbf{r}(x,t)$ may be represented by the sum of the products between natural eigen–modes

$$\boldsymbol{\varphi}_i(x) = \begin{bmatrix} \phi_y & \phi_z & \phi_\theta \end{bmatrix}_i^T \tag{4.6}$$

and unknown exclusively time dependent functions $\eta_i(t)$, i.e.

$$\mathbf{r}(x,t) = \sum_{i=1}^{N_{\mathrm{mod}}} \begin{bmatrix} \phi_y(x) \\ \phi_z(x) \\ \phi_\theta(x) \end{bmatrix}_i \cdot \eta_i(t) = \mathbf{\Phi}(x) \cdot \mathbf{\eta}(t) \tag{4.7}$$

where N_{mod} is the number of modes that has been deemed necessary for a sufficiently accurate or representative solution.

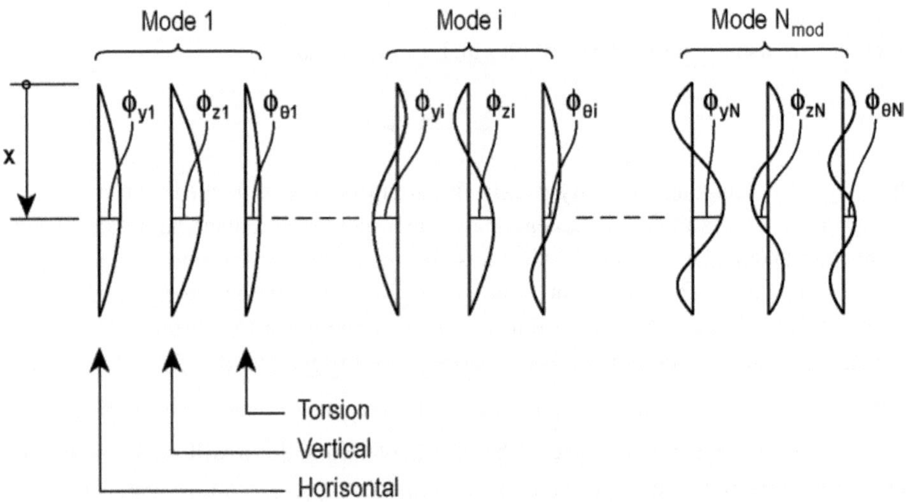

Fig. 4.3 *Mode shapes*

The mode shape matrix $\mathbf{\Phi}(x)$ and the vector $\mathbf{\eta}(t)$ that contains the so-called generalised coordinates are defined by

$$\left.\begin{aligned} \mathbf{\Phi}(x) &= \left[\boldsymbol{\varphi}_1(x) \quad \cdots \quad \boldsymbol{\varphi}_i(x) \quad \cdots \quad \boldsymbol{\varphi}_{N_{\mathrm{mod}}}(x) \right] \\ \mathbf{\eta}(t) &= \left[\eta_1(t) \quad \ldots\ldots \quad \eta_i(t) \quad \ldots\ldots \quad \eta_{N_{\mathrm{mod}}}(t) \right]^T \end{aligned}\right\} \tag{4.8}$$

The introduction of Eq. 4.7 into the equilibrium equations, followed by consecutive weighing with each (orthogonal) mode shape and span-wise integration will then render N_{mod} equivalent modal equilibrium conditions

$$\tilde{\mathbf{M}}_0 \cdot \ddot{\mathbf{\eta}} + \tilde{\mathbf{C}}_0 \cdot \dot{\mathbf{\eta}} + \tilde{\mathbf{K}}_0 \cdot \mathbf{\eta} = \tilde{\mathbf{Q}}_{tot} \tag{4.9}$$

where the zero index indicate that they contain structural properties in vacuum or in still air, and where the modal mass, damping and stiffness matrices are given by

$$\left.\begin{aligned}\tilde{\mathbf{M}}_0 &= diag\left[\tilde{M}_i\right]\\ \tilde{\mathbf{C}}_0 &= diag\left[\tilde{C}_i\right]\\ \tilde{\mathbf{K}}_0 &= diag\left[\tilde{K}_i\right]\end{aligned}\right\} \quad \text{where} \quad \begin{cases}\tilde{M}_i = \int\limits_L \left(\boldsymbol{\varphi}_i^T \cdot \mathbf{M}_0 \cdot \boldsymbol{\varphi}_i\right) dx\\ \tilde{C}_i = 2\tilde{M}_i \omega_i \zeta_i\\ \tilde{K}_i = \omega_i^2 \tilde{M}_i\end{cases} \tag{4.10}$$

The modal load vector in Eq. 4.9 is given by

$$\tilde{\mathbf{Q}}_{tot} = \begin{bmatrix}\tilde{Q}_1 & \cdots & \tilde{Q}_i & \cdots & \tilde{Q}_{N_{mod}}\end{bmatrix}_{tot}^T \tag{4.11}$$

where:

$$\tilde{Q}_{i_{tot}} = \int\limits_{L_{exp}} \left(\boldsymbol{\varphi}_i^T \cdot \mathbf{q}_{tot}\right) dx \tag{4.12}$$

In Eq. 4.10 ω_i are the eigen-frequencies and ζ_i are the damping ratios, each associated with the corresponding eigen-mode. It is in the following assumed that the structural damping ratios ζ_i are known quantities, chosen from experimental experience or an acknowledged code of practice, and that a pertinent mode shape variation has been adopted (e.g. a Rayleigh type of frequency dependency). The three by three mass matrix

$$\mathbf{M}_0 = diag\left[m_y(x) \quad m_z(x) \quad m_\theta(x)\right] \tag{4.13}$$

contains the cross sectional mass properties associated with the y, z and θ degrees of freedom, all taken with respect to the shear centre. (It may often be more convenient to calculate modal mass matrix $\tilde{\mathbf{M}}_0$ in Eq. 4.9 directly from the nodal mass lumping used in the preceding finite element eigen–value solution and the corresponding eigen-vectors, instead of the formal calculation procedures indicated above, as these already contain all the structural properties that are necessary for such a calculation.)

The cross sectional load vector $\mathbf{q}_{tot}$ contains the total drag, lift and moment loads per unit length (see Fig. 4.1) including flow induced as well as motion induces loads, i.e.

$$\mathbf{q}_{tot} = \begin{bmatrix}q_y & q_z & q_\theta\end{bmatrix}_{tot}^T \tag{4.14}$$

The symbolic integration limits L and L_{exp} indicate integration over the entire structure or over the wind exposed part of the structure. The modal matrices $\tilde{\mathbf{M}}_0$, $\tilde{\mathbf{C}}_0$ and $\tilde{\mathbf{K}}_0$ on the left hand side of Eq. 4.9 are all diagonal due to the orthogonal properties of the eigen-modes. However, we shall later see that motion induced parts of the load will be moved to the left hand side of the modal equilibrium equation, thus rendering non-diagonal mass, damping and stiffness matrices for the combined flow and structural system. For educational reasons the development of the necessary theory is divided into three parts, depending on the complexity of the problem. The first part of the presentation is dealing with the situation that the relevant eigen-frequencies are well separated and each mode only contains one component. The next is dealing with the same situation but now with each mode containing all three components. The final presentation is considering the situation that a full multi-mode investigation is required.

4.2 Single mode single component response calculations

In this first section it is assumed that the eigen-frequencies are well spaced out on the frequency axis. Furthermore, the cross sectional shear centre is assumed to coincide (or nearly coincide) with the centroid and there are no other significant sources of mechanical or flow induced coupling between the three displacement components. These assumptions imply that coupling between modes may be ignored, and that each mode shape only contains one component, i.e. any of the N_{mod} mode shapes is purely horizontal, vertical or torsion. The response covariance between modes will then be zero, and thus, the variance of the total dynamic horizontal, vertical or torsion displacement response can be obtained as the sum of contributions from each mode, i.e. the variance of a displacement component is the sum of all variance contributions from excited modes containing displacement components exclusively in the y, z or θ direction (see Eq. 2.27). E.g. σ_y^2 is the sum of all variances associated with the relevant number of modes containing only horizontal displacements, and so on. Thus,

$$\begin{bmatrix} \sigma_y^2 \\ \sigma_z^2 \\ \sigma_\theta^2 \end{bmatrix} = \begin{bmatrix} \sum\limits_{i_y} \sigma_{i_y}^2 \\ \sum\limits_{i_z} \sigma_{i_z}^2 \\ \sum\limits_{i_\theta} \sigma_{i_\theta}^2 \end{bmatrix} \tag{4.15}$$

Given an arbitrary horizontal, vertical or torsion mode shape $\phi_i(x)$ with eigen–frequency ω_i and damping ratio ζ_i, the time domain displacement response contribution of this mode is simply

$$r_i(x,t) = \phi_i(x) \cdot \eta_i(t) \tag{4.16}$$

As mentioned above, it is assumed that the corresponding instantaneous cross sectional load contains the sum of flow induced and motion induced contributions. Thus, the total load per unit length (horizontal, vertical or torsion) is given by

$$q_{tot} = q(x,t) + q_{ae}(x,t,r,\dot{r},\ddot{r}) \tag{4.17}$$

where $q(x,t)$ is the flow induced part and $q_{ae}(x,t,r,\dot{r},\ddot{r})$ is the additional load associated with interaction between flow and structural motion. The modal time domain equilibrium equation for mode number i is then given by

$$\tilde{M}_i \cdot \ddot{\eta}_i(t) + \tilde{C}_i \cdot \dot{\eta}_i(t) + \tilde{K}_i \cdot \eta_i(t) = \tilde{Q}_i(t) + \tilde{Q}_{ae_i}(t,\eta_i,\dot{\eta}_i,\ddot{\eta}_i) \tag{4.18}$$

where

$$\left.\begin{array}{l} \begin{bmatrix} \tilde{M}_i \\ \tilde{C}_i \\ \tilde{K}_i \end{bmatrix} = \begin{bmatrix} \int_L \phi_i^2 m dx \\ 2\tilde{M}_i \omega_i \zeta_i \\ \omega_i^2 \tilde{M}_i \end{bmatrix} \\ \begin{bmatrix} \tilde{Q}_i(t) \\ \tilde{Q}_{ae_i}(t,\eta,\dot{\eta},\ddot{\eta}) \end{bmatrix} = \int_{L_{\exp}} \phi_i \cdot \begin{bmatrix} q \\ q_{ae} \end{bmatrix} dx \end{array}\right\} \tag{4.19}$$

$L_{\exp}$ is the flow exposed part of the structure and $\tilde{Q}_{ae_i}(t,\eta_i,\dot{\eta}_i,\ddot{\eta}_i)$ is the modal motion induced load. It should be noted that structural mass $m(x)$ in the equation above will either be translational or rotational (with respect to the shear centre), depending on whether the mode shape involves displacements in the y or z directions or if it involves pure torsion. Transition into the frequency domain is obtained by taking the Fourier transform on either side of Eq. 4.18. Thus,

$$\left(-\tilde{M}_i\omega^2 + \tilde{C}_i i\omega + \tilde{K}_i\right) \cdot a_{\eta_i}(\omega) = a_{\tilde{Q}_i}(\omega) + a_{\tilde{Q}ae_i}(\omega,\eta_i,\dot{\eta}_i,\ddot{\eta}_i) \tag{4.20}$$

where a_{η_i}, $a_{\tilde{Q}_i}$ and $a_{\tilde{Q}ae_i}$ are the Fourier amplitudes of $\eta_i(t)$, $\tilde{Q}_i(t)$ and $\tilde{Q}_{ae_i}(t)$, respectively. (Index i is the mode shape number and the symbol i is the imaginary unit $i = \sqrt{-1}$.) It is now assumed that the Fourier amplitude of the modal motion induced load contains three cross sectional terms k_{ae}, c_{ae} and m_{ae}, proportional to and in phase with structural displacement, velocity and acceleration, i.e. it is assumed that

$$a_{\tilde{Q}ae_i} = \left(-\tilde{M}_{ae_i}\omega^2 + \tilde{C}_{ae_i} i\omega + \tilde{K}_{ae_i}\right)\cdot a_{\eta_i} \tag{4.21}$$

where

$$\begin{bmatrix} \tilde{M}_{ae_i} \\ \tilde{C}_{ae_i} \\ \tilde{K}_{ae_i} \end{bmatrix} = \int_{L_{\exp}} \phi_i^2 \begin{bmatrix} m_{ae} \\ c_{ae} \\ k_{ae} \end{bmatrix} dx \tag{4.22}$$

and where k_{ae}, c_{ae} and m_{ae} are known constants. Introducing $a_{\tilde{Q}ae_i}$ from Eq. 4.21 (as well as $\tilde{C}_i = 2\tilde{M}_i\omega_i\zeta_i$ and $\tilde{K}_i = \omega_i^2\tilde{M}_i$ from Eq. 4.10) into Eq. 4.20, gathering all motion dependent terms on the left hand side and dividing throughout the equation by $\tilde{K}_i$, the following is obtained

$$a_{\eta_i}(\omega) = \frac{\hat{H}_i(\omega)}{\tilde{K}_i}\cdot a_{\tilde{Q}_i}(\omega) \tag{4.23}$$

where

$$\hat{H}_i(\omega) = \left[1 - \frac{\tilde{K}_{ae_i}}{\omega_i^2\tilde{M}_i} - \left(1 - \frac{\tilde{M}_{ae_i}}{\tilde{M}_i}\right)\cdot\left(\frac{\omega}{\omega_i}\right)^2 + 2i\left(\zeta_i - \frac{\tilde{C}_{ae_i}}{2\omega_i\tilde{M}_i}\right)\cdot\frac{\omega}{\omega_i}\right]^{-1} \tag{4.24}$$

is the non-dimensional modal frequency-response-function.

Introducing $\mu_{ae_i} = \tilde{M}_{ae_i}/\tilde{M}_i$, $\kappa_{ae_i} = \tilde{K}_{ae_i}/\omega_i^2\tilde{M}_i$ and $\zeta_{ae_i} = \tilde{C}_{ae_i}/\left(2\omega_i\tilde{M}_i\right)$, then

$$\hat{H}_i(\omega) = \left[1 - \kappa_{ae_i} - \left(1 - \mu_{ae_i}\right)\cdot\left(\frac{\omega}{\omega_i}\right)^2 + 2i\left(\zeta_i - \zeta_{ae_i}\right)\cdot\frac{\omega}{\omega_i}\right]^{-1} \tag{4.25}$$

The single–sided spectrum of $\eta_i(t)$ is given by

$$S_{\eta_i}(\omega) = \lim_{T\to\infty}\frac{1}{\pi T}\cdot\left(a_{\eta_i}^* \cdot a_{\eta_i}\right) = \frac{\left|\hat{H}_i(\omega)\right|^2}{\tilde{K}_i^2}\cdot\lim_{T\to\infty}\frac{1}{\pi T}\cdot\left(a_{\tilde{Q}_i}^* \cdot a_{\tilde{Q}_i}\right) \tag{4.26}$$

$$\Rightarrow S_{\eta_i}(\omega) = \frac{\left|\hat{H}_i(\omega)\right|^2}{\tilde{K}_i^2}\cdot S_{\tilde{Q}_i}(\omega) \tag{4.27}$$

where it has been introduced that the single-sided spectrum of the modal loading is defined by

$$S_{\tilde{Q}_i}(\omega) = \lim_{T \to \infty} \frac{1}{\pi T} \cdot \left(a^*_{\tilde{Q}_i} \cdot a_{\tilde{Q}_i} \right) \tag{4.28}$$

This will render the displacement response at a position where $\phi_i(x) = 1$. The response at an arbitrary position x_r (e.g. where ϕ has its maximum) may simply be obtained by recognizing that due to linearity the Fourier amplitude at x_r is given by

$$a_{r_i}(\omega) = \phi_i(x_r) \cdot a_{\eta_i}(\omega) \tag{4.29}$$

and thus, the response spectrum for the displacement response at $x = x_r$ is given by

$$S_{r_i}(x_r, \omega) = \frac{\phi_i^2(x_r)}{\tilde{K}_i^2} \cdot \left| \hat{H}_i(\omega) \right|^2 \cdot S_{\tilde{Q}_i}(\omega) \tag{4.30}$$

In structural engineering it has been customary to split the response calculations into a background and a resonant part as illustrated in Fig. 4.4.

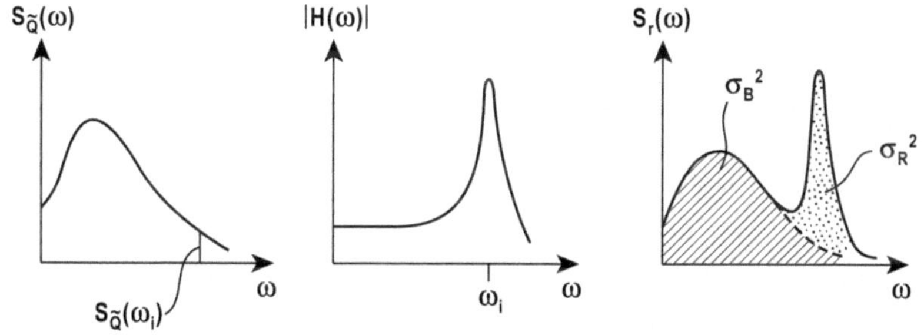

Fig. 4.4 *Frequency domain spectra and transfer function*

The motivation behind this is that static and quasi-static load effects are more accurately determined directly from time invariant equilibrium conditions. This is particularly important for the determination of cross sectional force resultants (or stresses), as shown in chapter 7. The variance of the displacement response in Eq. 4.30 split into a background and a resonant part is given by

$$\left.\begin{aligned}\sigma_{\tilde{r}_i}^2(x_r) &= \frac{\phi_i^2(x_r)}{\tilde{K}_i^2}\int_0^\infty \left|\hat{H}_i(\omega)\right|^2 \cdot S_{\tilde{Q}_i}(\omega)d\omega \\ &\approx \frac{\phi_i^2(x_r)}{\tilde{K}_i^2}\cdot\left[\left|\hat{H}_i(0)\right|^2\cdot\int_0^\infty S_{\tilde{Q}_i}(\omega)d\omega + S_{\tilde{Q}_i}(\omega_i)\cdot\int_0^\infty\left|\hat{H}_i(\omega)\right|^2 d\omega\right]\end{aligned}\right\} \tag{4.31}$$

It is in the following taken for granted that

$$\hat{H}_i(0) = \left[1-\kappa_{ae_i}\right]^{-1} = 1 \tag{4.32}$$

i.e. that $\tilde{K}_{ae_i} = 0$ at $\omega = 0$. (This is an obvious assumption as the structure is not in motion at $\omega = 0$.) Introducing

$$\left.\begin{aligned}&\int_0^\infty S_{\tilde{Q}_i}(\omega)d\omega = \sigma_{\tilde{Q}_i}^2 \\ &\int_0^\infty \left|\hat{H}_i(\omega)\right|^2 d\omega = \frac{\pi\omega_i}{4\left(1-\kappa_{ae_i}\right)\zeta_{tot_i}}\end{aligned}\right\} \tag{4.33}$$

where $\zeta_{tot_i} = \zeta_i - \zeta_{ae_i}$, the following is obtained

$$\sigma_{\tilde{r}_i}^2 \approx \sigma_{B_i}^2 + \sigma_{R_i}^2 = \left[\frac{\phi_i(x_r)}{\tilde{K}_i}\right]^2 \cdot \left[\sigma_{\tilde{Q}_i}^2 + \frac{\pi\omega_i S_{\tilde{Q}_i}(\omega_i)}{4\left(1-\kappa_{ae_i}\right)\zeta_{tot_i}}\right] \tag{4.34}$$

4.3 *Single mode three component response calculations*

In this second approach it is assumed that the eigen-frequencies are still well spaced out on the frequency axis, but that each mode shape contain three components, i.e. the displacements ϕ_y and ϕ_z, and the rotation ϕ_θ, as illustrated in Fig. 4.5. Adopting the same assumptions regarding motion induced load effects as presented in chapter 4.2 above, the total cross sectional load is given by

$$\mathbf{q}_{tot} = \mathbf{q}(x,t) + \mathbf{q}_{ae}(x,t,\mathbf{r},\dot{\mathbf{r}},\ddot{\mathbf{r}}) \tag{4.36}$$

where:

$$\mathbf{q}(x,t) = \begin{bmatrix} q_y & q_z & q_\theta \end{bmatrix}^T \tag{4.37}$$

is the flow induced part of the load, and

$$\mathbf{q}_{ae}(x,t,\mathbf{r},\dot{\mathbf{r}},\ddot{\mathbf{r}}) = \begin{bmatrix} q_y & q_z & q_\theta \end{bmatrix}_{ae}^T \tag{4.38}$$

is the motion induced part.

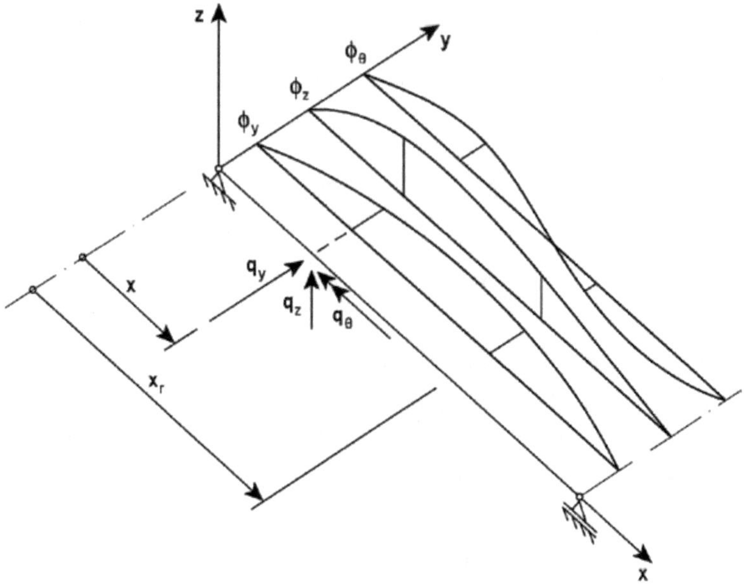

Fig. 4.5 *Single mode shape containing three components*

Adopting the same assumptions regarding motion induced load effects as presented in chapter 4.2 above, the total cross sectional load is given by

$$\mathbf{q}_{tot} = \mathbf{q}(x,t) + \mathbf{q}_{ae}(x,t,\mathbf{r},\dot{\mathbf{r}},\ddot{\mathbf{r}}) \tag{4.36}$$

where: $$\mathbf{q}(x,t) = \begin{bmatrix} q_y & q_z & q_\theta \end{bmatrix}^T \tag{4.37}$$

is the flow induced part of the load, and

$$\mathbf{q}_{ae}(x,t,\mathbf{r},\dot{\mathbf{r}},\ddot{\mathbf{r}}) = \begin{bmatrix} q_y & q_z & q_\theta \end{bmatrix}_{ae}^T \tag{4.38}$$

is the motion induced part. The time domain modal equilibrium condition given in Eq. 4.18 still holds, but for the expressions for $\tilde{M}$, $\tilde{C}$ and $\tilde{K}$ it is necessary to turn to Eq. 4.10, i.e. $\tilde{M}_i = \int_L \left(\boldsymbol{\varphi}_i^T \cdot \mathbf{M}_0 \cdot \boldsymbol{\varphi}_i\right) dx$, $\tilde{C}_i = 2\tilde{M}_i \omega_i \zeta_i$ and $\tilde{K}_i = \omega_i^2 \tilde{M}_i$ while the modal flow induced part of the load is given by (see Eq. 4.12)

$$\tilde{Q}_i(t) = \int_{L_{\exp}} \boldsymbol{\varphi}_i^T(x) \cdot \mathbf{q}(x,t)\, dx \tag{4.39}$$

The Fourier transform in Eq. 4.20 as well as the assumption regarding $a_{\tilde{Q}ae_i}$ in Eq. 4.21 are also still valid, but again, modal motion induced mass, damping and stiffness are now given by

$$\begin{bmatrix} \tilde{M}_{ae} \\ \tilde{C}_{ae} \\ \tilde{K}_{ae} \end{bmatrix}_i = \int_{L_{\exp}} \begin{bmatrix} \boldsymbol{\varphi}_i^T \cdot \mathbf{M}_{ae} \cdot \boldsymbol{\varphi}_i \\ \boldsymbol{\varphi}_i^T \cdot \mathbf{C}_{ae} \cdot \boldsymbol{\varphi}_i \\ \boldsymbol{\varphi}_i^T \cdot \mathbf{K}_{ae} \cdot \boldsymbol{\varphi}_i \end{bmatrix} dx \tag{4.40}$$

where $\mathbf{M}_{ae}$, $\mathbf{C}_{ae}$ and $\mathbf{K}_{ae}$ are three by three coefficient matrices associated with the motion induced part of the loading. To justify a mode by mode approach it is necessary to avoid the introduction of any motion induced coupling between modes, and therefore $\mathbf{M}_{ae}$, $\mathbf{C}_{ae}$ and $\mathbf{K}_{ae}$ must in this particular case be diagonal, i.e.

$$\left.\begin{aligned} \mathbf{M}_{ae} &= diag\begin{bmatrix} m_y & m_z & m_\theta \end{bmatrix}_{ae} \\ \mathbf{C}_{ae} &= diag\begin{bmatrix} c_y & c_z & c_\theta \end{bmatrix}_{ae} \\ \mathbf{K}_{ae} &= diag\begin{bmatrix} k_y & k_z & k_\theta \end{bmatrix}_{ae} \end{aligned}\right\} \tag{4.41}$$

Thus, altogether nine frequency domain motion dependent coefficients are required. In wind engineering $\mathbf{M}_{ae}$ is most often negligible.

Modally we are still dealing with a single-degree-of-freedom system, and thus, Eqs. 4.24 – 4.27 are valid. Linearity implies that the Fourier amplitudes of the displacement components at an arbitrary position x_r are given by

$$\mathbf{a}_{r_i}(x_r,\omega) = \begin{bmatrix} a_{r_y} \\ a_{r_z} \\ a_{r_\theta} \end{bmatrix} = \begin{bmatrix} \phi_y(x_r) \\ \phi_z(x_r) \\ \phi_\theta(x_r) \end{bmatrix}_i \cdot a_{\eta_i}(\omega) = \boldsymbol{\varphi}_{r_i}(x_r) \cdot a_\eta(\omega) \tag{4.42}$$

The cross spectral density matrix of the three components is then

$$\begin{aligned} \mathbf{S}_i(x_r,\omega) &= \lim_{T\to\infty} \frac{1}{\pi T}\left(\mathbf{a}_{r_i}^* \cdot \mathbf{a}_{r_i}^T\right) \\ &= \lim_{T\to\infty} \frac{1}{\pi T}\left\{\left(\boldsymbol{\varphi}_{r_i} \cdot a_{\eta_i}\right)^* \cdot \left(\boldsymbol{\varphi}_{r_i} \cdot a_{\eta_i}\right)^T\right\} = \boldsymbol{\varphi}_{r_i} \cdot \lim_{T\to\infty} \frac{1}{\pi T}\left(a_{\eta_i}^* \cdot a_{\eta_i}\right) \cdot \boldsymbol{\varphi}_{r_i}^T \end{aligned} \tag{4.43}$$

from which the following is obtained:

$$\mathbf{S}_i(x_r,\omega)=\begin{bmatrix}\ddots & & \iddots\\ & S_{nm}(x_r,\omega) & \\ \iddots & & \ddots\end{bmatrix}_i=\boldsymbol{\varphi}_{r_{\tilde{i}}}(x_r)\cdot S_{\eta_i}(\omega)\cdot\boldsymbol{\varphi}_{r_{\tilde{i}}}^T(x_r) \tag{4.44}$$

where: $\left.\begin{matrix}n\\m\end{matrix}\right\}=r_y,r_z,r_\theta$ and $S_{\eta_i}(\omega)=\lim\limits_{T\to\infty}\frac{1}{\pi T}\left(a_{\eta_i}^{*}\cdot a_{\eta_i}\right)$ is given in Eqs. 4.26-4.27, i.e.

$$\mathbf{S}_i(x_r,\omega)=\boldsymbol{\varphi}_{r_{\tilde{i}}}(x_r)\cdot\frac{\left|\hat{H}_i(\omega)\right|^2}{\tilde{K}_i^2}\cdot S_{\tilde{Q}_i}(\omega)\cdot\boldsymbol{\varphi}_{r_{\tilde{i}}}^T(x_r) \tag{4.45}$$

The response covariance matrix is obtained by the frequency domain integration of $\mathbf{S}_i(x_r,\omega)$, and thus

$$\int_0^\infty\mathbf{S}_i(x_r,\omega)d\omega=\begin{bmatrix}\sigma_{r_yr_y}^2(x_r) & Cov_{r_yr_z}(x_r) & Cov_{r_yr_\theta}(x_r)\\ Cov_{r_zr_y}(x_r) & \sigma_{r_zr_z}^2(x_r) & Cov_{r_zr_\theta}(x_r)\\ Cov_{r_\theta r_y}(x_r) & Cov_{r_\theta r_z}(x_r) & \sigma_{r_\theta r_\theta}^2(x_r)\end{bmatrix}_i \tag{4.46}$$

However, the three components of each mode shape are fully correlated and therefore all cross-covariance coefficients that may be extracted from Eq. 4.46 are equal to unity. Thus, it is only the terms on the diagonal of Eq, 4.46 that are of any interest, and then the calculations simplify into

$$\mathbf{S}_i(x_r,\omega)=\begin{bmatrix}S_{r_yr_y}\\ S_{r_zr_z}\\ S_{r_\theta r_\theta}\end{bmatrix}_i=\begin{bmatrix}\phi_y^2(x_r)\\ \phi_z^2(x_r)\\ \phi_\theta^2(x_r)\end{bmatrix}_i\cdot\frac{\left|\hat{H}_i(\omega)\right|^2}{\tilde{K}_i^2}\cdot S_{\tilde{Q}_i}(\omega) \tag{4.47}$$

and

$$\mathbf{Var}_i(x_r)=\begin{bmatrix}\sigma_{r_yr_y}^2\\ \sigma_{r_zr_z}^2\\ \sigma_{r_\theta r_\theta}^2\end{bmatrix}_i=\int_0^\infty\mathbf{S}_i(x_r,\omega)d\omega \tag{4.48}$$

The total response may be obtained by adding up variance contributions from all modes, i.e.

$$\mathbf{Var}(x_r)=\begin{bmatrix}\sigma_{r_yr_y}^2(x_r)\\ \sigma_{r_zr_z}^2(x_r)\\ \sigma_{r_\theta r_\theta}^2(x_r)\end{bmatrix}=\sum_{i=1}^{N_{mod}}\mathbf{Var}_i \tag{4.49}$$

4.4 General multi-mode response calculations

In the final section of this chapter it is assumed that a full multi-mode approach is required. The basic assumptions from chapter 4.1 are that

$$\mathbf{r}(x,t) = \mathbf{\Phi}(x) \cdot \boldsymbol{\eta}(t) \tag{4.50}$$

where
$$\begin{cases} \mathbf{r}(x,t) = \begin{bmatrix} r_y & r_z & r_\theta \end{bmatrix}^T \\ \mathbf{\Phi}(x) = \begin{bmatrix} \boldsymbol{\varphi}_1 \dots & \boldsymbol{\varphi}_i & \dots \boldsymbol{\varphi}_{N_{\text{mod}}} \end{bmatrix} \\ \boldsymbol{\eta}(x) = \begin{bmatrix} \eta_1 \dots & \eta_i & \dots \eta_{N_{\text{mod}}} \end{bmatrix}^T \end{cases} \tag{4.51}$$

$\boldsymbol{\varphi}_i(x) = \begin{bmatrix} \phi_y & \phi_z & \phi_\theta \end{bmatrix}_i^T$ and where N_{mod} is the number of modes chosen to be included in the calculations. Still adopting the assumptions regarding motion induced load effects as presented in chapter 4.2, the cross sectional load is

$$\mathbf{q}_{tot} = \mathbf{q}(x,t) + \mathbf{q}_{ae}(x,t,\mathbf{r},\dot{\mathbf{r}},\ddot{\mathbf{r}}) \tag{4.52}$$

where
$$\begin{cases} \mathbf{q}(x,t) = \begin{bmatrix} q_y & q_z & q_\theta \end{bmatrix}^T \\ \mathbf{q}_{ae}(x,t,\mathbf{r},\dot{\mathbf{r}},\ddot{\mathbf{r}}) = \begin{bmatrix} q_y & q_z & q_\theta \end{bmatrix}_{ae}^T \end{cases} \tag{4.53}$$

Thus, the time domain modal equilibrium equation is given by (see also Eq. 4.9)

$$\tilde{\mathbf{M}}_0 \cdot \ddot{\boldsymbol{\eta}}(t) + \tilde{\mathbf{C}}_0 \cdot \dot{\boldsymbol{\eta}}(t) + \tilde{\mathbf{K}}_0 \cdot \boldsymbol{\eta}(t) = \tilde{\mathbf{Q}}(t) + \tilde{\mathbf{Q}}(t,\eta,\dot{\eta},\ddot{\eta}) \tag{4.54}$$

where $\tilde{\mathbf{M}}_0$, $\tilde{\mathbf{C}}_0$ and $\tilde{\mathbf{K}}_0$ are N_{mod} by N_{mod} diagonal matrices defined in Eq. 4.10, and the modal N_{mod} by one flow induced load vector is given by

$$\tilde{\mathbf{Q}}(t) = \begin{bmatrix} \tilde{Q}_1 & \dots & \tilde{Q}_i & \dots & \tilde{Q}_{N_{\text{mod}}} \end{bmatrix}^T \tag{4.55}$$

Where
$$\tilde{Q}_i = \int_{L_{\exp}} \left(\boldsymbol{\varphi}_i^T \cdot \mathbf{q} \right) dx \tag{4.56}$$

Taking the Fourier transform on either side of Eq. 4.54

$$\left(-\tilde{\mathbf{M}}_0 \omega^2 + \tilde{\mathbf{C}}_0 i\omega + \tilde{\mathbf{K}}_0 \right) \cdot \mathbf{a}_\eta(\omega) = \mathbf{a}_{\tilde{Q}}(\omega) + \mathbf{a}_{\tilde{Q}_{ae}}(\omega,\eta,\dot{\eta},\ddot{\eta}) \tag{4.57}$$

where:

$$\left\{\begin{aligned}\mathbf{a}_{\eta} &= \begin{bmatrix} a_{\eta_1} \dots & a_{\eta_i} & \dots a_{\eta_N} \end{bmatrix}^T \\ \mathbf{a}_{\tilde{Q}} &= \begin{bmatrix} a_{\tilde{Q}_1} \dots & a_{\tilde{Q}_i} & \dots a_{\tilde{Q}_N} \end{bmatrix}^T\end{aligned}\right. \tag{4.58}$$

Since the assumption of a modal frequency domain motion induced load proportional to and in phase with structural displacement, velocity and acceleration is adopted, then $\mathbf{a}_{\tilde{Q}_{ae}}$ is given by

$$\mathbf{a}_{\tilde{Q}ae} = \left(-\tilde{\mathbf{M}}_{ae}\omega^2 + \tilde{\mathbf{C}}_{ae} i\omega + \tilde{\mathbf{K}}_{ae}\right) \cdot \mathbf{a}_{\eta} \tag{4.59}$$

where $\tilde{\mathbf{M}}_{ae}$, $\tilde{\mathbf{C}}_{ae}$ and $\tilde{\mathbf{K}}_{ae}$ are N_{mod} by N_{mod} matrices

$$\tilde{\mathbf{M}}_{ae} = \begin{bmatrix} \ddots & & ⋰ \\ & \tilde{M}_{ae_{ij}} & \\ ⋰ & & \ddots \end{bmatrix} \quad \tilde{\mathbf{C}}_{ae} = \begin{bmatrix} \ddots & & ⋰ \\ & \tilde{C}_{ae_{ij}} & \\ ⋰ & & \ddots \end{bmatrix} \quad \tilde{\mathbf{K}}_{ae} = \begin{bmatrix} \ddots & & ⋰ \\ & \tilde{K}_{ae_{ij}} & \\ ⋰ & & \ddots \end{bmatrix} \tag{4.60}$$

whose elements on row i column j are given by

$$\begin{bmatrix} \tilde{M}_{ae_{ij}} \\ \tilde{C}_{ae_{ij}} \\ \tilde{K}_{ae_{ij}} \end{bmatrix} = \int_{L_{\exp}} \begin{bmatrix} \boldsymbol{\varphi}_i^T \cdot \mathbf{M}_{ae} \cdot \boldsymbol{\varphi}_j \\ \boldsymbol{\varphi}_i^T \cdot \mathbf{C}_{ae} \cdot \boldsymbol{\varphi}_j \\ \boldsymbol{\varphi}_i^T \cdot \mathbf{K}_{ae} \cdot \boldsymbol{\varphi}_j \end{bmatrix} dx \tag{4.61}$$

where $\mathbf{M}_{ae}$, $\mathbf{C}_{ae}$ and $\mathbf{K}_{ae}$ are three by three motion dependent cross sectional load coefficient matrices

$$\mathbf{M}_{ae} = \begin{bmatrix} \ddots & & ⋰ \\ & m_{nm} & \\ ⋰ & & \ddots \end{bmatrix} \quad \mathbf{C}_{ae} = \begin{bmatrix} \ddots & & ⋰ \\ & c_{nm} & \\ ⋰ & & \ddots \end{bmatrix} \quad \mathbf{K}_{ae} = \begin{bmatrix} \ddots & & ⋰ \\ & k_{nm} & \\ ⋰ & & \ddots \end{bmatrix} \tag{4.62}$$

and where: $\left.\begin{matrix} n \\ m \end{matrix}\right\} = y, z, \theta$.

Example 4.2

The modal quantities given in Eq. 4.61 may be obtained from a fully expanded vector format given by:

$$\begin{aligned}
\tilde{M}_{ae_{ij}} &= (\phi_{y_i}^T m_{yy}\phi_{y_j} + \phi_{z_i}^T m_{zy}\phi_{y_j} + \phi_{\theta_i}^T m_{\theta y}\phi_{y_j} + \phi_{y_i}^T m_{yz}\phi_{z_j} + \phi_{z_i}^T m_{zz}\phi_{z_j} + \phi_{\theta_i}^T m_{\theta z}\phi_{z_j} \\
&\quad + \phi_{y_i}^T m_{y\theta}\phi_{\theta_j} + \phi_{z_i}^T m_{z\theta}\phi_{\theta_j} + \phi_{\theta_i}^T m_{\theta\theta}\phi_{\theta_j}) \cdot \Delta x \\
\tilde{C}_{ae_{ij}} &= (\phi_{y_i}^T c_{yy}\phi_{y_j} + \phi_{z_i}^T c_{zy}\phi_{y_j} + \phi_{\theta_i}^T c_{\theta y}\phi_{y_j} + \phi_{y_i}^T c_{yz}\phi_{z_j} + \phi_{z_i}^T c_{zz}\phi_{z_j} + \phi_{\theta_i}^T c_{\theta z}\phi_{z_j} \\
&\quad + \phi_{y_i}^T c_{y\theta}\phi_{\theta_j} + \phi_{z_i}^T c_{z\theta}\phi_{\theta_j} + \phi_{\theta_i}^T c_{\theta\theta}\phi_{\theta_j}) \cdot \Delta x \\
\tilde{K}_{ae_{ij}} &= (\phi_{y_i}^T k_{yy}\phi_{y_j} + \phi_{z_i}^T k_{zy}\phi_{y_j} + \phi_{\theta_i}^T k_{\theta y}\phi_{y_j} + \phi_{y_i}^T k_{yz}\phi_{z_j} + \phi_{z_i}^T k_{zz}\phi_{z_j} + \phi_{\theta_i}^T k_{\theta z}\phi_{z_j} \\
&\quad + \phi_{y_i}^T k_{y\theta}\phi_{\theta_j} + \phi_{z_i}^T k_{z\theta}\phi_{\theta_j} + \phi_{\theta_i}^T k_{\theta\theta}\phi_{\theta_j}) \cdot \Delta x
\end{aligned}$$

where Δx is the spanwise mesh separation (above assumed constant). If the coefficients vary along the span, their numerical values need to be given on the diagonal of an N by N matrix, where N is the number of nodes.

Thus, for a full description of the motion induced load effects altogether twenty–seven motion dependent load coefficients are required. First Eqs. 4.59 is introduced into 4.57 and all terms associated with structural motion are gathered on the left hand side

$$\left[-\left(\tilde{\mathbf{M}}_0 - \tilde{\mathbf{M}}_{ae}\right)\omega^2 + \left(\tilde{\mathbf{C}}_0 - \tilde{\mathbf{C}}_{ae}\right)i\omega + \left(\tilde{\mathbf{K}}_0 - \tilde{\mathbf{K}}_{ae}\right)\right] \cdot \mathbf{a}_\eta(\omega) = \mathbf{a}_{\tilde{Q}}(\omega) \tag{4.63}$$

and then the result is pre-multiplied with $\tilde{\mathbf{K}}_0^{-1}$, recalling that

$$\left.\begin{aligned}
\tilde{\mathbf{K}}_0 &= diag\left[\omega_i^2 \tilde{M}_i\right] \\
\tilde{\mathbf{C}}_0 &= diag\left[2\tilde{M}_i \omega_i \zeta_i\right]
\end{aligned}\right\} \tag{4.64}$$

It is convenient to introduce a reduced modal load vector

$$\mathbf{a}_{\hat{Q}}(\omega) = \tilde{\mathbf{K}}_0^{-1} \cdot \mathbf{a}_{\tilde{Q}}(\omega) = \begin{bmatrix} \cdots & \dfrac{\int_{L_{\exp}} \boldsymbol{\varphi}_i^T(x) \cdot \mathbf{a}_q(x,\omega)dx}{\omega_i^2 \tilde{M}_i} & \cdots \end{bmatrix}^T \tag{4.65}$$

where $\mathbf{a}_q(x,\omega) = \begin{bmatrix} a_{q_y} & a_{q_z} & a_{q_\theta} \end{bmatrix}^T$. The following is then obtained

$$\mathbf{a}_\eta(\omega) = \hat{\mathbf{H}}_\eta(\omega) \cdot \mathbf{a}_{\hat{Q}}(\omega) \tag{4.66}$$

where

$$\hat{\mathbf{H}}_\eta(\omega) = \left\{ \mathbf{I} - \tilde{\mathbf{K}}_0^{-1}\tilde{\mathbf{K}}_{ae} - \left(diag\left[\frac{1}{\omega_i^2}\right] - \tilde{\mathbf{K}}_0^{-1}\tilde{\mathbf{M}}_{ae} \right)\omega^2 + \left(diag\left[\frac{2\zeta_i}{\omega_i}\right] - \tilde{\mathbf{K}}_0^{-1}\tilde{\mathbf{C}}_{ae} \right) i\omega \right\}^{-1} \tag{4.67}$$

is the non-dimensional frequency-response-matrix, and $\mathbf{I}$ is the identity matrix (N_{mod} by N_{mod}). It is convenient to define the following N_{mod} by N_{mod} matrices

$$\left.\begin{aligned} \boldsymbol{\mu}_{ae} &= diag\left[\omega_i^2\right] \cdot \left(\tilde{\mathbf{K}}_0^{-1} \cdot \tilde{\mathbf{M}}_{ae}\right) \\ \boldsymbol{\kappa}_{ae} &= \tilde{\mathbf{K}}_0^{-1} \cdot \tilde{\mathbf{K}}_{ae} \\ \boldsymbol{\zeta}_{ae} &= \frac{1}{2} \cdot diag\left[\omega_i\right] \cdot \left(\tilde{\mathbf{K}}_0^{-1} \cdot \tilde{\mathbf{C}}_{ae}\right) \end{aligned}\right\} \tag{4.68}$$

as well as introducing $\boldsymbol{\zeta} = diag\left[\zeta_i\right]$. The non-dimensional frequency-response-matrix is then given by

$$\hat{\mathbf{H}}_\eta(\omega) = \left\{ \mathbf{I} - \boldsymbol{\kappa}_{ae} - \left(\omega \cdot diag\left[\frac{1}{\omega_i}\right] \right)^2 \cdot \left(\mathbf{I} - \boldsymbol{\mu}_{ae}\right) + 2i\omega \cdot diag\left[\frac{1}{\omega_i}\right] \cdot \left(\boldsymbol{\zeta} - \boldsymbol{\zeta}_{ae}\right) \right\}^{-1} \tag{4.69}$$

By combination of Eqs. 4.60, 4.61 and 4.64, then the content of

$$\boldsymbol{\mu}_{ae} = \begin{bmatrix} \ddots & & \iddots \\ & \mu_{ae_{ij}} & \\ \iddots & & \ddots \end{bmatrix} \quad \boldsymbol{\kappa}_{ae} = \begin{bmatrix} \ddots & & \iddots \\ & \kappa_{ae_{ij}} & \\ \iddots & & \ddots \end{bmatrix} \quad \boldsymbol{\zeta}_{ae} = \begin{bmatrix} \ddots & & \iddots \\ & \zeta_{ae_{ij}} & \\ \iddots & & \ddots \end{bmatrix} \tag{4.70}$$

are given by

$$\mu_{ae_{ij}} = \frac{\tilde{M}_{ae_{ij}}}{\tilde{M}_i} = \frac{\int_{L_{\exp}} \left(\boldsymbol{\varphi}_i^T \mathbf{M}_{ae} \boldsymbol{\varphi}_j \right) dx}{\tilde{M}_i} \tag{4.71}$$

$$\kappa_{ae_{ij}} = \frac{\tilde{K}_{ae_{ij}}}{\omega_i^2 \tilde{M}_i} = \frac{\int_{L_{\exp}} \left(\boldsymbol{\varphi}_i^T \mathbf{K}_{ae} \boldsymbol{\varphi}_j \right) dx}{\omega_i^2 \tilde{M}_i} \tag{4.72}$$

$$\zeta_{ae_{ij}} = \frac{\tilde{C}_{ae_{ij}}}{2\omega_i \tilde{M}_i} = \frac{\int_{L_{\exp}} \left(\boldsymbol{\varphi}_i^T \mathbf{C}_{ae} \boldsymbol{\varphi}_j \right) dx}{2\omega_i \tilde{M}_i} \tag{4.73}$$

Returning to Eq. 4.66, the response spectral density matrix (N_{mod} by N_{mod} and containing single-sided spectra) is obtained from the basic definition of spectra as expressed from the Fourier amplitudes, and thus, the following development applies:

$$\begin{aligned} \mathbf{S}_\eta(\omega) &= \lim_{T\to\infty} \frac{1}{\pi T} \left(\mathbf{a}_\eta^* \cdot \mathbf{a}_\eta^T \right) = \lim_{T\to\infty} \frac{1}{\pi T} \left[\left(\hat{\mathbf{H}}_\eta \mathbf{a}_{\hat{Q}} \right)^* \cdot \left(\hat{\mathbf{H}}_\eta \mathbf{a}_{\hat{Q}} \right)^T \right] \\ &= \hat{\mathbf{H}}_\eta^* \cdot \left[\lim_{T\to\infty} \frac{1}{\pi T} \left(\mathbf{a}_{\hat{Q}}^* \cdot \mathbf{a}_{\hat{Q}}^T \right) \right] \cdot \hat{\mathbf{H}}_\eta^T = \hat{\mathbf{H}}_\eta^* \cdot \mathbf{S}_{\hat{Q}} \cdot \hat{\mathbf{H}}_\eta^T \end{aligned} \tag{4.74}$$

where $\mathbf{S}_{\hat{Q}}$ is an N_{mod} by N_{mod} normalised modal load matrix

$$\begin{aligned} \mathbf{S}_{\hat{Q}}(\omega) &= \lim_{T\to\infty} \frac{1}{\pi T} \left(\mathbf{a}_{\hat{Q}}^* \cdot \mathbf{a}_{\hat{Q}}^T \right) = \lim_{T\to\infty} \frac{1}{\pi T} \left(\begin{bmatrix} a_{\hat{Q}_1}^* \\ \vdots \\ a_{\hat{Q}_i}^* \\ \vdots \\ a_{\hat{Q}_{N_{\text{mod}}}}^* \end{bmatrix} \cdot \begin{bmatrix} a_{\hat{Q}_1} & \cdots & a_{\hat{Q}_j} & \cdots & a_{\hat{Q}_{N_{\text{mod}}}} \end{bmatrix} \right) \\ \Rightarrow \mathbf{S}_{\hat{Q}}(\omega) &= \begin{bmatrix} \ddots & & \iddots \\ & S_{\hat{Q}_i \hat{Q}_j}(\omega) & \\ \iddots & & \ddots \end{bmatrix} \end{aligned} \tag{4.75}$$

whose elements on row i column j are given by

$$
\begin{aligned}
S_{\hat{Q}_i\hat{Q}_j}(\omega) &= \lim_{T\to\infty}\frac{1}{\pi T}\left[a^*_{\hat{Q}_i}(\omega)\cdot a_{\hat{Q}_j}(\omega)\right] \\
&= \lim_{T\to\infty}\frac{1}{\pi T}\left(\frac{\int\limits_{L_{\exp}} \boldsymbol{\varphi}_i^T(x)\mathbf{a}_q^*(x,\omega)dx}{\omega_i^2\tilde{M}_i}\cdot\frac{\int\limits_{L_{\exp}}\left[\boldsymbol{\varphi}_j^T(x)\mathbf{a}_q(x,\omega)\right]^T dx}{\omega_j^2\tilde{M}_j}\right) \\
&= \lim_{T\to\infty}\frac{1}{\pi T}\left\{\frac{\iint\limits_{L_{\exp}}\left[\boldsymbol{\varphi}_i^T(x_1)\mathbf{a}_q^*(x_1,\omega)\right]\cdot\left[\boldsymbol{\varphi}_j^T(x_2)\mathbf{a}_q(x_2,\omega)\right]^T dx_1dx_2}{\left(\omega_i^2\tilde{M}_i\right)\cdot\left(\omega_j^2\tilde{M}_j\right)}\right\} \\
&= \frac{\iint\limits_{L_{\exp}}\boldsymbol{\varphi}_i^T(x_1)\cdot\lim\limits_{T\to\infty}\frac{1}{\pi T}\left[\mathbf{a}_q^*(x_1,\omega)\cdot\mathbf{a}_q^T(x_2,\omega)\right]\cdot\boldsymbol{\varphi}_j(x_2)dx_1dx_2}{\left(\omega_i^2\tilde{M}_i\right)\cdot\left(\omega_j^2\tilde{M}_j\right)}
\end{aligned}
\tag{4.76}
$$

Thus, the elements of $\mathbf{S}_{\hat{Q}}(\omega)$ are given by

$$
\Rightarrow S_{\hat{Q}_i\hat{Q}_j}(\omega) = \frac{\iint\limits_{L_{\exp}}\boldsymbol{\varphi}_i^T(x_1)\cdot\mathbf{S}_{qq}(\Delta x,\omega)\cdot\boldsymbol{\varphi}_j(x_2)dx_1dx_2}{\left(\omega_i^2\tilde{M}_i\right)\cdot\left(\omega_j^2\tilde{M}_j\right)} \tag{4.77}
$$

where $\Delta x = |x_1 - x_2|$, and where $\mathbf{S}_{qq}(\Delta x,\omega)$ is the spectral density matrix of cross sectional loads, i.e. $\mathbf{S}_{qq}(\Delta x,\omega) = \lim\limits_{T\to\infty}(1/\pi T)\left[\mathbf{a}_q^*(x_1,\omega)\cdot\mathbf{a}_q^T(x_2,\omega)\right]$

$$
\Rightarrow \mathbf{S}_{qq}(\Delta x,\omega) = \lim_{T\to\infty}\frac{1}{\pi T}\begin{bmatrix} a^*_{q_y}\cdot a_{q_y} & a^*_{q_y}\cdot a_{q_z} & a^*_{q_y}\cdot a_{q_\theta} \\ a^*_{q_z}\cdot a_{q_y} & a^*_{q_z}\cdot a_{q_z} & a^*_{q_z}\cdot a_{q_\theta} \\ a^*_{q_\theta}\cdot a_{q_y} & a^*_{q_\theta}\cdot a_{q_z} & a^*_{q_\theta}\cdot a_{q_\theta}\end{bmatrix} = \begin{bmatrix} S_{q_yq_y} & S_{q_yq_z} & S_{q_yq_\theta} \\ S_{q_zq_y} & S_{q_zq_z} & S_{q_zq_\theta} \\ S_{q_\theta q_y} & S_{q_\theta q_z} & S_{q_\theta q_\theta}\end{bmatrix} \tag{4.78}
$$

Extracting from the mode shape matrix $\boldsymbol{\Phi} = \left[\boldsymbol{\varphi}_1 \;\; \dots \;\; \boldsymbol{\varphi}_i \;\; \dots \;\; \boldsymbol{\varphi}_N\right]$ (see Eq. 4.8) a three by N_{mod} matrix associated with a chosen span-wise position x_r

$$\begin{aligned}\boldsymbol{\Phi}_r(x_r) &= \begin{bmatrix}\boldsymbol{\varphi}_1(x_r) & \ldots & \boldsymbol{\varphi}_i(x_r) & \ldots & \boldsymbol{\varphi}_N(x_r)\end{bmatrix} \\ &= \begin{bmatrix}\begin{bmatrix}\phi_y(x_r)\\ \phi_z(x_r)\\ \phi_\theta(x_r)\end{bmatrix}_1 & \ldots & \begin{bmatrix}\phi_y(x_r)\\ \phi_z(x_r)\\ \phi_\theta(x_r)\end{bmatrix}_i & \ldots & \begin{bmatrix}\phi_y(x_r)\\ \phi_z(x_r)\\ \phi_\theta(x_r)\end{bmatrix}_N\end{bmatrix}\end{aligned} \tag{4.79}$$

then the three by three cross spectral density matrix of the unknown modal displacements r_y, r_z and r_θ at $x = x_r$

$$\mathbf{S}_{rr}(x_r,\omega) = \begin{bmatrix} S_{r_y r_y} & S_{r_y r_z} & S_{r_y r_\theta} \\ S_{r_z r_y} & S_{r_z r_z} & S_{r_z r_\theta} \\ S_{r_\theta r_y} & S_{r_\theta r_z} & S_{r_\theta r_\theta} \end{bmatrix} \tag{4.80}$$

is given by

$$\mathbf{S}_{rr}(x_r,\omega) = \boldsymbol{\Phi}_r(x_r)\cdot\mathbf{S}_\eta(\omega)\cdot\boldsymbol{\Phi}_r^T(x_r) \tag{4.81}$$

where $\mathbf{S}_\eta(\omega)$ is given in Eq. 4.74, i.e.:

$$\mathbf{S}_{rr}(x_r,\omega) = \boldsymbol{\Phi}_r(x_r)\cdot\left[\hat{\mathbf{H}}_\eta^*(\omega)\cdot\mathbf{S}_{\hat{Q}}(\omega)\cdot\hat{\mathbf{H}}_\eta^T(\omega)\right]\cdot\boldsymbol{\Phi}_r^T(x_r) \tag{4.82}$$

This equation is applicable to any linear load on a line–like structure. If all mechanical properties of the structure are known, then an eigen–value analysis will provide the basic input to $\hat{\mathbf{H}}_\eta$ and $\boldsymbol{\Phi}_r$. What then remains is the set-up of $\mathbf{S}_{\hat{Q}}$ and the motion induced contributions to $\hat{\mathbf{H}}_\eta$. This is shown in chapters 5 and 6.

Chapter 5

WIND AND MOTION INDUCED LOADS

5.1 *The buffeting theory*

The buffeting wind load on structures includes the part of the total load that may be ascribed to the velocity fluctuations in the oncoming flow, $U\left(x_f,y_f,z_f,t\right)=V\left(x_f,y_f,z_f\right)+u\left(x_f,y_f,z_f,t\right)$, $v\left(x_f,z_f,t\right)$ and $w\left(x_f,z_f,t\right)$, as well as any motion induced contributions. The theory presented below was first developed by A.G. Davenport [13, 14]. In the following it is a line like horizontal bridge type of structure that is considered. It is taken for granted that its z_f–position in the flow prior to any loading is constant along the entire span, that the wind field is stationary and homogeneous and that the main flow direction is perpendicular to the span-wise x -axis of the structure, in which case x_f is constant and y_f may be exchanged by x . It is then only the velocity fluctuations in the along wind and the across wind vertical directions expressed in structural axis that are of interest, i.e. the components $U(x,t)=V+u(x,t)$ and $w(x,t)$. The theory may readily be applied to a vertical (tower) type of structure, in which case any z_f -variation needs to be included and the w component must be replaced by the v component (but maintaining all other notations shown in Fig. 5.1 below). The basic assumptions behind the buffeting theory are that the load may be calculated from the instantaneous velocity pressure and the appropriate load coefficients that have been obtained from static tests, and that linearization of any fluctuating parts will render results with sufficient accuracy. Thus, the load may be calculated from an interpretation of the instantaneous relative velocity vector and the corresponding flow incidence dependent drag, lift and moment coefficients that are usually applied to calculate mean static load effects. It is taken for granted that structural displacements and cross sectional rotations are small. Furthermore, it is a requirement for linearization of load components that $u(x,t)$ and $w(x,t)$ are small as compared to V. The situation is illustrated in Fig. 5.1.

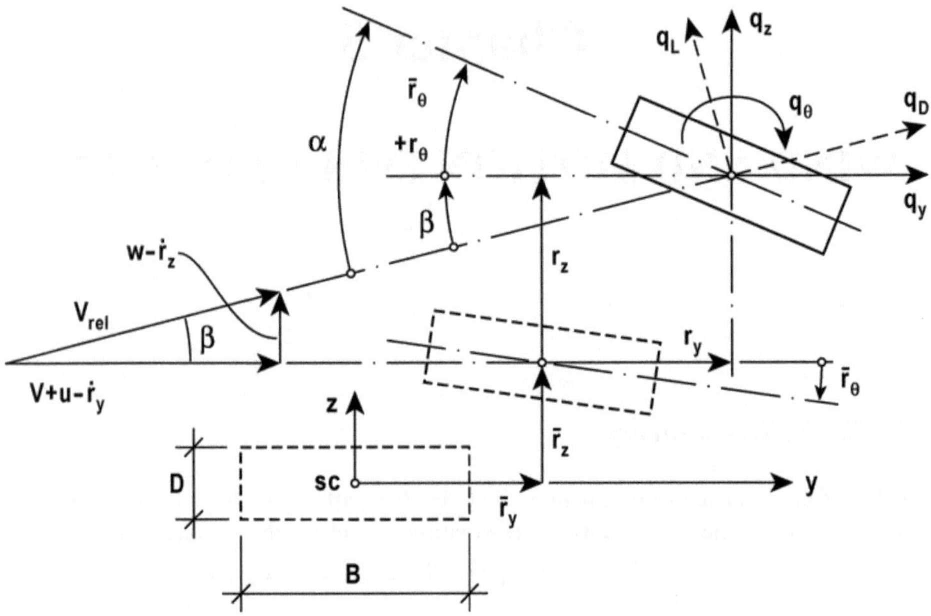

Fig. 5.1 *Instantaneous flow and displacement quantities*

As can be seen, the usual assumption that any fluctuating quantity can be split into a time invariant mean part and a zero mean fluctuating part is adopted (as previously mentioned in chapter 1.3). Thus, the cross section at an arbitrary position along the span is first given the displacements $\bar{r}_y(x)$, $\bar{r}_z(x)$ and $\bar{r}_\theta(x)$. In this position the wind velocity vector is $V+u(x,t)$ in the along wind horizontal direction and $w(x,t)$ in the vertical across wind direction. It is about this position that the structure oscillates. The cross section is then given the additional dynamic displacements $r_y(x,t)$, $r_z(x,t)$ and $r_\theta(x,t)$. In this position the instantaneous cross sectional drag, lift and moment forces in flow axes are by definition given by

$$\begin{bmatrix} q_D(x,t) \\ q_L(x,t) \\ q_M(x,t) \end{bmatrix} = \frac{1}{2}\rho V_{rel}^2 \cdot \begin{bmatrix} D \cdot C_D(\alpha) \\ B \cdot C_L(\alpha) \\ B^2 \cdot C_M(\alpha) \end{bmatrix} \tag{5.1}$$

where V_{rel} is the instantaneous relative wind velocity and α is the corresponding angle of flow incidence. Transformation into structural axis is given by

$$\mathbf{q}_{tot}(x,t) = \begin{bmatrix} q_y \\ q_z \\ q_\theta \end{bmatrix}_{tot} = \begin{bmatrix} \cos\beta & -\sin\beta & 0 \\ \sin\beta & \cos\beta & 0 \\ 0 & 0 & 1 \end{bmatrix} \cdot \begin{bmatrix} q_D \\ q_L \\ q_M \end{bmatrix} \tag{5.2}$$

where:

$$\beta = \arctan\left(\frac{w - \dot{r}_z}{V + u - \dot{r}_y}\right) \tag{5.3}$$

The first linearization involves the assumption that the fluctuating flow components $u(x,t)$ and $w(x,t)$ are small as compared to V, and that structural displacements (as well as cross sectional rotation) are also small. Then $\cos\beta \approx 1$ and $\sin\beta \approx \tan\beta \approx \beta \approx (w - \dot{r}_z)/(V + u - \dot{r}_y) \approx (w - \dot{r}_z)/V$, and thus

$$\left.\begin{aligned} V_{rel}^2 &= \left(V + u - \dot{r}_y\right)^2 + \left(w - \dot{r}_z\right)^2 \approx V^2 + 2Vu - 2V\dot{r}_y \\ \alpha &= \bar{r}_\theta + r_\theta + \beta \approx \bar{r}_\theta + r_\theta + \frac{w}{V} - \frac{\dot{r}_z}{V} \end{aligned}\right\} \tag{5.4}$$

The second linearization involves the flow incidence dependent load coefficients. As illustrated in Fig. 5.2, the nonlinear variation of the load coefficient curves is replaced by the following linear approximation

$$\begin{bmatrix} C_D(\alpha) \\ C_L(\alpha) \\ C_M(\alpha) \end{bmatrix} = \begin{bmatrix} C_D(\bar{\alpha}) \\ C_L(\bar{\alpha}) \\ C_M(\bar{\alpha}) \end{bmatrix} + \alpha_f \cdot \begin{bmatrix} C_D'(\bar{\alpha}) \\ C_L'(\bar{\alpha}) \\ C_M'(\bar{\alpha}) \end{bmatrix} \tag{5.5}$$

where $\bar{\alpha}$ and α_f are the mean value and the fluctuating part of the angle of incidence, and where C_D', C_L' and C_M' are the slopes of the load coefficient curves at $\bar{\alpha}$.

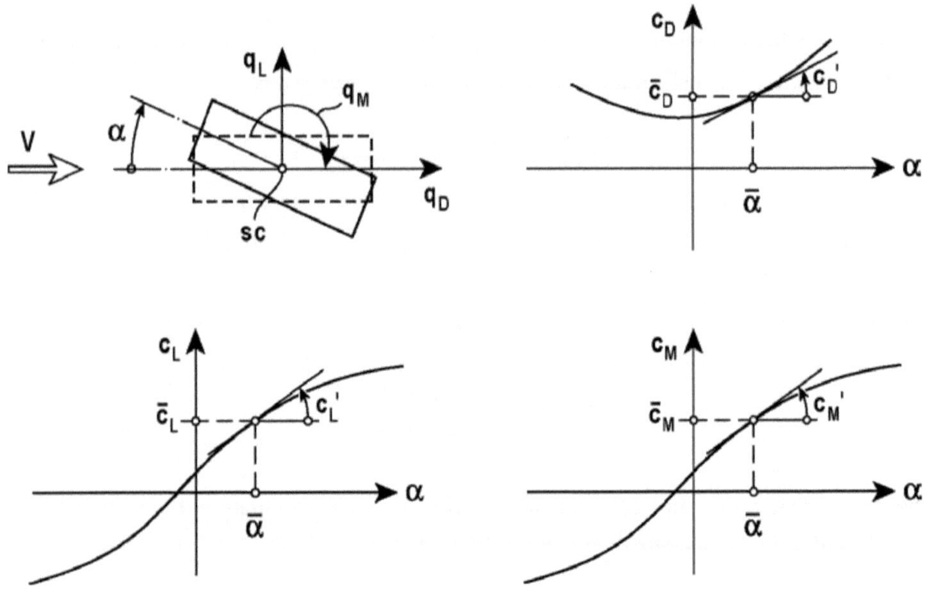

Fig. 5.2 *Load coefficients obtained from static tests*

It follows from Eq. 5.4 that $\bar{\alpha} = \bar{r}_\theta$ and $\alpha_f = r_\theta + w/V - \dot{r}_z/V$. For simplicity the following notation is introduced

$$\begin{bmatrix} C_D(\bar{\alpha}) \\ C_L(\bar{\alpha}) \\ C_M(\bar{\alpha}) \end{bmatrix} = \begin{bmatrix} \bar{C}_D \\ \bar{C}_L \\ \bar{C}_M \end{bmatrix} \quad \text{and} \quad \begin{bmatrix} C'_D(\bar{\alpha}) \\ C'_L(\bar{\alpha}) \\ C'_M(\bar{\alpha}) \end{bmatrix} = \begin{bmatrix} C'_D \\ C'_L \\ C'_M \end{bmatrix} \tag{5.6}$$

Combining Eqs. 5.2 – 5.6

$$\begin{bmatrix} q_y \\ q_z \\ q_\theta \end{bmatrix}_{tot} = \rho V \left(\frac{V}{2} + u - \dot{r}_y \right) \left\{ \begin{bmatrix} D\bar{C}_D \\ B\bar{C}_L \\ B^2\bar{C}_M \end{bmatrix} + \left(r_\theta + \frac{w}{V} - \frac{\dot{r}_z}{V} \right) \begin{bmatrix} DC'_D \\ BC'_L \\ B^2C'_M \end{bmatrix} + \frac{w - \dot{r}_z}{V} \begin{bmatrix} -B\bar{C}_L \\ D\bar{C}_D \\ 0 \end{bmatrix} \right\} \tag{5.7}$$

and discarding higher order terms (i.e. terms containing the product of quantities that have been assumed small) the following is obtained

$$\mathbf{q}_{tot}(x,t) = \begin{bmatrix} \bar{q}_y(x) \\ \bar{q}_z(x) \\ \bar{q}_\theta(x) \end{bmatrix} + \begin{bmatrix} q_y(x,t) \\ q_z(x,t) \\ q_\theta(x,t) \end{bmatrix} = \bar{\mathbf{q}} + \mathbf{B}_q \cdot \mathbf{v} + \mathbf{C}_{ae} \cdot \dot{\mathbf{r}} + \mathbf{K}_{ae} \cdot \mathbf{r} \tag{5.8}$$

where

$$\mathbf{v}(x,t) = \begin{bmatrix} u & w \end{bmatrix}^T \tag{5.9}$$

$$\mathbf{r}(x,t) = \begin{bmatrix} r_y & r_z & r_\theta \end{bmatrix}^T \tag{5.10}$$

$$\bar{\mathbf{q}}(x) = \begin{bmatrix} \bar{q}_y \\ \bar{q}_z \\ \bar{q}_\theta \end{bmatrix} = \frac{\rho V^2 B}{2} \begin{bmatrix} (D/B)\bar{C}_D \\ \bar{C}_L \\ B\bar{C}_M \end{bmatrix} = \frac{\rho V^2 B}{2} \cdot \hat{\mathbf{b}}_q \tag{5.11}$$

$$\mathbf{B}_q(x) = \frac{\rho VB}{2} \begin{bmatrix} 2(D/B)\bar{C}_D & \left((D/B)C_D' - \bar{C}_L\right) \\ 2\bar{C}_L & \left(C_L' + (D/B)\bar{C}_D\right) \\ 2B\bar{C}_M & BC_M' \end{bmatrix} = \frac{\rho VB}{2} \cdot \hat{\mathbf{B}}_q \tag{5.12}$$

$$\mathbf{C}_{ae}(x) = -\frac{\rho VB}{2} \begin{bmatrix} 2(D/B)\bar{C}_D & \left((D/B)C_D' - \bar{C}_L\right) & 0 \\ 2\bar{C}_L & \left(C_L' + (D/B)\bar{C}_D\right) & 0 \\ 2B\bar{C}_M & BC_M' & 0 \end{bmatrix} \tag{5.13}$$

$$\mathbf{K}_{ae}(x) = \frac{\rho V^2 B}{2} \begin{bmatrix} 0 & 0 & (D/B)C_D' \\ 0 & 0 & C_L' \\ 0 & 0 & BC_M' \end{bmatrix} \tag{5.14}$$

It is seen that the total load vector comprises a time invariant mean (static) part

$$\bar{\mathbf{q}}(x) = \begin{bmatrix} \bar{q}_y \\ \bar{q}_z \\ \bar{q}_\theta \end{bmatrix} = \frac{\rho V^2 B}{2} \cdot \hat{\mathbf{b}}_q \tag{5.15}$$

and a fluctuating (dynamic) part

$$\mathbf{q}(x,t) = \begin{bmatrix} q_y \\ q_z \\ q_\theta \end{bmatrix} = \mathbf{B}_q \cdot \mathbf{v} + \mathbf{C}_{ae} \cdot \dot{\mathbf{r}} + \mathbf{K}_{ae} \cdot \mathbf{r} \tag{5.16}$$

where $\mathbf{B}_q \cdot \mathbf{v}$ is the dynamic loading associated with turbulence (u and w) in the oncoming flow, while $\mathbf{C}_{ae} \cdot \dot{\mathbf{r}}$ and $\mathbf{K}_{ae} \cdot \mathbf{r}$ are motion induced loads associated with structural velocity and displacement. It is seen that linearity has been obtained, and thus, the theory is applicable in time domain as well as in frequency domain. The frequency domain amplitudes of the dynamic load are obtained by taking the Fourier transform throughout Eq. 5.16. Thus,

$$\mathbf{a}_q = \mathbf{B}_q \cdot \mathbf{a}_v + \left(i\omega \mathbf{C}_{ae} + \mathbf{K}_{ae}\right) \cdot \mathbf{a}_r \tag{5.17}$$

where:

$$\left.\begin{aligned} \mathbf{a}_q(x,\omega) &= \begin{bmatrix} a_{q_y} & a_{q_z} & a_{q_\theta} \end{bmatrix}^T \\ \mathbf{a}_r(x,\omega) &= \begin{bmatrix} a_{r_y} & a_{r_z} & a_{r_\theta} \end{bmatrix}^T \\ \mathbf{a}_v(x,\omega) &= \begin{bmatrix} a_u & a_w \end{bmatrix}^T \end{aligned}\right\} \tag{5.18}$$

and where i is the imaginary unit. Taking it for granted that the theory will primarily be applied in a modal frequency domain approach it is favourable to introduce two major improvements. First, for the purpose of frequency domain calculations it has been suggested to include frequency dependent flow induced dynamic loads, i.e. to replace $\mathbf{B}_q(x)$ in Eq. 5.12 with

$$\mathbf{B}_q(x,\omega) = \frac{\rho VB}{2} \begin{bmatrix} 2(D/B)\bar{C}_D A_{yu} & \left((D/B)C_D' - \bar{C}_L\right)A_{yw} \\ 2\bar{C}_L A_{zu} & \left(C_L' + (D/B)\bar{C}_D\right)A_{zw} \\ 2B\bar{C}_M A_{\theta u} & BC_M' A_{\theta w} \end{bmatrix} \tag{5.19}$$

where:

$$A_{mn}(\omega) \qquad \begin{cases} m = y,z,\theta \\ n = u,w \end{cases} \tag{5.20}$$

are the so-called cross sectional admittance functions. They are frequency dependent functions characteristic to the cross section in question.

In general, they may be determined from section model wind tunnel experiments, either directly from pressure tap measurements around the periphery of the cross section, or from time series of drag, lift and moment forces on the model that are otherwise used to determine mean load coefficients, in which case it is necessary to assume that the length scales of the fluctuating forces are identical to the appropriate length scales of the turbulent flow components. Cross sectional admittance functions have been theoretically developed for a thin airfoil by Sears [15], but since Sears solution is complex and contain cumbersome Bessel functions, approximate expressions, usually of the following type have been suggested (first by Liepmann [16])

$$A_{mn}(\omega) = \frac{1}{\left(1 + a_{mn} B\omega/V\right)^{b_{mn}}} \qquad \begin{cases} m = y,z,\theta \\ n = u,w \end{cases} \tag{5.21}$$

where a_{mn} and b_{mn} are cross sectional dependent constants. As can be seen,

$$\left.\begin{aligned} & A_{mn}(\omega) \le 1 \\ & A_{mn}(\omega = 0) = 1 \\ & \lim_{\omega \to \infty} A_{mn}(\omega) = 0 \end{aligned}\right\} \tag{5.22}$$

and thus, its main effect is to filter off load contributions at high frequencies. (Other expressions may be expected for complex cross sections.) The second major improvement to the frequency domain application of the buffeting theory is to replace the content of $\mathbf{C}_{ae}$ and $\mathbf{K}_{ae}$ with the so-called aerodynamic derivatives. That is dealt with in the next chapter.

5.2 *Aerodynamic derivatives*

As derived from the buffeting theory $\mathbf{C}_{ae}$ and $\mathbf{K}_{ae}$ are given in Eqs. 5.13 and 5.14. They are three by three matrices containing all the eighteen coefficients that are required for a full frequency domain description of motion induced dynamic forces associated with structural velocity and displacement. The modal frequency domain counterparts to $\mathbf{C}_{ae}$ and $\mathbf{K}_{ae}$ are first fully presented in Eq. 4.62 in chapter 4.4. (Basic assumptions are given in Eq. 4.59. $\mathbf{M}_{ae}$ is in the following considered negligible.) The essential theory presented below was first developed in the field of aeronautics and later made applicable to bridges by Scanlan & Tomko [17]. Following their notations, rather than the more general use of symbols shown in chapter 4.4, the frequency domain versions of $\mathbf{C}_{ae}$ and $\mathbf{K}_{ae}$ are given by

$$\mathbf{C}_{ae} = \begin{bmatrix} P_1 & P_5 & P_2 \\ H_5 & H_1 & H_2 \\ A_5 & A_1 & A_2 \end{bmatrix} \quad \text{and} \quad \mathbf{K}_{ae} = \begin{bmatrix} P_4 & P_6 & P_3 \\ H_6 & H_4 & H_3 \\ A_6 & A_4 & A_3 \end{bmatrix} \tag{5.23}$$

The coefficients contained in $\mathbf{C}_{ae}$ and $\mathbf{K}_{ae}$ are then functions of the frequency of motion, the mean wind velocity and the type of cross section (and to some extent the initial or mean angle of incidence and the turbulence properties in the oncoming flow). Usually, they have been experimentally determined in wind tunnel aeroelastic section model tests, limited to vertical and torsion displacements. Since their main use lies in the detection of unstable motion at high wind velocities, the primary modal mass and stiffness properties of the section model will intentionally only contain the eigen-frequencies associated with the most onerous modes with respect to unstable structural oscillations. For a plate-like bridge cross section this is usually the lowest mode in torsion together with the shape-wise similar and lowest vertical mode. (Shape-wise similarity is required because the effect of aerodynamic coupling between the two modes is often important.) Since the along wind motion is absent in the section model, all terms associated with this direction must either be disregarded or taken from the quasi static buffeting theory (see Eqs. 5.25 and 5.26). The tests may be performed in three alternative ways. The original procedure was to extract the motion induced forces from the changes in resonance frequency and damping properties in transient (i.e. decay) recordings at various wind velocities under the conditions of pure vertical motion, pure torsion and finally combined vertical and torsion (see appendix C). Another procedure is to perform ambient vibration tests, again at various wind velocities, and use the theory of system identification to extract the sought flow-structure interaction properties. The third procedure is to use a section model that undergoes forced oscillations at various frequencies, amplitudes and wind velocities. From such a steady-state situation cross sectional forces are measured by pressure tap recordings on the surface of the model hull. Subtraction of the forces at zero motion will then render net motion induced effects. The method of forced oscillations is demanding and generally not in use. Thus, the frequency at which the aerodynamic derivatives are determined will most often be associated with the mass and stiffness properties of the relevant section model, as well as the motion induced forces themselves at various mean wind velocities. I.e., the aerodynamic derivatives will be associated with the eigen-frequencies of the chosen set of section model mode shapes, and thus, they will be functions of the reduced velocity $\hat{V} = V/(B\omega_i)$. For the purpose of full scale calculations the similarity requirements between model scale and full scale conditions must be fulfilled, and thus, the aerodynamic derivatives will have to be extracted as functions of $\hat{V} = V/(B\omega_i)$.

Furthermore, it has been considered convenient to normalise $\mathbf{C}_{ae}$ and $\mathbf{K}_{ae}$ with $\rho B^2\omega_i/2$ and $\rho B^2\omega_i^2/2$, where ω_i is the in-wind (mean wind velocity dependent) resonance frequency associated with the mode shape (number i) from which they have been extracted. Thus,

$$\mathbf{C}_{ae} = \frac{\rho B^2}{2}\cdot\omega_i(V)\cdot\hat{\mathbf{C}}_{ae} \qquad \text{and} \qquad \mathbf{K}_{ae} = \frac{\rho B^2}{2}\cdot\left[\omega_i(V)\right]^2\cdot\hat{\mathbf{K}}_{ae} \tag{5.24}$$

where

$$\hat{\mathbf{C}}_{ae} = \begin{bmatrix} P_1^* & P_5^* & BP_2^* \\ H_5^* & H_1^* & BH_2^* \\ BA_5^* & BA_1^* & B^2A_2^* \end{bmatrix} \qquad \text{and} \qquad \hat{\mathbf{K}}_{ae} = \begin{bmatrix} P_4^* & P_6^* & BP_3^* \\ H_6^* & H_4^* & BH_3^* \\ BA_6^* & BA_4^* & B^2A_3^* \end{bmatrix} \tag{5.25}$$

It is the non–dimensional coefficients P_k^*, H_k^*, A_k^*, $k = 1-6$ that are usually called aerodynamic derivatives. The values that emerge from the buffeting theory are obtained by comparison to Eqs. 5.13 and 5.14, rendering quasi-static aerodynamic derivartives

$$\begin{bmatrix} P_1^* & H_1^* & A_1^* \\ P_2^* & H_2^* & A_2^* \\ P_3^* & H_3^* & A_3^* \\ P_4^* & H_4^* & A_4^* \\ P_5^* & H_5^* & A_5^* \\ P_6^* & H_6^* & A_6^* \end{bmatrix} = \begin{bmatrix} -2\bar{C}_D\dfrac{D}{B}\dfrac{V}{B\omega_i(V)} & -\left(C_L' + \bar{C}_D\dfrac{D}{B}\right)\dfrac{V}{B\omega_i(V)} & -C_M'\dfrac{V}{B\omega_i(V)} \\ 0 & 0 & 0 \\ C_D'\dfrac{D}{B}\left(\dfrac{V}{B\omega_i(V)}\right)^2 & C_L'\left(\dfrac{V}{B\omega_i(V)}\right)^2 & C_M'\left(\dfrac{V}{B\omega_i(V)}\right)^2 \\ 0 & 0 & 0 \\ \left(\bar{C}_L - C_D'\dfrac{D}{B}\right)\dfrac{V}{B\omega_i(V)} & -2\bar{C}_L\dfrac{V}{B\omega_i(V)} & -2\bar{C}_M\dfrac{V}{B\omega_i(V)} \\ 0 & 0 & 0 \end{bmatrix} \tag{5.26}$$

As shown in Eq. 5.26, the aerodynamic derivatives will be functions of the reduced velocity $V/\left[\omega_i(V)B\right]$. It should be noted that in the determination of the reduced velocity [or the non-dimensional resonance frequency $\hat{\omega}_i = B\omega_i(V)/V$] the resonance frequency $\omega_i(V)$ is a function of the mean wind velocity, V. To start off with, i.e. at $V = 0$, $\omega_i(V = 0)$ is the eigen-frequency in still air conditions. It is then only dependent on the relevant structural properties. At $V \neq 0$ the aerodynamic derivatives contained in $\mathbf{K}_{ae}$ will have the effect of changing the total stiffness of the combined structure and flow system. This implies that the resonance frequency at $V \neq 0$ is different from the initial value that was determined at $V = 0$ (or in vacuum). In general the consequence of this effect is that any response calculation involving the aerodynamic derivatives contained in $\mathbf{K}_{ae}$ will demand iterations. However, under normal circumstances the effects of $\mathbf{K}_{ae}$ will only be of significant importance in the velocity region at or immediately below an instability limit. At a characteristic mean wind velocity well below such an instability limit it is usually the aerodynamic derivatives contained in $\mathbf{C}_{ae}$ that play the leading role, and the effects of the changes of ω_i with increasing V to the determination of the aerodynamic derivatives are most often only of minor importance, especially as compared to other uncertainties in the theory (see further discussion in chapters 6.3 and 8). On the other hand, at or in the vicinity of an instability limit the flow induced changes to the resonance frequency will in most cases be of great importance, and thus, for the determination of an instability limit this effect can usually not be ignored (see chapter 8).

Aerodynamic derivatives for an ideal flat plate type of cross section were first developed by Theodorsen [28]. They are given by:

$$\begin{bmatrix} H_1^* & A_1^* \\ H_2^* & A_2^* \\ H_3^* & A_3^* \\ H_4^* & A_4^* \end{bmatrix} = \begin{bmatrix} -2\pi F\hat{V}_i & -\frac{\pi}{2}F\hat{V}_i \\ \frac{\pi}{2}\left(1+F+4G\hat{V}_i\right)\hat{V}_i & -\frac{\pi}{8}\left(1-F-4G\hat{V}_i\right)\hat{V}_i \\ 2\pi\left(F\hat{V}_i - G/4\right)\hat{V}_i & \frac{\pi}{2}\left(F\hat{V}_i - G/4\right)\hat{V}_i \\ \frac{\pi}{2}\left(1+4G\hat{V}_i\right) & \frac{\pi}{2}G\hat{V}_i \end{bmatrix} \tag{5.27}$$

where $\hat{V}_i = V/\left[B\omega_i(V)\right]$ is the reduced velocity, and

$$\left.\begin{aligned} F\left(\frac{\hat{\omega}_i}{2}\right) &= \frac{J_1\cdot(J_1+Y_0)+Y_1\cdot(Y_1-J_0)}{(J_1+Y_0)^2+(Y_1-J_0)^2} \\ G\left(\frac{\hat{\omega}_i}{2}\right) &= -\frac{J_1\cdot J_0+Y_1\cdot Y_0}{(J_1+Y_0)^2+(Y_1-J_0)^2} \end{aligned}\right\} \tag{5.28}$$

are the real and imaginary parts of the so-called Theodorsen's circulatory function. Their content $J_n(\hat{\omega}_i/2)$ and $Y_n(\hat{\omega}_i/2)$, $n = 0$ or 1, are first and second kinds of Bessel functions with order n, and $\hat{\omega}_i$ is the non-dimensional resonance frequency, i.e. $\hat{\omega}_i = B\omega_i(V)/V = \hat{V}^{-1}$. The flat plate aerodynamic derivatives given in Eq. 5.27 are plotted in Fig. 5.3. (The division of $\hat{\omega}$ with 2 in Eq. 5.28 stems from Theodorsen's choice of frequency normalization with $B/2$ rather than B which is chosen herein.).

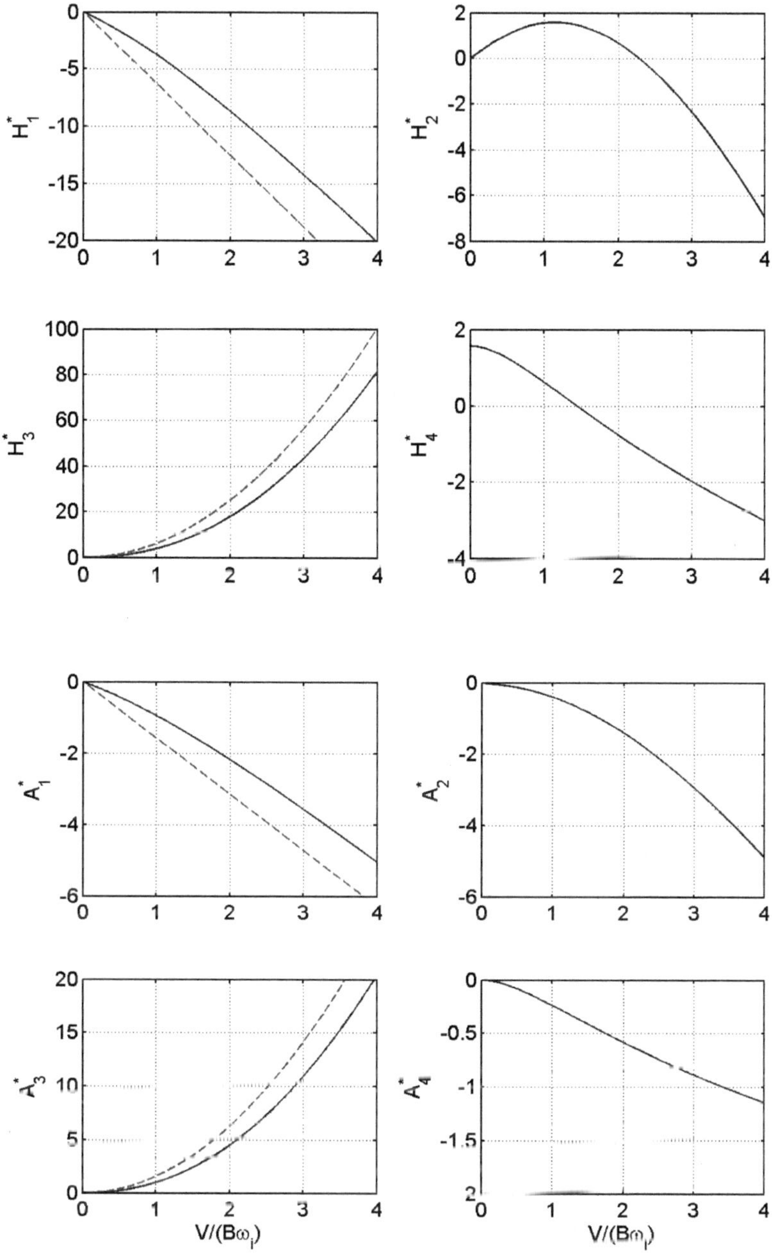

Fig. 5.3 *Flat plate aerodynamic derivatives (broken lines are the quasi- static values)*

The aerodynamic derivatives for a flat plate that emerge from the buffeting theory (i.e. quasi-static values) are obtained from Eq. 5.26 by the introduction of

$$\begin{bmatrix} \bar{C}_D & C'_D \\ \bar{C}_L & C'_L \\ \bar{C}_M & C'_M \end{bmatrix} = \begin{bmatrix} 0 & 0 \\ 0 & 2\pi \\ 0 & \pi/2 \end{bmatrix} \tag{5.29}$$

Thus, for a flat plate the non-zero quasi-static aerodynamic derivatives are given by

$$\begin{bmatrix} H_1^* & A_1^* \\ H_3^* & A_3^* \end{bmatrix} = \begin{bmatrix} -2\pi\hat{V}_i & -\frac{\pi}{2}\hat{V}_i \\ 2\pi\hat{V}_i^2 & \frac{\pi}{2}\hat{V}_i^2 \end{bmatrix} \tag{5.30}$$

5.3 *Vortex shedding*

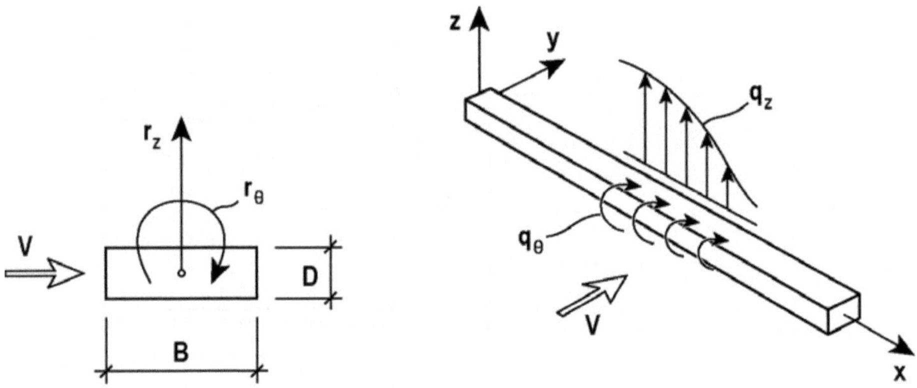

Fig. 5.4 *Relevant displacement components and vortex shedding forces*

When the air flow is met by a solid bridge or tower type of structure flow separation will occur on the surface of the structure causing vortices to be shed alternately on either side of the structure.

Assuming that along wind load effects may be disregarded, these vortices give rise to fluctuating across wind forces q_z and cross sectional torsion moment q_θ, accompanied by fluctuating displacements r_z or r_θ, as shown in Fig. 5.4. Harmful vortex induced vibrations may particularly occur in cases of resonance.

Experimental investigations of the single point q_m process ($m = z$ or θ) on stiff models where $r_m \approx 0$ show that fluctuating loads are more or less narrow banded centred at a vortex shedding frequency f_s, as illustrated in Fig. 5.5.a. The properties of the shedding frequency are characteristic to the cross section of the line-like structure. It is proportional to the mean wind velocity V and inversely proportional to the across wind width D. Thus,

$$f_s = St \cdot \frac{V}{D} \tag{5.31}$$

where St is the Strouhal number, which is available for a good number of typical structural cross sections in the literature. Two-dimensional investigations also show that q_m has a more or less random distribution in the span-wise direction, as illustrated on the right hand side of Fig. 5.4. and indicated by the decaying co-spectrum in Fig. 5.5.b.

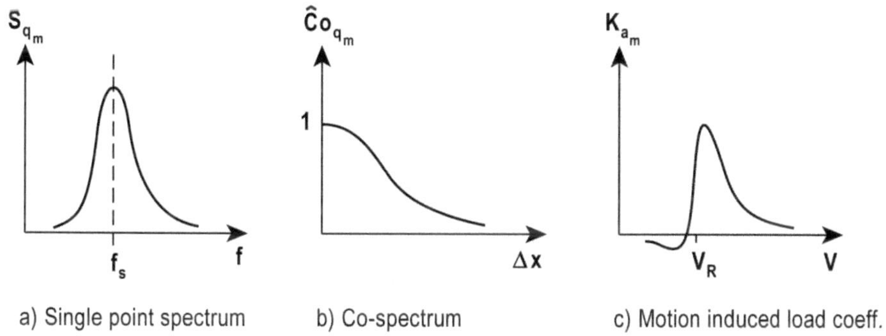

Fig. 5.5 *Load characteristics associated with vortex shedding*

Turning to a flexible structure it is assumed that the properties of f_s are maintained, i.e. that Eq. 5.31 still holds. The situation is illustrated in Fig. 5.6. Assuming that V is slowly increasing (from zero), then f_s will increase accordingly, and resonance will first occur when f_s becomes equal to the lowest eigen-frequency with respect to vibrations in the across wind direction or torsion. Further increase of V will cause resonance to occur when f_s is equal to the next eigen frequency, and so on. Theoretically, resonance will occur when f_s is equal to any eigen-frequency f_i. According to Eq. 5.31, the event that $f_s = f_i$ will occur when the mean wind velocity has a value given by

$$V_{R_i} = f_i D / St \tag{5.32}$$

Thus, there is a resonance velocity for every eigen-frequency associated with vibrations in the across wind direction or in torsion. Experiments show that when resonance occurs the flow and the oscillating structure will interact, and for a certain range of ensuing wind velocity settings f_s will deviate from Eq. 5.31 and stay equal or close to f_i, as shown on the upper right hand side of Fig. 5.6. This is what is usually called lock-in. Such vortex shedding induced interaction is accompanied by two important load effects. At lock-in the fluctuating load becomes better correlated in the span-wise direction, but what is more important is that a significant motion induced part is added. However, these effects are self-destructive in the sense that they diminish when fluctuating structural displacements become large. Thus, vortex shedding induced vibrations are self-limiting, as illustrated on the diagram on the lower right hand side in Fig. 5.6.

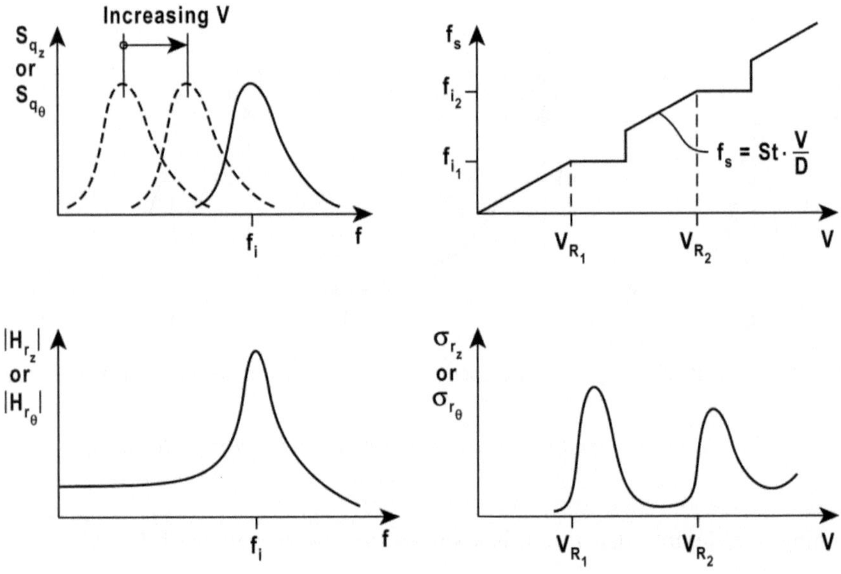

Fig. 5.6 *Response characteristics associated with vortex shedding*

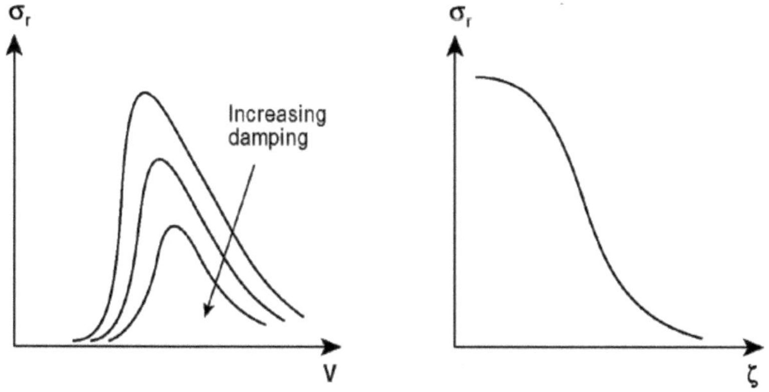

Fig. 5.7 *Vortex shedding induced response characteristics at different levels of structural eigen-damping*

Experiments show that vortex induced vibrations are greatly affected by the damping properties of the structure, as shown on the left hand side diagram in Fig. 5.7. The motion induced self-limiting and damping dependent nature of vortex induced vibrations is further illustrated on the diagram to the right in Fig. 5.7. Thus, it has been customary to ascribe the motion induced part of the load to the structural displacements and the velocity, i.e. $q_m = q_m(x,t,r_m,\dot{r}_m)$, where $m = z$ or θ.

Extensive research has been carried out on the investigation of vortex shedding induced vibrations. In the following it is the theory first developed by Vickery & Basu [18, 19] that will be presented. The motivation is convenience, as it is the only comprehensive stochastic frequency domain theory currently available, rendering a solution at any setting of the mean wind velocity. An alternative approach applicable at resonance has been presented by Ruscheweyh [20].

In the theory as developed by Vickery & Basu the description of the net motion–independent cross sectional load spectra and corresponding co-spectra are shape-wise shown in Fig. 5.5. Mathematically they are given by

$$\begin{bmatrix} S_{q_z}(\omega) \\ S_{q_\theta}(\omega) \end{bmatrix} = \frac{\left(\frac{1}{2}\rho V^2\right)^2}{\sqrt{\pi}\cdot\omega_s} \cdot \begin{bmatrix} \dfrac{\left(B\cdot\hat{\sigma}_{q_z}\right)^2}{b_z}\cdot\exp\left\{-\left(\dfrac{1-\omega/\omega_s}{b_z}\right)^2\right\} \\ \dfrac{\left(B^2\cdot\hat{\sigma}_{q_\theta}\right)^2}{b_\theta}\cdot\exp\left\{-\left(\dfrac{1-\omega/\omega_s}{b_\theta}\right)^2\right\} \end{bmatrix} \tag{5.33}$$

and

$$\hat{C}o_{q_m}(\Delta x) = \cos\left(\frac{2}{3}\frac{\Delta x}{\lambda_m D}\right)\cdot\exp\left[-\left(\frac{\Delta x}{3\lambda_m D}\right)^2\right] \tag{5.34}$$

where $m = z$ or θ, $\omega_s = 2\pi f_s$, $\hat{\sigma}_{q_m}$ is the non-dimensional root mean square lift or torsion moment coefficient, b_m is a non-dimensional load spectrum band width parameter, λ_m is a non-dimensional coherence length scale and Δx is span–wise separation. [By substituting $\frac{\Delta x}{\lambda D} = \alpha$, $a = 1/3$ and $b = 2/3$, and using the known integral $\int_0^\infty \cos(b\alpha)\cdot\exp\left[-(a\alpha)^2\right]d\alpha = \frac{\sqrt{\pi}}{2a}\cdot\exp\left[-(b/2a)^2\right]$ it may readily be shown that $\int_0^\infty \hat{C}o_{q_m}(\Delta x)d(\Delta x) = \frac{3\sqrt{\pi}}{2}\cdot e^{-1}\cdot\lambda D \approx 0.9778\cdot\lambda D \approx \lambda D$.]

In general, $\hat{\sigma}_{q_z}$ increases with increasing bluffness of the cross section, b_z attains values between 0.1 and 0.3, while λ_z is typically in the order of 2 to 5. Similar properties may be expected of q_θ.

For the description of the characteristic motion induced load effects at "lock-in" Vickery & Basu [18, 19] have suggested that this may be accounted for by a negative motion dependent aerodynamic modal damping ratio, ζ_{ae_i}, such that the total modal damping ratio associated with mode i is given by

$$\zeta_{tot_i} = \zeta_i - \zeta_{ae_i} \tag{5.35}$$

This is equivalent to the introduction of motion dependent aerodynamic derivatives as described in chapter 5.2 above. Adopting the notation given in Eqs. 5.24 and 5.25, it is the aerodynamic derivatives H_1^* and A_2^* that are responsible for aerodynamic damping exclusively effective in the across wind vertical (z) direction or in torsion (θ). Assuming that in the vicinity of a distinct vortex shedding type of response all other motion induced effects may be ignored, then

$$\mathbf{C}_{ae} \approx \frac{\rho B^2}{2}\omega_i(V)\begin{bmatrix} 0 & 0 & 0 \\ 0 & H_1^* & 0 \\ 0 & 0 & B^2 A_2^* \end{bmatrix} \quad \text{and} \quad \mathbf{K}_{ae} \approx 0 \tag{5.36}$$

where

$$H_1^* = K_{a_z}\left[1-\left(\frac{\sigma_z}{a_z D}\right)^2\right] \quad \text{and} \quad A_2^* = K_{a_\theta}\left[1-\left(\frac{\sigma_\theta}{a_\theta}\right)^2\right] \tag{5.37}$$

and where K_{a_z} and K_{a_θ} are the velocity dependent damping coefficients equivalent to those defined by Vickery & Basu [18, 19]. (However, if appropriate experimental evidence is available, there is no reason why $\mathbf{C}_{ae}$ and $\mathbf{K}_{ae}$ should not be full three by three matrices, also in the region of distinct vortex shedding excitation.) Assuming that $\omega_i(V) \approx \omega_i(V=0)$, then the aerodynamic damping term in Eq. 5.35 may be taken from Eq. 4.73, and thus,

$$\left.\begin{aligned} \zeta_{ae_i} &= \frac{\tilde{C}_{ae_i}}{2\omega_i \tilde{M}_i} = \frac{\int_{L_{\exp}} \boldsymbol{\varphi}_i^T \cdot \mathbf{C}_{ae} \cdot \boldsymbol{\varphi}_i dx}{2\omega_i \tilde{m}_i \int_L \boldsymbol{\varphi}_i^T \cdot \boldsymbol{\varphi}_i dx} \\ \Rightarrow \zeta_{ae_i} &= \frac{\rho B^2}{4\tilde{m}_i} \cdot \frac{\int_{L_{\exp}} \left(H_1^* \phi_z^2 + B^2 A_2^* \phi_\theta^2 \right) dx}{\int_L \left(\phi_y^2 + \phi_z^2 + \phi_\theta^2 \right) dx} \end{aligned}\right\} \tag{5.38}$$

where

$$\tilde{m}_i = \frac{\tilde{M}_i}{\int_L \boldsymbol{\varphi}_i^T \cdot \boldsymbol{\varphi}_i dx} = \frac{\tilde{M}_i}{\int_L \left(\phi_y^2 + \phi_z^2 + \phi_\theta^2 \right) dx} \tag{5.39}$$

are the evenly distributed and modally equivalent masses associated with mode i. K_{a_m} ($m = z$ or θ) are the coefficients that account for the accelerating part of the motion induced load when V is close to V_{R_i}. Apart from being cross sectional characteristics, they are functions of V and the resonance frequency of the mode in question (see right hand side diagram in Fig. 5.5). $a_z D$ and a_θ are quantities associated with the self-limiting nature of vortex shedding, i.e. they represent upper displacement or rotation limits at which the aerodynamic damping becomes insignificant.

It should be noted that in Eq. 5.37 the damping coefficients are defined such that consistency is obtained with the general definition of aerodynamic derivatives in Eqs. 5.24 and 5.25 rather than the definition adopted by Vickery & Basu [18, 19]. [Thus, the K_{a_z} values given by Vickery & Basu in references [18, 19] are applicable in the expressions given above if they are multiplied by $4(D/B)^2$. Vickery & Basu have not given any recommendations regarding the K_{a_θ} coefficient.]

It should also be noted that vortex shedding effects are to some extent dependent on the Reynolds number ($\mathrm{Re} = VD/\upsilon$ where $\upsilon = 1.5 \cdot 10^{-5}$ m^2/s is the kinematic viscosity of air) and of the turbulence properties in the oncoming flow. Information about these effects is presented by Simui & Scanlan [4] and by Dyrbye & Hansen [5]. For a tubular cross section the Reynolds number effect is to change the point of flow separation, thus changing the Strouhal number as well as the load intensity. The main effect of turbulence is to broaden the band-width and disturb the size and coherence of the pressure fluctuations on the surface of the structure. Most structures are more prone to vortex induced oscillations in smooth flow.

Above, only the effects in the across wind direction and torsion have been included. In general, vortex shedding will also generate more or less narrow–banded load fluctuations in the along wind direction, but at a frequency twice that which occurs in the across wind direction and for most bridges at an insignificant load intensity.

Chapter 6

WIND INDUCED STATIC AND DYNAMIC RESPONSE CALCULATIONS

6.1 *Introduction*

The wind induced dynamic response calculations dealt with in this chapter focus on structural displacements for a line-like type of bridge structure. The calculations of corresponding cross sectional forces are shown in chapter 7. The problem at hand is illustrated in Fig. 6.1.

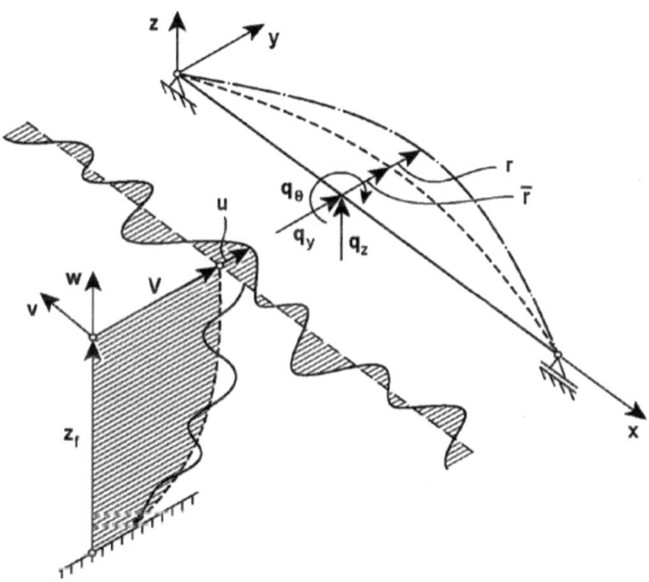

Fig. 6.1 *Simple bridge structural system subject to fluctuating wind field*

As shown, the velocity components in a wind field vary in time and space. When a structure is subject to such a fluctuating wind field, the passing of the flow will generate fluctuating drag, lift and moment loads on the structure, which in turn will cause the structure to oscillate. The time domain chain of events is illustrated in Fig. 6.2.a. From a design point of view the main focus is on quantifying the maximum value of the response that is most critical with respect to structural safety.

The flow is in general assumed Gaussian, stationary and homogeneous over a certain short term period T (e.g. 10 min), i.e. the response calculations are performed for a chosen design weather condition that is stable in time and space. If the mathematical transfer from flow properties to forces is linear and the structure is linear elastic, then the assumption of Gaussian and stationary properties also holds for any structural response quantity. Thus, any response quantity (e.g. a displacement) may be described by its mean value and probability density distribution, as shown to the right in Fig. 6.2.a. Its maximum value at position x_r is then given by

$$r_{\max}(x_r) = \bar{r}(x_r) + k_p \cdot \sigma_r(x_r) \tag{6.1}$$

where $\bar{r}(x_r)$ is the mean value, k_p is the peak factor that depends on the type of process (see chapter 2.4) and $\sigma_r(x_r)$ is the standard deviation of the fluctuating part of the response.

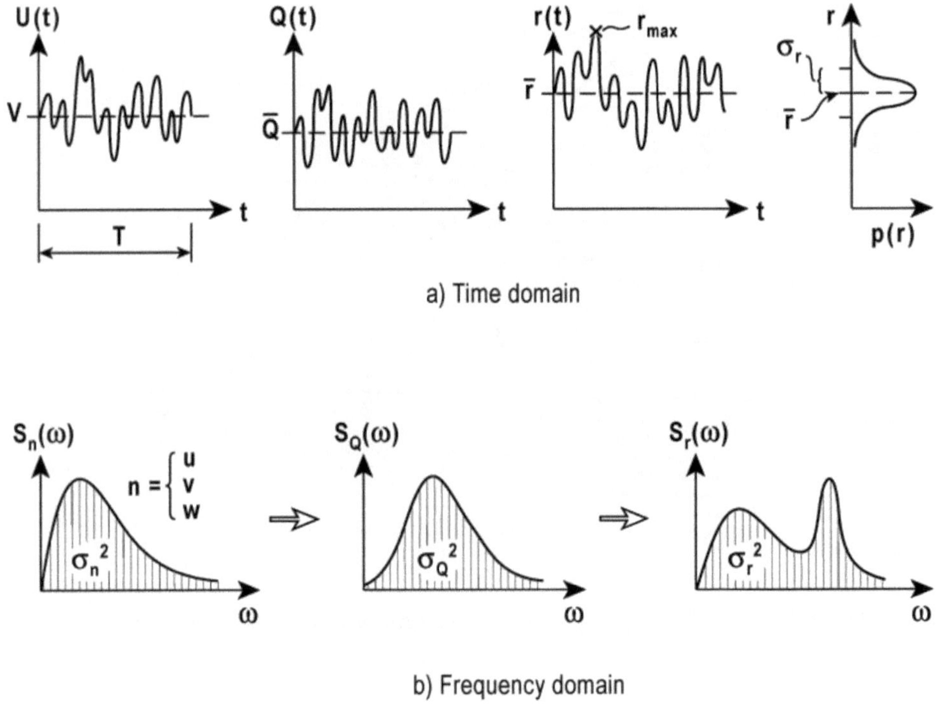

Fig. 6.2 *Time and frequency domain representations*

The mean value $\bar{r}(x_r)$ may be obtained from simple static equilibrium conditions. The standard deviation of the fluctuating part of the response $\sigma_r(x_r)$ may either be obtained from a time domain integration of the dynamic load effects from the fluctuating flow field and possible vortex shedding, or from a modal approach in frequency domain. The former alternative is computationally a demanding task, as it requires the time domain simulation of a wind field that is usually broad banded and spatially un-correlated (such a simulation procedure is shown in appendix A). In the following it is the alternative of a modal frequency domain approach that is presented. As illustrated in Fig. 6.2.b, the main steps involve the transfer from a wind field cross-spectral density via a corresponding modal load spectrum to the final sought response spectrum. The area under the response spectrum is then the variance σ_r^2 of the response.

As shown in Eq. 1.6 (and illustrated in Fig. 1.3.b) the cross sectional displacement at a position x_r that has been chosen for the relevant response calculation is in general a vector containing three components: r_y in the along wind horizontal direction, r_z in the across wind (for a bridge) vertical direction and the cross sectional rotation r_θ. Since these describe a combined cross sectional displacement in a plane perpendicular to the span, the peak factor in Eq. 6.1 is equally applicable to each of the components, and thus

$$\mathbf{r}_{\max}(x_r) = \begin{bmatrix} r_y \\ r_z \\ r_\theta \end{bmatrix}_{\max} = \begin{bmatrix} \bar{r}_y \\ \bar{r}_z \\ \bar{r}_\theta \end{bmatrix} + k_p \cdot \begin{bmatrix} \sigma_{r_y} \\ \sigma_{r_z} \\ \sigma_{r_\theta} \end{bmatrix} \tag{6.2}$$

As mentioned above (and further discussed in chapter 1), the wind induced response of a slender structure is assumed stationary, and then the total response may be split into a mean (static) and a fluctuating (dynamic) part. What can in general be expected in the case of a slender structure is illustrated in Fig. 6.3. The static part is proportional to the mean velocity pressure, i.e. to the mean wind velocity squared, until motion induced forces may reduce the total stiffness of the combined structure and flow system, after which the static response may approach an instability limit (torsion divergence). The dynamic part of the response may conveniently be separated into three mean wind velocity regions. Vortex shedding effects will usually occur at fairly low mean wind velocities, buffeting will usually be the dominant effect in an intermediate velocity region, while at high wind velocities motion induced load effects may entirely govern the response. Such a partition should not be taken literally, as there are no tight borders. E.g., important motion induced load effects may also occur in what seems like a typical vortex shedding or buffeting behaviour. In the vicinity of a certain limiting (critical) mean wind velocity the response curve may increase rapidly, i.e. the structure shows signs of unstable behaviour in the sense that a small increase of V implies a large increase of static or dynamic response, indicating an upper stability limit (V_{cr}).

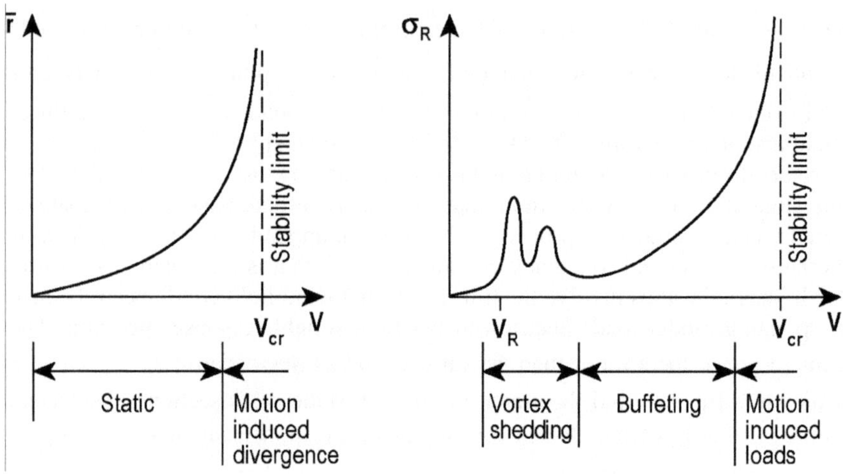

Fig. 6.3 *Typical response variation with mean wind velocity*

Finally, a comment regarding the use of aerodynamic derivatives is appropriate. As discussed in chapter 5.2, motion induced forces may change the combined flow and structural stiffness (as well as damping), and thus, the current resonance frequencies are functions of the mean wind velocity. In the dynamic response calculations below this effect is fully included in the relevant frequency-response-functions. However, in the quantification of aerodynamic derivatives and their contribution to total stiffness and damping it is assumed that the effect of changing resonance frequencies may be ignored. For the response calculations in this chapter motion induced load effects may then be taken at a reduced velocity $V/(\omega_i B)$ where ω_i is the predetermined resonance frequency based on structural properties alone and at $V = 0$. Otherwise, iterations are required. Thus, it is assumed that the response calculations are not taken in close vicinity to a motion induced instability limit. However, in the determination of an instability limit as shown in chapter 8, this effect can not be ignored, and ω_i will be taken at the relevant critical wind velocity, V_{cr}. Thus, the determination of V_{cr} in chapter 8 will demand iterations.

6.2 The mean value of the response

The mean value of the response is the load effects of the mean flow induced load as defined in Eq. 5.11. It may readily be calculated according to standard static equilibrium type of procedures in structural mechanics. Such procedures are in general mathematically formulated within a finite element type of description where the solution strategy is based on the displacement method, i.e. for a chosen discrete model containing N number of nodes the mean displacement vector $\bar{\mathbf{r}}$ is obtained from

$$\mathbf{K} \cdot \bar{\mathbf{r}} = \bar{\mathbf{R}} \tag{6.3}$$

where $\mathbf{K}$ is the static stiffness matrix and $\bar{\mathbf{R}}$ is the mean load vector. A line like structure will in general be modelled by beam or beam-column type of elements, in which case there will usually be six degrees of freedom in each node (as illustrated in Fig. 6.4.a). Thus, $\bar{\mathbf{r}}$ and $\bar{\mathbf{R}}$ are $6 \cdot N$ by one vectors and $\mathbf{K}$ is a $6 \cdot N$ by $6 \cdot N$ matrix. Herein, the establishment of $\mathbf{K}$ and the ensuing strategy for the calculation of $\bar{\mathbf{r}}$ will not be further pursued. However, the establishment of $\bar{\mathbf{R}}$ is presented below.

Let us consider a typical finite element type of modelling with six load components in each node. According to Eq. 5.11 the mean value of the evenly distributed load on an element is given by

$$\bar{\mathbf{q}}(x) = \begin{bmatrix} \bar{q}_y \\ \bar{q}_z \\ \bar{q}_\theta \end{bmatrix} = \frac{\rho V^2}{2} \cdot \mathbf{b}_q \qquad \text{where} \qquad \mathbf{b}_q = \begin{bmatrix} D\bar{C}_D \\ B\bar{C}_L \\ B^2\bar{C}_M \end{bmatrix} \tag{6.4}$$

At an arbitrary node p the load contribution from an adjoining element m (see Fig. 6.4.b and c) is then

$$\bar{\mathbf{Q}}_{pm} = \begin{bmatrix} \bar{Q}_y \\ \bar{Q}_z \\ \bar{Q}_\theta \end{bmatrix} = \bar{\mathbf{q}}_m(x) \cdot \frac{L_m}{2} = \left(\frac{\rho V^2}{2} \right)_p \cdot \frac{L_m}{2} \cdot \mathbf{b}_{qm} \tag{6.5}$$

where L_m is the element length, $\mathbf{b}_{qm}$ is the $\mathbf{b}_q$ vector that contains the properties associated with element m , and where it has for simplicity been assumed that the nodal discretisation is such that $\bar{\mathbf{q}}$ may be taken constant within the length of the element (otherwise, $L_m/2$ may be replaced by the result of a simple span-wise integration).

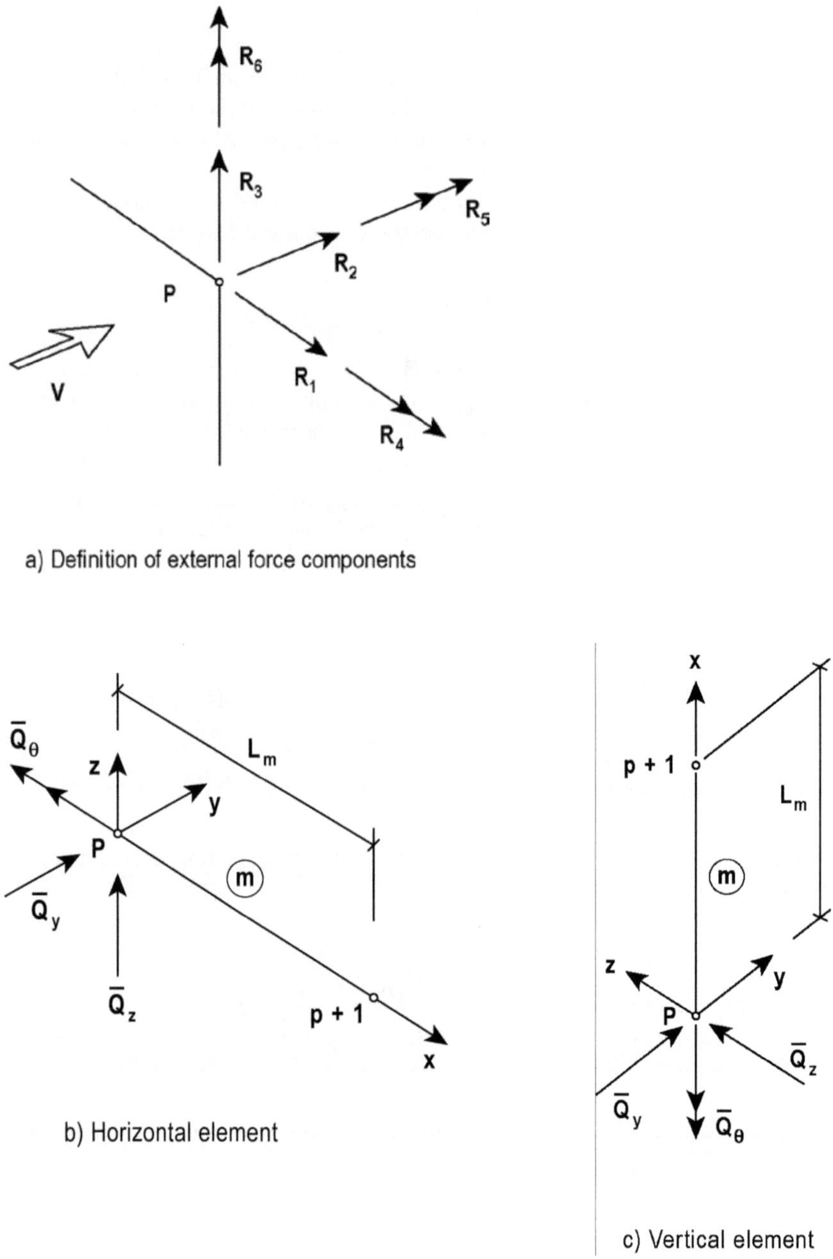

Fig. 6.4 *Wind induced mean load components*

Comparing the situation in Fig. 6.4.b and c to the general definition of external load components in Fig. 6.4.a, it is then seen that the contribution from $\bar{\mathbf{Q}}_{p_m}$ to the load vector is

$$\bar{\mathbf{R}}_{p_m} = \begin{bmatrix} R_1 & R_2 & R_3 & R_4 & R_5 & R_6 \end{bmatrix}_{p_m}^T = \begin{bmatrix} 0 & \bar{Q}_y & \bar{Q}_z & -\bar{Q}_\theta & 0 & 0 \end{bmatrix}_{p_m}^T \tag{6.6}$$

if m is horizontal, and

$$\bar{\mathbf{R}}_{p_m} = \begin{bmatrix} R_1 & R_2 & R_3 & R_4 & R_5 & R_6 \end{bmatrix}_{p_m}^T = \begin{bmatrix} -\bar{Q}_z & \bar{Q}_y & 0 & 0 & 0 & -\bar{Q}_\theta \end{bmatrix}_{p_m}^T \tag{6.7}$$

if m is vertical. Thus,

$$\bar{\mathbf{R}}_{p_m} = \boldsymbol{\theta}_m \cdot \bar{\mathbf{Q}}_{p_m} = \left(\frac{\rho V^2}{2} \right)_p \cdot \frac{L_m}{2} \cdot \boldsymbol{\theta}_m \cdot \mathbf{b}_{q_m} \tag{6.8}$$

where

$$\boldsymbol{\theta}_m = \begin{bmatrix} 0 & 0 & 0 \\ 1 & 0 & 0 \\ 0 & 1 & 0 \\ 0 & 0 & -1 \\ 0 & 0 & 0 \\ 0 & 0 & 0 \end{bmatrix} \text{ if } m \text{ is horizontal, and } \boldsymbol{\theta}_m = \begin{bmatrix} 0 & -1 & 0 \\ 1 & 0 & 0 \\ 0 & 0 & 0 \\ 0 & 0 & 0 \\ 0 & 0 & 0 \\ 0 & 0 & -1 \end{bmatrix} \text{ if } m \text{ is vertical.}$$

The six by one load vector $\bar{\mathbf{R}}_p$ in node p is then given by the sum of the contributions from all adjoining elements, i.e.

$$\bar{\mathbf{R}}_p = \begin{bmatrix} R_1 & R_2 & R_3 & R_4 & R_5 & R_6 \end{bmatrix}_p^T = \sum_m \bar{\mathbf{R}}_{p_m} \tag{6.9}$$

and the total $6 \cdot N$ by one load vector is given by: $\bar{\mathbf{R}} = \begin{bmatrix} \mathbf{R}_1 & \cdots & \bar{\mathbf{R}}_p & \cdots & \bar{\mathbf{R}}_N \end{bmatrix}^T$.

6.3 Buffeting response

As previously discussed in chapter 4, for practical reasons it is in the following distinguished between three cases. First a case of single mode single component response will be shown. This will render a suitable solution if eigen-frequencies are well separated and there is insignificant structural or flow induced coupling between horizontal, vertical and torsion displacement components. Second, a case of single mode three component response will be shown. This is a suitable solution strategy if there is significant structural or flow induced coupling between any of the three displacement components, and if eigen-frequencies are still well separated. Finally, a full multi mode approach is presented.

The buffeting load is given in chapter 5.1. As shown in Eq. 5.8 (see also Eqs. 5.15 and 5.16), it comprises a time invariant mean part $\bar{\mathbf{q}}(x)$, previously dealt with in chapter 6.2 above, and a fluctuating part

$$\mathbf{q}(x,t) = \mathbf{B}_q \cdot \mathbf{v} + \mathbf{C}_{ae} \cdot \dot{\mathbf{r}} + \mathbf{K}_{ae} \cdot \mathbf{r} \tag{6.10}$$

that contains a flow induced contribution $\mathbf{B}_q \cdot \mathbf{v}$ and two motion induced parts $\mathbf{C}_{ae} \cdot \dot{\mathbf{r}}$ and $\mathbf{K}_{ae} \cdot \mathbf{r}$. The content of Eq. 6.10 is defined in Eqs. 5.9 – 5.14. It is applicable in time domain as well as in frequency domain. Improved frequency domain counterparts to $\mathbf{B}_q$, $\mathbf{C}_{ae}$ and $\mathbf{K}_{ae}$ are given in Eqs. 5.19, 5.24 and 5.25. As shown in chapter 4.2 – 4.4, in a modal frequency domain solution the flow induced part of the load (i.e. the modal versions of $\mathbf{C}_{ae} \cdot \dot{\mathbf{r}}$ and $\mathbf{K}_{ae} \cdot \mathbf{r}$) are moved to the left hand side of the equilibrium equation and included in the modal frequency-response-function. Thus, the development of a modal buffeting load needs only consideration of the flow induced part $\mathbf{B}_q \cdot \mathbf{v}$, while the motion induced parts need consideration in the development of the modal frequency-response-function.

Single mode single component buffeting response calculations

The response spectrum of an arbitrary displacement component at span–wise position x_r due to excitation in a corresponding mode shape number i is given in Eqs. 4.28 – 4.30. The variance of the displacement response at x_r is then obtained by frequency domain integration, i.e.

$$\sigma_{\tilde{r}}^2(x_r) = \int_0^\infty S_{\tilde{r}}(x_r,\omega)d\omega = \frac{\phi_i^2(x_r)}{\tilde{K}_i^2} \cdot \int_0^\infty \left|\hat{H}_i(\omega)\right|^2 \cdot S_{\tilde{Q}_i}(\omega)d\omega \tag{6.11}$$

where

$$S_{\tilde{Q}_i}(\omega) = \lim_{T\to\infty} \frac{1}{\pi T}\left(a_{\tilde{Q}_i}^* \cdot a_{\tilde{Q}_i}\right) \tag{6.12}$$

and $a_{\tilde{Q}_i}$ is the Fourier amplitude of the appropriate flow induced modal loading component q_y, q_z or q_θ. The modal stiffness $\tilde{K}_i$ and the modal frequency-response-function $\hat{H}_i(\omega)$ are defined in Eqs. 4.19 and 4.24. As shown in Eq. 4.24, any motion induced load effects are included in $\hat{H}_i(\omega)$.

Let us for simplicity consider the displacement response in the along wind horizontal direction r_y at x_r, and develop its variance contribution from one of the predominantly y–modes, $\boldsymbol{\varphi}_i \approx \begin{bmatrix}\phi_y & 0 & 0\end{bmatrix}^T$, with corresponding eigen-frequency $\omega_i = \omega_y$ (e.g. the contribution from the y-mode with lowest eigen-frequency). The flow induced modal load is then given by (see Eqs. 4.19 and 5.12)

$$\begin{aligned} \tilde{Q}_y(t) &= \int_{L_{exp}} \phi_y(x)\cdot q_y(x,t)dx \\ &= \frac{\rho VB}{2}\cdot \int_{L_{exp}} \phi_y(x)\cdot\left[2\frac{D}{B}\bar{C}_D\cdot u(x,t) + \left(\frac{D}{B}C_D' - \bar{C}_L\right)\cdot w(x,t)\right]dx \end{aligned} \tag{6.13}$$

where L_{exp} is the flow exposed part of the structure. Taking the Fourier transform on either side renders

$$\begin{aligned} a_{\tilde{Q}_y}(\omega) &= \int_{L_{exp}} \phi_y(x)\cdot a_{q_y}(x,\omega)dx \\ &= \frac{\rho VB}{2}\cdot \int_{L_{exp}} \phi_y(x)\cdot\left[2\frac{D}{B}\bar{C}_D\cdot a_u(x,\omega) + \left(\frac{D}{B}C_D' - \bar{C}_L\right)\cdot a_w(x,\omega)\right]dx \end{aligned} \tag{6.14}$$

and thus, the modal load spectrum is given by

$$S_{\tilde{Q}_y}(\omega) = \left(\frac{\rho VB}{2}\right)^2 \lim_{T\to\infty} \frac{1}{\pi T} \left\{ \int_{L_{exp}} \phi_y \left[2\frac{D}{B}\bar{C}_D a_u^* + \left(\frac{D}{B}C_D' - \bar{C}_L\right) a_w^* \right] dx \right\}$$
$$\cdot \left\{ \int_{L_{exp}} \phi_y \left[2\frac{D}{B}\bar{C}_D a_u + \left(\frac{D}{B}C_D' - \bar{C}_L\right) a_w \right] dx \right\} \tag{6.15}$$

Acknowledging that

$$S_{mn}(\Delta x,\omega) = \lim_{T\to\infty} \frac{1}{\pi T} \left[a_m^*(x_1,\omega) \cdot a_n(x_2,\omega) \right] \quad \text{where} \quad \left.\begin{matrix} m \\ n \end{matrix}\right\} = u,w \tag{6.16}$$

and assuming that the cross spectra between flow components are negligible, i.e. that

$$S_{uw}(\Delta x,\omega) = S_{wu}(\Delta x,\omega) \approx 0 \tag{6.17}$$

then

$$S_{\tilde{Q}_y}(\omega) = \left[\frac{\rho V^2 B}{2} \cdot J_y(\omega) \right]^2 \tag{6.18}$$

where

$$J_y^2(\omega) = \iint_{L_{exp}} \phi_y(x_1) \cdot \phi_y(x_2) \cdot \left\{ \left(2\frac{D}{B}\bar{C}_D I_u \right)^2 \frac{S_{uu}(\Delta x,\omega)}{\sigma_u^2} \right.$$
$$\left. + \left[\left(\frac{D}{B}C_D' - \bar{C}_L \right) I_w \right]^2 \frac{S_{ww}(\Delta x,\omega)}{\sigma_w^2} \right\} dx_1 dx_2 \tag{6.19}$$

is the joint acceptance function containing the span-wise statistical averaging of variance contributions from the fluctuating u and w flow components. I_u and I_w are the corresponding turbulence intensities and $\Delta x = |x_1 - x_2|$ is the spatial (span-wise) separation. Combining Eqs. 6.11 and 6.18, using $\tilde{K}_y = \omega_y^2 \tilde{M}_y$, and introducing the modally equivalent and evenly distributed mass

$$\tilde{m}_y = \tilde{M}_y \Big/ \int_L \phi_y^2 dx = \int_L m_y \phi_y^2 dx \Big/ \int_L \phi_y^2 dx \tag{6.20}$$

then the following expression is obtained for the standard deviation of the dynamic response in the along wind y direction

$$\sigma_y(x_r) = |\phi_y(x_r)| \cdot \frac{\rho B^3}{2\tilde{m}_y} \cdot \left(\frac{V}{B\omega_y}\right)^2 \cdot \left[\int_0^\infty |\hat{H}_y(\omega)|^2 \cdot \hat{J}_y^2(\omega) d\omega\right]^{1/2} \tag{6.21}$$

where

$$\hat{J}_y(\omega) = J_y / \int_L \phi_y^2 dx \tag{6.22}$$

The non-dimensional frequency response function is given in Eq. 4.25. Neglecting any aerodynamic mass effects and introducing the notation given in Eqs. 5.24 and 5.25, it is then given by

$$\hat{H}_y(\omega) = \left[1 - \kappa_{ae_y} - \left(\frac{\omega}{\omega_y}\right)^2 + 2i\left(\zeta_y - \zeta_{ae_y}\right) \cdot \frac{\omega}{\omega_y}\right]^{-1} \tag{6.23}$$

where

$$\begin{bmatrix} \kappa_{ae_y} \\ \zeta_{ae_y} \end{bmatrix} = \begin{bmatrix} \dfrac{\tilde{K}_{ae_y}}{\omega_y^2 \tilde{M}_y} \\ \dfrac{\tilde{C}_{ae_y}}{2\omega_y \tilde{M}_y} \end{bmatrix} = \begin{bmatrix} \dfrac{\frac{\rho B^2}{2} \omega_y^2 P_4^* \int_{L_{\exp}} \phi_y^2 dx}{\omega_y^2 \tilde{m}_y \int_L \phi_y^2 dx} \\ \dfrac{\frac{\rho B^2}{2} \omega_y P_1^* \int_{L_{\exp}} \phi_y^2 dx}{2\omega_y \tilde{m}_y \int_L \phi_y^2 dx} \end{bmatrix} = \frac{\rho B^2}{\tilde{m}_y} \cdot \frac{\int_{L_{\exp}} \phi_y^2 dx}{\int_L \phi_y^2 dx} \cdot \begin{bmatrix} \frac{1}{2} P_4^* \\ \frac{1}{4} P_1^* \end{bmatrix} \tag{6.24}$$

Similarly, the standard deviation of the dynamic response in the z direction and in torsion are given by

$$\sigma_z(x_r) = |\phi_z(x_r)| \cdot \frac{\rho B^3}{2\tilde{m}_z} \cdot \left(\frac{V}{B\omega_z}\right)^2 \cdot \left[\int_0^\infty |\hat{H}_z(\omega)|^2 \cdot \hat{J}_z^2(\omega) d\omega\right]^{1/2} \tag{6.25}$$

$$\sigma_\theta(x_r) = |\phi_\theta(x_r)| \cdot \frac{\rho B^4}{2\tilde{m}_\theta} \cdot \left(\frac{V}{B\omega_\theta}\right)^2 \cdot \left[\int_0^\infty |\hat{H}_\theta(\omega)|^2 \cdot \hat{J}_\theta^2(\omega) d\omega\right]^{1/2} \tag{6.26}$$

where:

$$\left.\begin{aligned}\tilde{m}_z &= \tilde{M}_z / \int_L \phi_z^2 dx = \frac{\int_L m_z \phi_z^2 dx}{\int_L \phi_z^2 dx}\\ \tilde{m}_\theta &= \tilde{M}_\theta / \int_L \phi_\theta^2 dx = \frac{\int_L m_\theta \phi_\theta^2 dx}{\int_L \phi_\theta^2 dx}\end{aligned}\right\} \tag{6.27}$$

and where the joint acceptance functions are given by

$$\begin{aligned}\hat{J}_z(\omega) = &\left(\iint_{L_{exp}} \phi_z(x_1)\cdot\phi_z(x_2)\cdot\left\{\left(2\bar{C}_L I_u\right)^2 \frac{S_{uu}(\Delta x,\omega)}{\sigma_u^2}\right.\right.\\ &\left.\left. +\left[\left(C_L' + \frac{D}{B}\bar{C}_D\right) I_w\right]^2 \frac{S_{ww}(\Delta x,\omega)}{\sigma_w^2}\right\} dx_1 dx_2 \right)^{1/2} \Bigg/ \int_L \phi_z^2 dx\end{aligned} \tag{6.28}$$

$$\begin{aligned}\hat{J}_\theta(\omega) = &\left(\iint_{L_{exp}} \phi_\theta(x_1)\cdot\phi_\theta(x_2)\cdot\left\{\left(2\bar{C}_M I_u\right)^2 \frac{S_{uu}(\Delta x,\omega)}{\sigma_u^2}\right.\right.\\ &\left.\left. +\left(C_M' I_w\right)^2 \frac{S_{ww}(\Delta x,\omega)}{\sigma_w^2}\right\} dx_1 dx_2 \right)^{1/2} \Bigg/ \int_L \phi_\theta^2 dx\end{aligned} \tag{6.29}$$

The corresponding frequency response functions (see Eqs. 4.24, 4.68 and 4.69) are given by

$$\left.\begin{aligned}\hat{H}_z(\omega) &= \left[1-\kappa_{ae_z} - \left(\frac{\omega}{\omega_z}\right)^2 + 2i\left(\zeta_z - \zeta_{ae_z}\right)\cdot\frac{\omega}{\omega_z}\right]^{-1}\\ \hat{H}_\theta(\omega) &= \left[1-\kappa_{ae_\theta} - \left(\frac{\omega}{\omega_\theta}\right)^2 + 2i\left(\zeta_\theta - \zeta_{ae_\theta}\right)\cdot\frac{\omega}{\omega_\theta}\right]^{-1}\end{aligned}\right\} \tag{6.30}$$

where:

$$\left.\begin{aligned}
\begin{bmatrix}\kappa_{ae_z}\\ \zeta_{ae_z}\end{bmatrix} &= \frac{\rho B^2}{\tilde{m}_z}\cdot\frac{\int_{L_{exp}}\phi_z^2 dx}{\int_L \phi_z^2 dx}\cdot\begin{bmatrix}\frac{1}{2}H_4^*\\ \frac{1}{4}H_1^*\end{bmatrix}\\
\begin{bmatrix}\kappa_{ae_\theta}\\ \zeta_{ae_\theta}\end{bmatrix} &= \frac{\rho B^4}{\tilde{m}_\theta}\cdot\frac{\int_{L_{exp}}\phi_\theta^2 dx}{\int_L \phi_\theta^2 dx}\cdot\begin{bmatrix}\frac{1}{2}A_3^*\\ \frac{1}{4}A_2^*\end{bmatrix}
\end{aligned}\right\} \tag{6.31}$$

Example 6.1

The volume integral in the joint acceptance functions above, e.g. as first defined in Eq. 6.19 or as normalised versions given in Eqs. 6.22, 6.28 and 6.29, may in general be expressed by

$$J_{r_m r_n}^2 = \int_0^{L_{exp}}\int_0^{L_{exp}} g_{r_m r_n}(x_1,x_2)\,dx_1 dx_2 \qquad \left.\begin{matrix}m\\ n\end{matrix}\right\} = y,z,\theta$$

where: $g_{r_m r_n}(x_1,x_2) = G_{r_m}(x_1)\cdot G_{r_n}(x_2)\cdot\psi_{kk}(\Delta x)$, $k = u,w$. It will in most cases demand a fine mesh, particularly in the region of small separation $\Delta x = |x_1 - x_2|$. The reason for this is that ψ_{kk}, is usually rather steep close to zero, and thus, $g_{r_m r_n}(x_1,x_2)$ will rapidly drop in the region close to a diagonal plane through $x_1 = x_2$. This difficulty may readily be overcome by adopting Dyrbye & Hansen's [21] following procedure for turning a volume integral back into two line integrals. The position coordinates x_1 and x_2 are interchangeable, and therefore $g_{r_m r_n}(x_1,x_2)$ will be symmetric about the plane through $x_1 = x_2$. Thus,

$$J_{r_m r_n}^2 = 2\int_0^{L_{exp}}\left[\int_{\Delta x}^{L_{exp}} g_{r_m r_n}(x_1,x_1-\Delta x)\,dx_1\right]d\Delta x$$

Introducing the notation $x_1 = x + \Delta x$ and

$$g_{r_m r_n}(x_1,x_1-\Delta x) = G_{r_m}(x_1)\cdot G_{r_n}(x_1-\Delta x)\cdot\psi_{kk}(\Delta x) = G_{r_m}(x+\Delta x)\cdot G_{r_n}(x)\cdot\psi_{kk}(\Delta x)$$

then the following is obtained:

$$J_{r_m r_n}^2 = 2\int_0^{L_{exp}}\left[\int_0^{L_{exp}-\Delta x} G_{r_m}(x+\Delta x)\cdot G_{r_n}(x)\,dx\right]\cdot\psi_{kk}(\Delta x)\,d\Delta x$$

It is usually convenient to introduce the normalised coordinate $\hat{x} = x/L_{exp}$ and separation $\Delta\hat{x} = \Delta x/L_{exp}$. Thus, in a normalised format the joint acceptance function is given by

$$J_{r_m r_n}^2 = 2L_{exp}^2\int_0^1\left[\int_0^{1-\Delta\hat{x}} G_{r_m}(\hat{x}+\Delta\hat{x})\,G_{r_n}(\hat{x})\,d\hat{x}\right]\cdot\psi_{kk}(\Delta\hat{x})\,d\Delta\hat{x}$$

Let for instance $G_{r_m}(x_1) = x_1/L_{exp}$ and $G_{r_n}(x_2) = x_2/L_{exp}$, then

$$J_{r_m r_n}^2 = 2L_{\exp}^2 \int_0^1 \left[\int_0^{1-\Delta\hat{x}} \frac{\hat{x}+\Delta\hat{x}}{L_{\exp}} \cdot \frac{\hat{x}}{L_{\exp}} d\hat{x} \right] \cdot \psi_{kk}(\Delta\hat{x}) d\Delta\hat{x} = \frac{1}{3} \int_0^1 \left[2 - 3(\Delta\hat{x}) + (\Delta\hat{x})^3 \right] \psi_{kk}(\Delta\hat{x}) d\Delta\hat{x}$$

The solutions to a good number of cases have been shown by Dyrbye & Hansen [21] and by Davenport [14], who has also developed simple approximate expressions.

The most common cases are graphically illustrated in appendix B.

Example 6.2

Let us consider a typical single mode single component situation, where the three modes k,m,n

$$\boldsymbol{\varphi}_k = \begin{bmatrix} \phi_y & 0 & 0 \end{bmatrix}^T \qquad \boldsymbol{\varphi}_m = \begin{bmatrix} 0 & \phi_z & 0 \end{bmatrix}^T \qquad \boldsymbol{\varphi}_n = \begin{bmatrix} 0 & 0 & \phi_\theta \end{bmatrix}^T$$

with corresponding eigen-frequencies $\omega_y, \omega_z, \omega_\theta$ have been singled out for a response calculation. Since the main girder cross section of many bridges are close to a flat plate, the load coefficient properties

$$\left.\begin{matrix} \bar{C}_D \\ C_L' \\ C_M' \end{matrix}\right\} \neq 0, \qquad \left.\begin{matrix} C_D' \\ \bar{C}_L \\ \bar{C}_M \end{matrix}\right\} \approx 0 \qquad \text{and} \qquad \frac{D}{B}\bar{C}_D \ll C_L'$$

are frequently encountered in bridge engineering. In that case

$$J_y^2(\omega) = \left(2\frac{D}{B}\bar{C}_D I_u \right)^2 \iint_{L_{\exp}} \phi_y(x_1) \cdot \phi_y(x_2) \cdot \frac{S_{uu}(\Delta x, \omega)}{\sigma_u^2} dx_1 dx_2$$

$$J_z^2(\omega) = \left(C_L' I_w \right)^2 \iint_{L_{\exp}} \phi_z(x_1) \cdot \phi_z(x_2) \cdot \frac{S_{ww}(\Delta x, \omega)}{\sigma_w^2} dx_1 dx_2$$

$$J_\theta^2(\omega) = \left(C_M' I_w \right)^2 \iint_{L_{\exp}} \phi_\theta(x_1) \cdot \phi_\theta(x_2) \cdot \frac{S_{ww}(\Delta x, \omega)}{\sigma_w^2} dx_1 dx_2$$

Introducing:

$$S_{uu}(\Delta x, \omega) = S_u(\omega) \cdot \hat{C}o_{uu}(\Delta x, \omega), \qquad S_{ww}(\Delta x, \omega) = S_w(\omega) \cdot \hat{C}o_{ww}(\Delta x, \omega),$$

$$\tilde{K}_n = \omega_n^2 \cdot \tilde{M}_n = \omega_n^2 \cdot \tilde{m}_n \int_L \phi_n^2 dx, \quad n = y,\ z \text{ or } \theta$$

and the non–dimensional joint acceptance functions

$$\hat{J}_y(\omega) = \left(\iint_{L_{\exp}} \phi_y(x_1) \cdot \phi_y(x_2) \cdot \hat{C}o_{uu}(\Delta x, \omega) dx_1 dx_2 \right)^{1/2} \Big/ \int_L \phi_y^2 dx$$

$$\hat{J}_z(\omega) = \left(\iint_{L_{\exp}} \phi_z(x_1) \cdot \phi_z(x_2) \cdot \hat{C}o_{ww}(\Delta x, \omega) dx_1 dx_2 \right)^{1/2} \Big/ \int_L \phi_z^2 dx$$

$$\hat{J}_\theta(\omega) = \left(\iint\limits_{L_{\exp}} \phi_\theta(x_1)\cdot\phi_\theta(x_2)\cdot\hat{C}o_{ww}(\Delta x,\omega)dx_1dx_2 \right)^{1/2} \Bigg/ \int\limits_L \phi_\theta^2 dx$$

then the r_y, r_z and r_θ response spectra are given by (see Eq. 4.30, 6.18, 6.19, 6.28 and 6.29)

$$S_{r_y}(\omega,x_r) = \left[\phi_y(x_r)\cdot\frac{\rho B^2 D}{\tilde{m}_y}\cdot\left(\frac{V}{B\omega_y}\right)^2 \cdot\bar{C}_D I_u\cdot\left|\hat{H}_y(\omega)\right|\cdot\hat{J}_y(\omega)\right]^2 \cdot\frac{S_u(\omega)}{\sigma_u^2}$$

$$S_{r_z}(\omega,x_r) = \left[\phi_z(x_r)\frac{\rho B^3}{2\tilde{m}_z}\cdot\left(\frac{V}{B\omega_z}\right)^2 \cdot C_L' I_w\cdot\left|\hat{H}_z(\omega)\right|\cdot\hat{J}_z(\omega)\right]^2 \cdot\frac{S_w(\omega)}{\sigma_w^2}$$

$$S_{r_\theta}(\omega,x_r) = \left[\phi_\theta(x_r)\frac{\rho B^4}{2\tilde{m}_\theta}\cdot\left(\frac{V}{B\omega_\theta}\right)^2 \cdot C_M' I_w\cdot\left|\hat{H}_\theta(\omega)\right|\cdot\hat{J}_\theta(\omega)\right]^2 \cdot\frac{S_w(\omega)}{\sigma_w^2}$$

Integrating across the entire frequency domain, the following response standard deviations are obtained:

$$\sigma_{r_y}(x_r) - \left|\phi_y(x_r)\right|\cdot\frac{\rho B^2 D}{\tilde{m}_y}\cdot\bar{C}_D I_u\cdot\left(\frac{V}{B\omega_y}\right)^2\cdot\left[\int_0^\infty \left|\hat{H}_y(\omega)\right|^2\cdot\frac{S_u(\omega)}{\sigma_u^2}\cdot\hat{J}_y^2(\omega)d\omega\right]^{1/2}$$

$$\sigma_{r_z}(x_r) = \left|\phi_z(x_r)\right|\cdot\frac{\rho B^3}{2\tilde{m}_z}\cdot C_L' I_w\cdot\left(\frac{V}{B\omega_z}\right)^2\cdot\left[\int_0^\infty \left|\hat{H}_z(\omega)\right|^2\cdot\frac{S_w(\omega)}{\sigma_w^2}\cdot\hat{J}_z^2(\omega)d\omega\right]^{1/2}$$

$$\sigma_{r_\theta}(x_r) = \left|\phi_\theta(x_r)\right|\cdot\frac{\rho B^4}{2\tilde{m}_\theta}\cdot C_M' I_w\cdot\left(\frac{V}{B\omega_\theta}\right)^2\cdot\left[\int_0^\infty \left|\hat{H}_\theta(\omega)\right|^2\cdot\frac{S_w(\omega)}{\sigma_w^2}\cdot\hat{J}_\theta^2(\omega)d\omega\right]^{1/2}$$

Let us focus exclusively on the response in the y (drag) direction, and consider a simply supported horizontal beam type of bridge with span $L = 500m$ that is elevated at a position $z_f = 50m$. Let us for simplicity assume that the relevant mode shape $\phi_y(x) = \sin(\pi x/L)$ and that $x_r = L/2$, in which case $\phi_y(x_r) = 1$. Let us also assume that the entire span is flow exposed, i.e. $L_{\exp} = L$, and adopt the following wind field properties:

1) the turbulence intensity $\quad I_u = \sigma_u/V = 0.15 \quad$ (see Eq. 3.14)

2) the integral length scale: $\quad {}^{xf}L_u = 100\cdot(z_f/10)^{0.3} = 162m \quad$ (see Eq. 3.36),

3) the auto spectral density: $\quad \dfrac{S_u(\omega)}{\sigma_u^2} = \dfrac{1.08\cdot{}^{xf}L_u/V}{\left(1+1.62\cdot\omega\cdot{}^{xf}L_u/V\right)^{5/3}} \quad$ (see Eq. 3.25)

4) the normalised co-spectrum: $\quad \hat{C}o_{uu}(\omega,\Delta x) = \exp(-C_{ux}\cdot\omega\cdot\Delta x/V) \quad$ (see Eq. 3.41)

where $C_{ux} = C_{uyf} = 9/(2\pi) \approx 1.4$.

Let us allot the following values to the remaining constants that are necessary for a numerical calculation of $\sigma_{r_y}(x_r = L/2)$:

ρ (kg/m^3)	$\bar{C}_D$	B (m)	D (m)	m_y (kg/m)	ω_y (rad/s)	ζ_y
1.25	0.7	20	4	10000	0.4	0.005

Since m_y is constant along the span, then the modally equivalent and evenly distributed mass $\tilde{m}_y = m_y$. Finally, let us adopt quasi–static values to the aerodynamic derivatives, in which case $\kappa_{ae_y} = 0$ and the aerodynamic damping ζ_{ae_y} is given by (see Eqs. 5.26 and 6.24)

$$\zeta_{ae_y} = \frac{\rho B^2}{4\tilde{m}_y} P_1^* = \frac{\rho B^2}{4\tilde{m}_y}\left(-2\bar{C}_D \frac{D}{B}\frac{V}{B\omega_y}\right) = -\frac{\rho D \bar{C}_D V}{2\tilde{m}_y \omega_y} \approx -4.375\cdot 10^{-4}\cdot V$$

The non-dimensional joint acceptance function $\hat{J}_y$ may readily be obtained by numerical calculations. However, as shown by Davenport [14], in many cases closed form solutions may be obtained. The situation that $\phi_y(x) = \sin(\pi x/L)$ and $\hat{C}o_{uu}(\omega,\Delta x)$ is a simple exponential function is such a case. Substituting $x_1 = x$, $x_2 = x + \Delta x$, $\hat{x} = \pi x/L$, $\Delta\hat{x} = \Delta x/L$ and $\hat{\omega} = C_{ux}\omega L/V$, then

$$\hat{J}_y^2 = \iint\limits_{L_{\exp}} \phi_y(x_1)\cdot\phi_y(x_2)\cdot\hat{C}o_{uu}(\Delta x,\omega)\,dx_1 dx_2 \bigg/ \left(\int_L \phi_y^2 dx\right)^2$$

$$= 2\int_0^{L_{\exp}}\left[\int_0^{L_{\exp}-\Delta x} \phi_y(x+\Delta x)\cdot\phi_y(x)\,dx\right]\hat{C}o_{uu}(\Delta x,\omega)\,d\Delta x \bigg/ \left(\int_0^L \phi_y^2 dx\right)^2$$

$$= 2\int_0^{L_{\exp}}\left[\int_0^{L_{\exp}-\Delta x} \sin\frac{\pi}{L}(x+\Delta x)\cdot\sin\frac{\pi}{L}x\,dx\right]\cdot\exp(-C_{ux}\omega\Delta x/V)\,d\Delta x \bigg/ \left(\int_0^L \sin^2\frac{\pi}{L}x\,dx\right)^2$$

Using that $\sin 2\alpha = 2\sin\alpha\cos\alpha$ and that $\sin(\alpha+\beta) = \sin\alpha\cdot\cos\beta + \cos\alpha\cdot\sin\beta$, then this may be expanded into

$$\hat{J}_y^2 = \frac{8}{L^2}\int_0^{L_{\exp}}\left[\int_0^{L_{\exp}-\Delta x}\left(\cos\frac{\pi}{L}\Delta x\cdot\sin^2\frac{\pi}{L}x + \frac{1}{2}\sin\frac{\pi}{L}\Delta x\cdot\sin\frac{2\pi}{L}x\right)dx\right]\cdot\exp\left(-\frac{C_{ux}\omega\Delta x}{V}\right)d\Delta x$$

$$= \frac{8}{\pi}\int_0^1\left(\cos\frac{\pi}{L}\Delta\hat{x}\int_0^{\pi(1-\Delta\hat{x})}\sin^2\hat{x}\,d\hat{x} + \frac{1}{2}\sin\frac{\pi}{L}\Delta\hat{x}\int_0^{\pi(1-\Delta\hat{x})}\sin 2\hat{x}\,d\hat{x}\right)\cdot\exp(-\hat{\omega}\Delta\hat{x})\,d\Delta\hat{x}$$

$$\hat{J}_y^2 = \frac{8}{\pi}\int_0^1\left\{\frac{\pi}{2}(1-\Delta\hat{x})\cos\pi\Delta\hat{x} - \frac{1}{4}\left[\cos\pi\Delta\hat{x}\cdot\sin 2\pi(1-\Delta\hat{x}) - \sin\pi\Delta\hat{x}\cdot\cos 2\pi(1-\Delta\hat{x})\right]\right.$$
$$\left. + \frac{1}{4}\sin\pi\Delta\hat{x}\right\}\exp(-\hat{\omega}\Delta\hat{x})\,d\Delta\hat{x}$$

Using that $\sin\alpha\cdot\cos\beta-\cos\alpha\cdot\sin\beta=\sin(\alpha-\beta)$ and $\sin(-\alpha+2\pi)=-\sin\alpha$ this simplifies into

$$\hat{J}_y^2=4\int_0^1\left[(1-\Delta\hat{x})\cos\pi\Delta\hat{x}+\frac{1}{\pi}\sin\pi\Delta\hat{x}\right]\exp(-\hat{\omega}\Delta\hat{x})d\Delta\hat{x}=4\left[\int_0^1\cos\pi\Delta\hat{x}\cdot\exp(-\hat{\omega}\Delta\hat{x})d\Delta\hat{x}\right.$$

$$\left.-\int_0^1\Delta\hat{x}\cdot\cos\pi\Delta\hat{x}\cdot\exp(-\hat{\omega}\Delta\hat{x})d\Delta\hat{x}+\frac{1}{\pi}\int_0^1\sin\pi\Delta\hat{x}\cdot\exp(-\hat{\omega}\Delta\hat{x})d\Delta\hat{x}\right]$$

$$=4\left\{\left[\frac{(1-\Delta\hat{x})\exp(-\hat{\omega}\Delta\hat{x})}{\hat{\omega}^2+\pi^2}\cdot(-\hat{\omega}\cos\pi\Delta\hat{x}+\pi\sin\pi\Delta\hat{x})\right]_0^1\right.$$

$$+\left[\frac{\exp(-\hat{\omega}\Delta\hat{x})}{\left(\hat{\omega}^2+\pi^2\right)^2}\left(\left(\hat{\omega}^2-\pi^2\right)\cos\pi\Delta\hat{x}-2\pi\sin\pi\Delta\hat{x}\right)\right]_0^1$$

$$\left.-\frac{1}{\pi}\left[\frac{\exp(-\hat{\omega}\Delta\hat{x})}{\hat{\omega}^2+\pi^2}\cdot(\hat{\omega}\sin\pi\Delta\hat{x}+\pi\cos\pi\Delta\hat{x})\right]_0^1\right\}$$

Thus, the following is obtained:

$$\Rightarrow\hat{J}_y^2(\hat{\omega})=4\cdot\psi(\hat{\omega})\qquad\text{where}\qquad\psi(\hat{\omega})=\left[\frac{\hat{\omega}}{\hat{\omega}^2+\pi^2}+2\pi^2\frac{1+\exp(-\hat{\omega})}{\left(\hat{\omega}^2+\pi^2\right)^2}\right]$$

The standard deviation of the dynamic response at $x_r=L/2$ is then given by

$$\sigma_{r_y}(L/2)=3.28\cdot10^{-4}\cdot V^2\left[\int_0^\infty\left|\hat{H}_y(\omega)\right|^2\cdot\frac{S_u(\omega)}{\sigma_u^2}\cdot\hat{J}_y^2(\hat{\omega})d\omega\right]^{1/2}$$

where $S_u(\omega)/\sigma_u^2$, $\hat{J}_y^2(\hat{\omega})$ and $\hat{\omega}$ are defined above, and where

$$\hat{H}_y(\omega)=\left[1-(\omega/0.4)^2+2i\left(0.005+4.375\cdot10^{-4}\cdot V\right)\cdot\omega/0.4\right]^{-1}$$

The chosen single point spectral density and corresponding normalised co–spectrum of the turbulent u component are shown on the top left and right hand side diagrams in Fig. 6.5. The non-dimensional frequency response function and the squared normalised joint acceptance functions are shown on the lower left and right hand side diagrams in Fig. 6.5. The response spectrum of the along wind r_y component at $x_r=L/2$ and $V=40\ m/s$ is shown in Fig. 6.6. As can be seen, it contains a broad banded background part and a narrow banded resonant part at $\omega=0.4\ rad/s$. The standard deviation of the dynamic response at $x_r=L/2$ is plotted versus the mean wind velocity in Fig. 6.7. [It should be noted that the effect of aerodynamic damping is considerable (see Example 6.3), and that the validity of the quasi-static theory may be limited.]

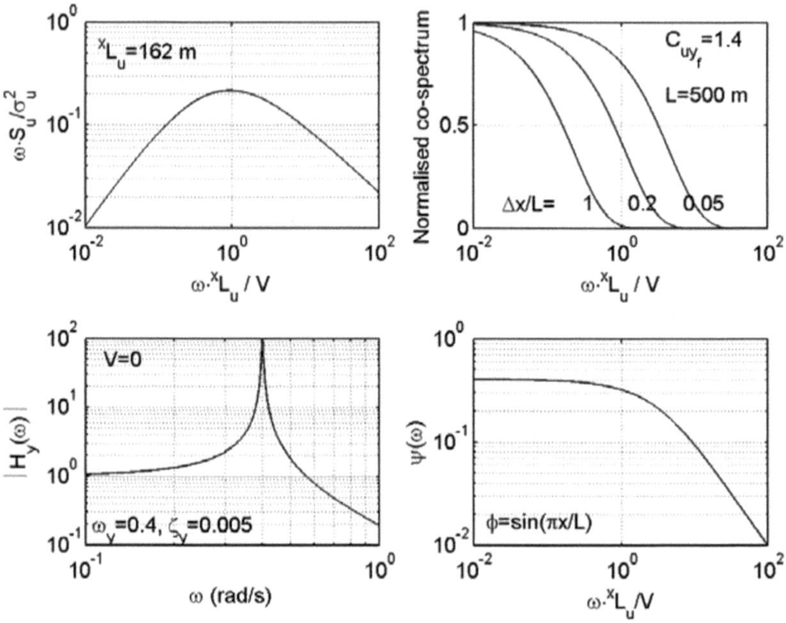

Fig. 6.5 *Top left and right: single point u spectrum and corresponding normalised co-spectrum, lower left and right: frequency response function and joint acceptance function*

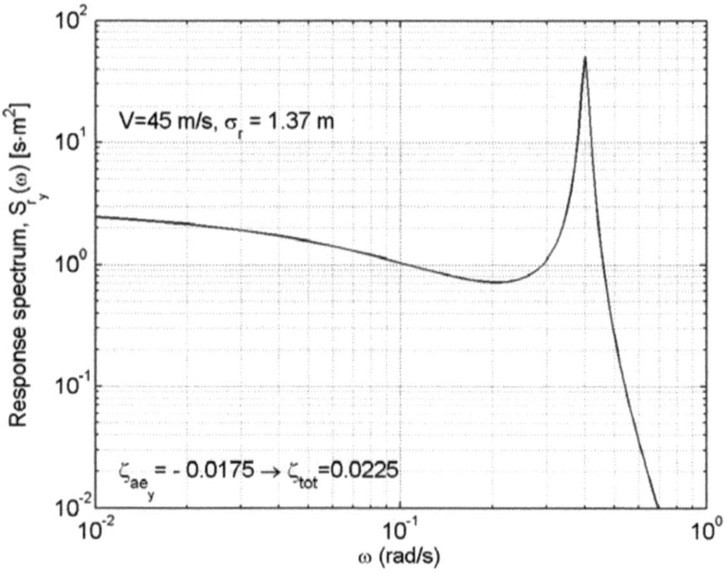

Fig. 6.6 *Response spectrum of r_y displacements at $x_r = L/2$ and $V = 45\ m/s$*

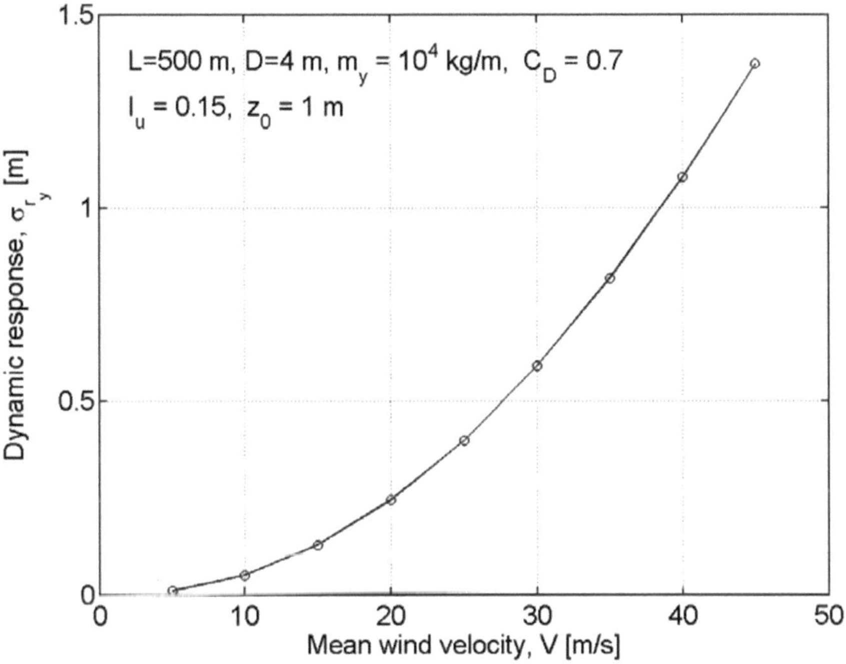

Fig. 6.7 *The standard deviation of the dynamic response at* $x_r = L/2$ *versus the mean wind velocity*

Single mode three component buffeting response calculations

The solution to this case is given in chapter 4.3, see Eqs. 4.47 and 4.48. What remains from the development in chapter 4.3 is to expand on the modal load spectrum $S_{\tilde{Q}_i}$ using the results from chapter 5.1. As shown above (see Eq. 6.4 and ensuing discussion), the flow induced buffeting part of the fluctuating load is

$$\begin{bmatrix} q_y(x,t) \\ q_z(x,t) \\ q_\theta(x,t) \end{bmatrix} = \mathbf{B}_q(x) \cdot \mathbf{v}(x,t) = (\rho V B/2) \cdot \hat{\mathbf{B}}_q \cdot \mathbf{v} \tag{6.32}$$

where $\hat{\mathbf{B}}_q$ and $\mathbf{v}$ are defined in Eqs. 5.9 and 5.12. Thus (see Eq. 4.39)

$$\tilde{Q}_i(t) = \frac{\rho VB}{2} \int_{L_{\exp}} \boldsymbol{\varphi}_i^T \cdot \left(\hat{\mathbf{B}}_q \cdot \mathbf{v}\right) dx \tag{6.33}$$

where $\boldsymbol{\varphi}_i(x) = \begin{bmatrix} \phi_y & \phi_z & \phi_\theta \end{bmatrix}_i^T$. The Fourier transform of Eq. 6.33

$$a_{\tilde{Q}_i}(\omega) = \frac{\rho VB}{2} \int_{L_{\exp}} \boldsymbol{\varphi}_i^T \cdot \left(\hat{\mathbf{B}}_q \cdot \mathbf{a}_v\right) dx \tag{6.34}$$

where

$$\mathbf{a}_v(x,\omega) = \begin{bmatrix} a_u & a_w \end{bmatrix}^T \tag{6.35}$$

contains the Fourier amplitudes of the u and w components. This will then render the following modal load spectrum

$$\begin{aligned}
S_{\tilde{Q}_i}(\omega) &= \lim_{T\to\infty} \frac{1}{\pi T}\left(a_{\tilde{Q}_i}^* \cdot a_{\tilde{Q}_i}\right) \\
&= \left(\frac{\rho VB}{2}\right)^2 \lim_{T\to\infty} \frac{1}{\pi T} \left\{ \left[\int_{L_{\exp}} \boldsymbol{\varphi}_i^T \cdot \left(\hat{\mathbf{B}}_q \cdot \mathbf{a}_v^*\right) dx \right] \cdot \left[\int_{L_{\exp}} \boldsymbol{\varphi}_i^T \cdot \left(\hat{\mathbf{B}}_q \cdot \mathbf{a}_v\right) dx \right]^T \right\} \\
&= \left(\frac{\rho VB}{2}\right)^2 \cdot \iint_{L_{\exp}} \boldsymbol{\varphi}_i^T(x_1) \cdot \left\{ \hat{\mathbf{B}}_q(x_1) \cdot \mathbf{S}_v(\Delta x,\omega) \cdot \hat{\mathbf{B}}_q^T(x_2) \right\} \cdot \boldsymbol{\varphi}_i(x_2) dx_1 dx_2
\end{aligned} \tag{6.36}$$

where

$$\mathbf{S}_v(\Delta x,\omega) = \lim_{T\to\infty} \frac{1}{\pi T}\left[\mathbf{a}_v^*(x_1,\omega) \cdot \mathbf{a}_v^T(x_2,\omega)\right] = \lim_{T\to\infty} \frac{1}{\pi T} \begin{bmatrix} a_u^* a_u & a_u^* a_w \\ a_w^* a_u & a_w^* a_w \end{bmatrix} = \begin{bmatrix} S_{uu} & S_{uw} \\ S_{wu} & S_{ww} \end{bmatrix} \tag{6.37}$$

This is greatly simplified if the cross spectra between flow components are negligible, i.e. $S_{uw} = S_{wu} \approx 0$, see Eq. 6.17. Then

$$S_{\tilde{Q}_i}(\omega) = \left[\frac{\rho V^2 B}{2} \cdot J_i(\omega)\right]^2 \tag{6.38}$$

where:

$$J_i^2 = \iint_{L_{\exp}} \boldsymbol{\varphi}_i^T(x_1) \cdot \left\{ \hat{\mathbf{B}}_q(x_1) \cdot \left[\mathbf{I}_v^2 \cdot \hat{\mathbf{S}}_v(\Delta x,\omega)\right] \cdot \hat{\mathbf{B}}_q^T(x_2) \right\} \cdot \boldsymbol{\varphi}_i(x_2) dx_1 dx_2 \tag{6.39}$$

is the joint acceptance function, and where

$$\left.\begin{aligned} \mathbf{I}_v &= diag\left[I_u \quad I_w\right] \\ \hat{\mathbf{S}}_v(\Delta x,\omega) &= diag\left[S_{uu}/\sigma_u^2 \quad S_{ww}/\sigma_w^2\right] \end{aligned}\right\} \tag{6.40}$$

Introducing the modal stiffness $\tilde{K}_i = \omega_i^2 \tilde{M}_i$ and defining

$$\tilde{m}_i = \tilde{M}_i / \int_L \left(\boldsymbol{\varphi}_i^T \cdot \boldsymbol{\varphi}_i\right) dx \tag{6.41}$$

then from Eqs. 4.47 and 4.48 the following standard deviations of displacement responses at x_r are obtained

$$\begin{bmatrix} \sigma_y \\ \sigma_z \\ \sigma_\theta \end{bmatrix}_i = \begin{bmatrix} \phi_y(x_r) \\ \phi_z(x_r) \\ \phi_\theta(x_r) \end{bmatrix}_i \frac{\rho B^3}{2\tilde{m}_i} \cdot \left(\frac{V}{B\omega_i}\right)^2 \cdot \left[\int_0^\infty \left|\hat{H}_i(\omega)\right|^2 \cdot \hat{J}_i^2(\omega) d\omega\right]^{1/2} \tag{6.42}$$

where

$$\hat{J}_i = \frac{J_i}{\int_L \boldsymbol{\varphi}_i^T \cdot \boldsymbol{\varphi}_i dx} \tag{6.43}$$

Again, neglecting any aerodynamic mass and introducing the notation given in Eqs. 4.25, 4.40 and 5.25, then the frequency response function is given by

$$\hat{H}_i(\omega) = \left[1 - \kappa_{ae_i} - \left(\frac{\omega}{\omega_i}\right)^2 + 2i\left(\zeta_i - \zeta_{ae_i}\right) \cdot \frac{\omega}{\omega_i}\right]^{-1} \tag{6.44}$$

where

$$\kappa_{ae_i} = \frac{\rho B^2}{2\tilde{m}_i} \cdot \frac{\int_{L_{\exp}} \left(\boldsymbol{\varphi}_i^T \hat{\mathbf{K}}_{ae} \boldsymbol{\varphi}_i\right) dx}{\int_L \left(\boldsymbol{\varphi}_i^T \boldsymbol{\varphi}_i\right) dx} \tag{6.45}$$

$$\zeta_{ae_i} = \frac{\rho B^2}{4\tilde{m}_i} \cdot \frac{\int_{L_{\exp}} \left(\boldsymbol{\varphi}_i^T \hat{\mathbf{C}}_{ae} \boldsymbol{\varphi}_i\right) dx}{\int_L \left(\boldsymbol{\varphi}_i^T \boldsymbol{\varphi}_i\right) dx} \tag{6.46}$$

As explained in chapter 4.3 (see Eq. 4.41), only diagonal $\hat{\mathbf{K}}_{ae}$ and $\hat{\mathbf{C}}_{ae}$ will maintain the presupposition that no modal coupling will occur. Flow induced coupling will occur if $\hat{\mathbf{K}}_{ae}$ and $\hat{\mathbf{C}}_{ae}$ are not diagonal.

Multi mode buffeting response calculations

The general solution to a multi mode approach is given by the three by three response matrix shown in Eqs. 4.80 – 4.82. The corresponding three by three response covariance matrix

$$\mathbf{Cov}_{rr}(x_r) = \begin{bmatrix} \sigma^2_{r_y r_y} & Cov_{r_y r_z} & Cov_{r_y r_\theta} \\ Cov_{r_z r_y} & \sigma^2_{r_z r_z} & Cov_{r_z r_\theta} \\ Cov_{r_\theta r_y} & Cov_{r_\theta r_z} & \sigma^2_{r_\theta r_\theta} \end{bmatrix} \tag{6.47}$$

which contains the variance of each response displacement component r_y, r_z and r_θ at $x = x_r$ on its diagonal and cross covariance on its off-diagonal terms, is obtained by frequency domain integration. Thus,

$$\mathbf{Cov}_{rr}(x_r) = \int_0^\infty \mathbf{S}_{rr}(x_r,\omega)d\omega = \mathbf{\Phi}_r(x_r)\left[\int_0^\infty \hat{\mathbf{H}}^*_\eta(\omega)\mathbf{S}_{\hat{Q}}(\omega)\hat{\mathbf{H}}^T_\eta(\omega)d\omega\right]\mathbf{\Phi}^T_r(x_r) \tag{6.48}$$

where $\hat{\mathbf{H}}_\eta(\omega)$ and $\mathbf{S}_{\hat{Q}}(\omega)$ are N_{mod} by N_{mod} matrices given in Eqs. 4.69 and 4.75, and $\mathbf{\Phi}_r(x_r)$ is a three by N_{mod} matrix defined in Eq. 4.79. What remains is to bring the results from chapter 5.1 into $\hat{\mathbf{H}}_\eta(\omega)$ and $\mathbf{S}_{\hat{Q}}(\omega)$. Disregarding any aerodynamic mass effects, the frequency response matrix $\hat{\mathbf{H}}_\eta(\omega)$ in Eq. 4.69 is reduced to

$$\hat{\mathbf{H}}_\eta(\omega) = \left\{\mathbf{I} - \boldsymbol{\kappa}_{ae} - \left(\omega \cdot diag\left[\frac{1}{\omega_i}\right]\right)^2 + 2i\omega \cdot diag\left[\frac{1}{\omega_i}\right]\cdot(\boldsymbol{\zeta} - \boldsymbol{\zeta}_{ae})\right\}^{-1} \tag{6.49}$$

where $\mathbf{I}$ is the identity matrix (N_{mod} by N_{mod}), and where $\boldsymbol{\zeta}$, $\boldsymbol{\zeta}_{ae}$ and $\boldsymbol{\kappa}_{ae}$ are defined in Eq. 4.68. By introducing the modal stiffness matrix $\tilde{\mathbf{K}}_0^{-1} = diag\left[1/\left(\omega_i^2 \tilde{M}_i\right)\right]$, the definition of $\tilde{m}_i$ in Eq. 6.41 and the notation in Eqs. 5.24 and 5.25, then the content of

$$\boldsymbol{\kappa}_{ae} = \begin{bmatrix} \ddots & & \cdot^{\cdot^{\cdot}} \\ & \kappa_{ae_{ij}} & \\ \cdot^{\cdot^{\cdot}} & & \ddots \end{bmatrix} \quad \text{and} \quad \boldsymbol{\zeta}_{ae} = \begin{bmatrix} \ddots & & \cdot^{\cdot^{\cdot}} \\ & \zeta_{ae_{ij}} & \\ \cdot^{\cdot^{\cdot}} & & \ddots \end{bmatrix} \tag{6.50}$$

are given by

$$\kappa_{ae_{ij}} = \frac{\tilde{K}_{ae_{ij}}}{\omega_i^2 \tilde{M}_i} = \frac{\rho B^2}{2\tilde{m}_i} \cdot \frac{\int_{L_{\exp}} \left(\boldsymbol{\varphi}_i^T \cdot \hat{\mathbf{K}}_{ae} \cdot \boldsymbol{\varphi}_j \right) dx}{\int_L \left(\boldsymbol{\varphi}_i^T \cdot \boldsymbol{\varphi}_i \right) dx} \tag{6.51}$$

$$\zeta_{ae_{ij}} = \frac{\omega_i}{2} \frac{\tilde{C}_{ae_{ij}}}{\omega_i^2 \tilde{M}_i} = \frac{\rho B^2}{4\tilde{m}_i} \cdot \frac{\int_{L_{\exp}} \left(\boldsymbol{\varphi}_i^T \cdot \hat{\mathbf{C}}_{ae} \cdot \boldsymbol{\varphi}_j \right) dx}{\int_L \left(\boldsymbol{\varphi}_i^T \cdot \boldsymbol{\varphi}_i \right) dx} \tag{6.52}$$

Fully expanded versions of these expressions are given by

$$\frac{\kappa_{ae_{ij}}}{\frac{\rho B^2}{2\tilde{m}_i}} = \left[\int_{L_{\exp}} \left(\phi_{y_i}\phi_{y_j}P_4^* + \phi_{z_i}\phi_{y_j}H_6^* + \phi_{\theta_i}\phi_{y_j}BA_6^* + \phi_{y_i}\phi_{z_j}P_6^* + \phi_{z_i}\phi_{z_j}H_4^* + \phi_{\theta_i}\phi_{z_j}BA_4^* \right.\right.$$
$$\left.\left. + \phi_{y_i}\phi_{\theta_j}BP_3^* + \phi_{z_i}\phi_{\theta_j}BH_3^* + \phi_{\theta_i}\phi_{\theta_j}B^2A_3^* \right) dx \right] \Big/ \left[\int_L \left(\phi_{y_i}^2 + \phi_{z_i}^2 + \phi_{\theta_i}^2 \right) dx \right] \tag{6.53}$$

$$\frac{\zeta_{ae_{ij}}}{\frac{\rho B^2}{4\tilde{m}_i}} = \left[\int_{L_{\exp}} \left(\phi_{y_i}\phi_{y_j}P_1^* + \phi_{z_i}\phi_{y_j}H_5^* + \phi_{\theta_i}\phi_{y_j}BA_5^* + \phi_{y_i}\phi_{z_j}P_5^* + \phi_{z_i}\phi_{z_j}H_1^* + \phi_{\theta_i}\phi_{z_j}BA_1^* \right.\right.$$
$$\left.\left. + \phi_{y_i}\phi_{\theta_j}BP_2^* + \phi_{z_i}\phi_{\theta_j}BH_2^* + \phi_{\theta_i}\phi_{\theta_j}B^2A_2^* \right) dx \right] \Big/ \left[\int_L \left(\phi_{y_i}^2 + \phi_{z_i}^2 + \phi_{\theta_i}^2 \right) dx \right] \tag{6.54}$$

As mentioned above, the normalised modal load matrix $\mathbf{S}_{\hat{Q}}$ (N_{mod} by N_{mod}) is given in Eq. 4.75. Its content $S_{\hat{Q}_i\hat{Q}_j}(\omega)$, containing the cross sectional load matrix $\mathbf{S}_{qq}(\Delta x, \omega)$, is defined in Eq. 4.77 (and 4.78). Based on the buffeting load expressions in chapter 5.1 it is now only $\mathbf{S}_{qq}$ that remains for further expansion. Recalling from Eq. 6.32 that the buffeting part of the cross sectional loading is $\begin{bmatrix} q_y & q_z & q_\theta \end{bmatrix}^T = \mathbf{B}_q \cdot \mathbf{v} = (\rho VB/2) \cdot \hat{\mathbf{B}}_q \cdot \mathbf{v}$, then its Fourier transform is

$$\mathbf{a}_q(x,\omega) = \begin{bmatrix} a_{q_y} \\ a_{q_z} \\ a_{q_\theta} \end{bmatrix} = (\rho VB/2) \cdot \hat{\mathbf{B}}_q \cdot \mathbf{a}_v \tag{6.55}$$

where

$$\mathbf{a}_v(x,\omega) = \begin{bmatrix} a_u & a_w \end{bmatrix}^T \tag{6.56}$$

The cross spectrum $\mathbf{S}_{qq}(\Delta x,\omega)$ is then given by

$$\begin{aligned} \mathbf{S}_{qq}(\Delta x,\omega) &= \lim_{T\to\infty} \frac{1}{\pi T}\left[\mathbf{a}_q^*(x_1,\omega) \cdot \mathbf{a}_q^T(x_2,\omega)\right] \\ &= \left(\frac{\rho VB}{2}\right)^2 \cdot \hat{\mathbf{B}}_q \cdot \lim_{T\to\infty} \frac{1}{\pi T}\left[\mathbf{a}_v^*(x_1,\omega) \cdot \mathbf{a}_v^T(x_2,\omega)\right] \cdot \hat{\mathbf{B}}_q^T \\ &= \left(\frac{\rho VB}{2}\right)^2 \cdot \hat{\mathbf{B}}_q \cdot \mathbf{S}_v(\Delta x,\omega) \cdot \hat{\mathbf{B}}_q^T \end{aligned} \tag{6.57}$$

where $\mathbf{S}_v(\Delta x,\omega)$ is defined in Eq. 6.37. Adopting the assumption that $S_{uw} = S_{wu} \approx 0$, see Eq. 6.17, and introducing Eq. 6.40, then the content of the normalised modal load matrix (N_{mod} by N_{mod})

$$\mathbf{S}_{\hat{Q}}(\omega) = \begin{bmatrix} \ddots & & \cdot^{\cdot^{\cdot}} \\ & S_{\hat{Q}_i\hat{Q}_j}(\omega) & \\ \cdot^{\cdot^{\cdot}} & & \ddots \end{bmatrix} \tag{6.58}$$

is given by

$$\begin{aligned} S_{\hat{Q}_i\hat{Q}_j}(\omega) &= \left(\frac{\rho V^2 B}{2}\right)^2 \cdot \frac{\iint_{L_{\exp}} \boldsymbol{\varphi}_i^T(x_1) \cdot \left\{\hat{\mathbf{B}}_q \cdot \left[\mathbf{I}_v^2 \cdot \hat{\mathbf{S}}_v(\Delta x,\omega)\right] \cdot \hat{\mathbf{B}}_q^T\right\} \cdot \boldsymbol{\varphi}_j(x_2)\, dx_1 dx_2}{\left(\omega_i^2 \tilde{M}_i\right) \cdot \left(\omega_j^2 \tilde{M}_j\right)} \\ &= \frac{\rho B^3}{2\tilde{m}_i} \cdot \frac{\rho B^3}{2\tilde{m}_j} \cdot \left(\frac{V}{B\omega_i}\right)^2 \cdot \left(\frac{V}{B\omega_j}\right)^2 \cdot \hat{J}_{ij}^2 \end{aligned} \tag{6.59}$$

Thus, in case of multi mode calculations there will be $N_{\text{mod}} \cdot N_{\text{mod}}$ such reduced joint acceptance functions $\hat{J}_{ij}^2$, each defined by

$$\hat{J}_{ij}^2 = \frac{\iint\limits_{L_{\exp}} \boldsymbol{\varphi}_i^T(x_1) \cdot \left\{ \hat{\mathbf{B}}_q \cdot \left[\mathbf{I}_v^2 \cdot \hat{\mathbf{S}}_v(\Delta x, \omega) \right] \cdot \hat{\mathbf{B}}_q^T \right\} \cdot \boldsymbol{\varphi}_j(x_2) dx_1 dx_2}{\left(\int\limits_L \boldsymbol{\varphi}_i^T \cdot \boldsymbol{\varphi}_i dx \right) \cdot \left(\int\limits_L \boldsymbol{\varphi}_j^T \cdot \boldsymbol{\varphi}_j dx \right)} \tag{6.60}$$

A fully expanded version of J_{ij}^2 is given by

$$\begin{aligned}
J_{ij}^2 = \iint\limits_{L_{\exp}} \Bigg\{ & \phi_{y_i}(x_1)\phi_{y_j}(x_2)\left[\left(2\frac{D}{B}\bar{C}_D \right)^2 I_u^2 \hat{S}_{uu} + \left(\frac{D}{B}C_D' - \bar{C}_L \right)^2 I_w^2 \hat{S}_{ww} \right] \\
& + \phi_{z_i}(x_1)\phi_{z_j}(x_2)\left[\left(2\bar{C}_L \right)^2 I_u^2 \hat{S}_{uu} + \left(C_L' + \frac{D}{B}\bar{C}_D \right)^2 I_w^2 \hat{S}_{ww} \right] \\
& + \phi_{\theta_i}(x_1)\phi_{\theta_j}(x_2)\left[\left(2B\bar{C}_M \right)^2 I_u^2 \hat{S}_{uu} + \left(BC_M' \right)^2 I_w^2 \hat{S}_{ww} \right] \\
& + \left[\phi_{y_i}(x_1)\phi_{z_j}(x_2) + \phi_{z_i}(x_1)\phi_{y_j}(x_2) \right]\left[4\frac{D}{B}\bar{C}_D\bar{C}_L I_u^2 \hat{S}_{uu} + \left(\frac{D}{B}C_D' - \bar{C}_L \right)\left(C_L' + \frac{D}{B}\bar{C}_D \right) I_w^2 \hat{S}_{ww} \right] \\
& + \left[\phi_{y_i}(x_1)\phi_{\theta_j}(x_2) + \phi_{\theta_i}(x_1)\phi_{y_j}(x_2) \right]\left[4\frac{D}{B}\bar{C}_D B\bar{C}_M I_u^2 \hat{S}_{uu} + \left(\frac{D}{B}C_D' - \bar{C}_L \right) BC_M' I_w^2 \hat{S}_{ww} \right] \\
& + \left[\phi_{z_i}(x_1)\phi_{\theta_j}(x_2) + \phi_{\theta_i}(x_1)\phi_{z_j}(x_2) \right]\left[4\bar{C}_L B\bar{C}_M I_u^2 \hat{S}_{uu} + \left(C_L' + \frac{D}{B}\bar{C}_D \right) BC_M' I_w^2 \hat{S}_{ww} \right] \Bigg\} dx_1 dx_2
\end{aligned} \tag{6.61}$$

and the corresponding reduced version is given by

$$\hat{J}_{ij}^2 = \frac{J_{ij}^2}{\left(\int\limits_L \left(\phi_{y_i}^2 + \phi_{z_i}^2 + \phi_{\theta_i}^2 \right) dx \right) \cdot \left(\int\limits_L \left(\phi_{y_j}^2 + \phi_{z_j}^2 + \phi_{\theta_j}^2 \right) dx \right)} \tag{6.62}$$

The reduced cross spectra $\hat{S}_{uu}$ and $\hat{S}_{ww}$ are defined by

$$\left. \begin{aligned} \hat{S}_{uu} &= S_{uu}(\Delta x, \omega) / \sigma_u^2 \\ \hat{S}_{ww} &= S_{ww}(\Delta x, \omega) / \sigma_w^2 \end{aligned} \right\} \tag{6.63}$$

where S_{uu} and S_{ww} are defined in Eq. 3.39. (A transition between spectral density descriptions using f rather than ω as the frequency variable is shown in Eq. 2.68.) Since spatial averaging will eliminate any complex parts of the cross spectra, Eq. 6.63 may for all practical purposes be replaced by

$$\left.\begin{aligned}\hat{S}_{uu} &= \mathrm{Re}\left[S_{uu}(\Delta x,\omega)\right]/\sigma_u^2 = \frac{S_u(\omega)}{\sigma_u^2}\cdot \hat{C}o_{uu}(\Delta x,\omega)\\ \hat{S}_{ww} &= \mathrm{Re}\left[S_{ww}(\Delta x,\omega)\right]/\sigma_w^2 = \frac{S_w(\omega)}{\sigma_w^2}\cdot \hat{C}o_{ww}(\Delta x,\omega)\end{aligned}\right\} \tag{6.64}$$

where $\hat{C}o_{uu}$ and $\hat{C}o_{ww}$ are the reduced u- and w- component co-spectra (see Eq. 3.40).

Example 6.3

Let us again (similar to example 6.2) consider a simply supported horizontal beam type of bridge with span $L = 500m$ that is elevated at a position $z_f = 50m$, but now we set out to calculate the dynamic response at $x_r = L/2$ associated with the two mode shapes

$$\boldsymbol{\varphi}_1 = \begin{bmatrix} 0 & \phi_{z_1} & 0 \end{bmatrix}^T \qquad \text{and} \qquad \boldsymbol{\varphi}_2 = \begin{bmatrix} 0 & 0 & \phi_{\theta_2} \end{bmatrix}^T$$

with corresponding eigen–frequencies $\omega_1 = 0.8$ and $\omega_2 = 2.0\ rad/s$. As can be seen, $\boldsymbol{\varphi}_1$ contains only the displacement component in the across wind vertical direction while $\boldsymbol{\varphi}_2$ only contains torsion. Let us for simplicity assume that $\phi_{z_1} = \phi_{\theta_2} = \sin \pi x/L$. Thus, the aim of this example is to calculate the corresponding dynamic response quantities $\sigma_{r_z r_z}$ and $\sigma_{r_\theta r_\theta}$ at $x_r = L/2$ and the covariance $Cov_{r_\theta r_z}$ between them. It is taken for granted that the chosen mean wind velocity settings are well below any instability limit, such that any changes to resonance frequencies may be ignored. Again, it is assumed that the cross section is close to a flat plate with the following static load coefficient properties:

$$\bar{C}_L = 0 \qquad C_L' = 5 \qquad \bar{C}_M = 0 \qquad C_M' = 1.5 \qquad \text{and} \qquad \frac{D}{B}\bar{C}_D \ll C_L'$$

(Quantifying the drag coefficient is obsolete since y direction response is not excited.)

Let us also assume that the entire span is flow exposed, i.e. $L_{\exp} = L$, and adopt the following wind field properties:

1) the turbulence intensity $I_w = \sigma_w / V = 0.08$ (see Eq. 3.14)

2) the integral length scales: ${}^{xf}L_u = 100 \cdot \left(\frac{z_f}{10} \right)^{0.3} = 162m$, ${}^{xf}L_w = \frac{{}^{xf}L_u}{12}$ (see Eq. 3.36),

3) the auto spectral density: $\frac{S_w(\omega)}{\sigma_w^2} = \frac{1.5 \cdot {}^{xf}L_w / V}{\left(1 + 2.25 \cdot \omega \cdot {}^{xf}L_w / V\right)^{5/3}}$ (see Eq. 3.25)

4) the normalised co-spectrum: $\hat{Co}_{ww}(\omega, \Delta x) = \exp\left(-C_{wx} \cdot \frac{\omega \cdot \Delta x}{V} \right)$ (see Eq. 3.41)

where $C_{wx} = C_{wy_f} = 6.5/(2\pi) \approx 1.0$.

Let us allot the following values to the remaining constants that are necessary for a numerical calculation of the relevant dynamic response quantities at $x_r = L/2$:

ρ (kg/m^3)	B (m)	D (m)	m_1 (kg/m)	m_2 (kgm^2/m)	ω_1 (rad/s)	ω_2 (rad/s)	ζ_1	ζ_2
1.25	20	4	10^4	$6 \cdot 10^5$	0.8	2.0	0.005	0.005

Since m_1 and m_2 are constant along the span, then the modally equivalent and evenly distributed masses $\tilde{m}_1 = m_1$ and $\tilde{m}_2 = m_2$. It should be noted that

$$\boldsymbol{\varphi}_1^T \cdot \boldsymbol{\varphi}_1 = \phi_{z_1}^2 = \sin^2 \pi x/L \qquad \text{and} \qquad \boldsymbol{\varphi}_2^T \cdot \boldsymbol{\varphi}_2 = \phi_{\theta_2}^2 = \sin^2 \pi x/L$$

and that $\int_0^L \phi_m \cdot \phi_n dx = \frac{L}{2}$ for any combination of $\left.\begin{matrix} m \\ n \end{matrix}\right\} = z_1$ or θ_2.

Finally, let us for simplicity adopt quasi-static values to the aerodynamic derivatives, except for A_2^* which is responsible for aerodynamic damping in torsion. Adopting $A_2^* = -\beta_M C_M' (V/B\omega_i)^2$ and $\beta_M = 0.2$ provides a good approximation to the flat plate properties. Thus, the aerodynamic derivatives associated with motion in the across wind vertical direction and torsion are given by (see Eq. 5.26):

$$\begin{bmatrix} H_1^* \\ H_2^* \\ H_3^* \end{bmatrix} = C_L' \cdot \begin{bmatrix} -\hat{V} \\ 0 \\ \hat{V}^2 \end{bmatrix} \qquad \begin{bmatrix} A_1^* \\ A_2^* \\ A_3^* \end{bmatrix} = C_M' \cdot \begin{bmatrix} -\hat{V} \\ -\beta_M \cdot \hat{V}^2 \\ \hat{V}^2 \end{bmatrix} \qquad \begin{bmatrix} H_4^* & A_4^* \\ H_5^* & A_5^* \\ H_6^* & A_6^* \end{bmatrix} = \mathbf{0}$$

where: $\hat{V} = V/(B\omega_i)$. The aerodynamic coefficients associated with changes in stiffness and damping are then given by (see Eq. 6.51 and 6.52, or the fully expanded versions in Eqs. 6.53 and 6.54):

$$\kappa_{ae_{ij}} = \frac{\rho B^2}{2\tilde{m}_i} \cdot \int_{L_{exp}} \left(\phi_{z_i} \phi_{\theta_j} B H_3^* + \phi_{\theta_i} \phi_{\theta_j} B^2 A_3^* \right) dx \Big/ \int_L \left(\phi_{z_i}^2 + \phi_{\theta_i}^2 \right) dx$$

$$\zeta_{ae_{ij}} = \frac{\rho B^2}{4\tilde{m}_i} \cdot \int_{L_{exp}} \left(\phi_{z_i} \phi_{z_j} H_1^* + \phi_{\theta_i} \phi_{z_j} B A_1^* + \phi_{\theta_i} \phi_{\theta_j} B^2 A_2^* \right) dx \Big/ \int_L \left(\phi_{z_i}^2 + \phi_{\theta_i}^2 \right) dx$$

where in this case i and j are equal to 1 or 2. Introducing the choice of aerodynamic derivatives given above, then:

$$\kappa_{ae11} = 0\,, \quad \kappa_{ae12} = \frac{\rho B^2}{2\tilde{m}_1} \cdot \frac{\int_{L_{\exp}} \phi_{z1}\phi_{\theta 2} B H_3^* dx}{\int_L \phi_{z1}^2 dx} = \frac{\rho B^3}{2\tilde{m}_1} \cdot H_3^* = \frac{\rho B^3}{2\tilde{m}_1} \cdot C_L' \cdot \left(\frac{V}{B\omega_1}\right)^2$$

$$\kappa_{ae21} = 0\,, \quad \kappa_{ae22} = \frac{\rho B^2}{2\tilde{m}_2} \cdot \frac{\int_{L_{\exp}} \phi_{\theta 2}^2 B^2 A_3^* dx}{\int_L \phi_{\theta 2}^2 dx} = \frac{\rho B^4}{2\tilde{m}_2} \cdot A_3^* = \frac{\rho B^4}{2\tilde{m}_2} \cdot C_M' \cdot \left(\frac{V}{B\omega_2}\right)^2$$

$$\zeta_{ae11} = \frac{\rho B^2}{4\tilde{m}_1} \cdot \frac{\int_{L_{\exp}} \phi_{z1}^2 H_1^* dx}{\int_L \phi_{z1}^2 dx} = \frac{\rho B^2}{4\tilde{m}_1} \cdot H_1^* = -\frac{\rho B^2}{4\tilde{m}_1} \cdot C_L' \cdot \frac{V}{B\omega_1}\,, \qquad \zeta_{ae12} = 0$$

$$\zeta_{ae21} = \frac{\rho B^2}{4\tilde{m}_2} \cdot \frac{\int_{L_{\exp}} \phi_{\theta 2}\phi_{z1} B A_1^* dx}{\int_L \phi_{\theta 2}^2 dx} = \frac{\rho B^3}{4\tilde{m}_2} \cdot A_1^* = -\frac{\rho B^3}{4\tilde{m}_2} \cdot C_M' \cdot \frac{V}{B\omega_2}$$

$$\zeta_{ae22} = \frac{\rho B^2}{4\tilde{m}_2} \cdot \frac{\int_{L_{\exp}} \phi_{\theta 2}^2 B^2 A_2^* dx}{\int_L \phi_{\theta 2}^2 dx} = \frac{\rho B^4}{4\tilde{m}_2} \cdot A_2^* = -\frac{\rho B^4}{4\tilde{m}_2} \cdot \beta_M C_M' \cdot \left(\frac{V}{B\omega_2}\right)^2$$

The non-dimensional frequency response function is then given by (see Eq. 6.49)

$$\hat{\mathbf{H}}_\eta(\omega) = \left\{\mathbf{I} - \boldsymbol{\kappa}_{ae} - \left(\omega \cdot diag\left[\frac{1}{\omega_i}\right]\right)^2 + 2i\omega \cdot diag\left[\frac{1}{\omega_i}\right] \cdot (\boldsymbol{\zeta} - \boldsymbol{\zeta}_{ae})\right\}^{-1} =$$

$$\left\{\begin{bmatrix} 1 & 0 \\ 0 & 1 \end{bmatrix} - \begin{bmatrix} 0 & \kappa_{ae12} \\ 0 & \kappa_{ae22} \end{bmatrix} - \omega^2 \begin{bmatrix} \omega_1^{-2} & 0 \\ 0 & \omega_2^{-2} \end{bmatrix} + 2i\omega \begin{bmatrix} \omega_1^{-1} & 0 \\ 0 & \omega_2^{-1} \end{bmatrix} \left(\begin{bmatrix} \zeta_1 & 0 \\ 0 & \zeta_2 \end{bmatrix} - \begin{bmatrix} \zeta_{ae11} & 0 \\ \zeta_{ae21} & \zeta_{ae22} \end{bmatrix}\right)\right\}^{-1}$$

where: $\kappa_{ae12} = 97.66 \cdot 10^{-4} \cdot V^2$, $\kappa_{ae22} = 1.563 \cdot 10^{-4} \cdot V^2$, $\zeta_{ae11} = -39.06 \cdot 10^{-4} \cdot V$, $\zeta_{ae21} = -1.563 \cdot 10^{-4} \cdot V$, $\zeta_{ae22} = -0.1563 \cdot 10^{-4} \cdot V^2$, and where all other quantities are given above. The aerodynamic stiffness and damping coefficients κ_{ae12}, κ_{ae21}, ζ_{ae11}, ζ_{ae21}, ζ_{ae22} are shown in Fig. 6.8. The absolute value of the determinant of the non–dimensional frequency response function (at $V = 0$) is shown in Fig. 6. 9 together with the single point spectral density and normalised co-spectrum of the wind turbulence w component.

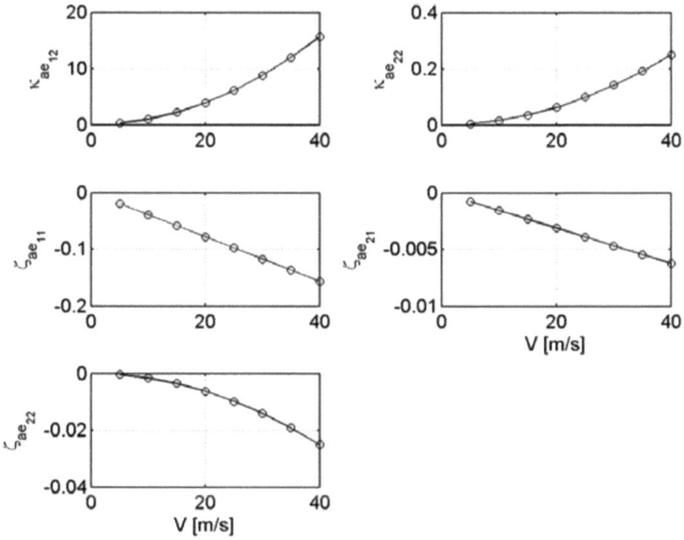

Fig. 6.8 *Aerodynamic stiffness and damping coefficients*

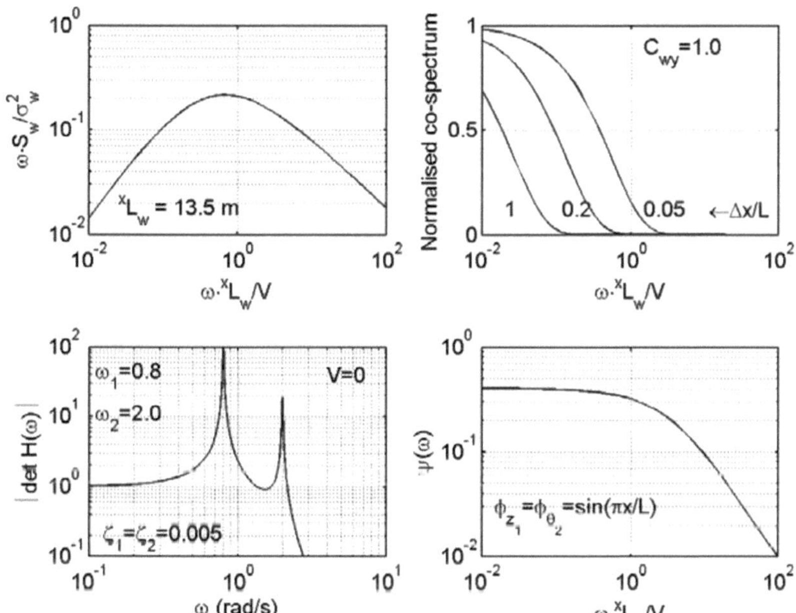

Fig. 6.9 *Top left and right hand side diagrams: w component spectral density and normalised co-spectrum, lower left: absolute value of the determinant of the non-dimensional frequency response function at* $V = 0$, *lower right: the joint acceptance function of normalized mode shapes.*

The content of the normalised modal load matrix

$$\mathbf{S}_{\hat{Q}}(\omega) = \begin{bmatrix} S_{\hat{Q}_1\hat{Q}_1} & S_{\hat{Q}_1\hat{Q}_2} \\ S_{\hat{Q}_2\hat{Q}_1} & S_{\hat{Q}_2\hat{Q}_2} \end{bmatrix}$$

is given in Eq. 6.59:

$$S_{\hat{Q}_i\hat{Q}_j}(\omega) = \frac{\rho B^3}{2\tilde{m}_i} \cdot \frac{\rho B^3}{2\tilde{m}_j} \cdot \left(\frac{V}{B\omega_i}\right)^2 \cdot \left(\frac{V}{B\omega_j}\right)^2 \cdot \hat{J}_{ij}^2$$

where the reduced joint acceptance function $\hat{J}_{ij}$ is given in Eq. 6.60. An expanded version of the joint acceptance function itself is given in in Eq. 6.61. Under the present circumstances it simplifies into

$$J_{11}^2 = \iint\limits_{L_{exp}} \phi_{z_1}(x_1) \cdot \phi_{z_1}(x_2) \cdot (C_L' I_w)^2 \cdot \frac{S_{ww}(\omega, \Delta x)}{\sigma_w^2} dx_1 dx_2$$

$$J_{12}^2 = \iint\limits_{L_{exp}} \phi_{z_1}(x_1) \cdot \phi_{\theta_2}(x_2) \cdot C_L' B C_M' I_w^2 \cdot \frac{S_{ww}(\omega, \Delta x)}{\sigma_w^2} dx_1 dx_2, \qquad J_{21}^2 = J_{12}^2$$

$$J_{22}^2 = \iint\limits_{L_{exp}} \phi_{\theta_2}(x_1) \cdot \phi_{\theta_2}(x_2) \cdot (B C_M' I_w)^2 \cdot \frac{S_{ww}(\omega, \Delta x)}{\sigma_w^2} dx_1 dx_2$$

Introducing $S_{ww}(\omega, \Delta x) = S_w(\omega) \cdot \hat{C}o_{ww}(\omega, \Delta x)$ and $I_w = \sigma_w / V$, then the content of the normalised modal load matrix is given by

$$S_{\hat{Q}_1\hat{Q}_1}(\omega) = \left(\frac{\rho V B C_L'}{2\omega_1^2 \tilde{m}_1} \cdot \hat{J}_{11}(\omega)\right)^2 S_w(\omega), \quad S_{\hat{Q}_1\hat{Q}_2}(\omega) = \left(\frac{\rho V B \sqrt{C_L' B C_M'}}{2\omega_1\omega_2\sqrt{\tilde{m}_1\tilde{m}_2}} \cdot \hat{J}_{21}(\omega)\right)^2 S_w(\omega),$$

$$S_{\hat{Q}_2\hat{Q}_1}(\omega) = S_{\hat{Q}_1\hat{Q}_2}(\omega) \qquad \text{and} \qquad S_{\hat{Q}_2\hat{Q}_2}(\omega) = \left(\frac{\rho V B^2 C_M'}{2\omega_2^2 \tilde{m}_2} \cdot \hat{J}_{22}(\omega)\right)^2 S_w(\omega)$$

where:

$$\hat{J}_{11}^2 = \iint\limits_{L_{exp}} \phi_{z_1}(x_1) \cdot \phi_{z_1}(x_2) \cdot \hat{C}o_{ww}(\omega, \Delta x) dx_1 dx_2 \Bigg/ \left(\int_L \phi_{z_1}^2 dx\right)^2$$

$$\hat{J}_{21}^2 = \iint\limits_{L_{exp}} \phi_{z_1}(x_1) \cdot \phi_{\theta_2}(x_2) \cdot \hat{C}o_{ww}(\omega, \Delta x) dx_1 dx_2 \Bigg/ \left(\int_L \phi_{z_1}^2 dx \cdot \int_L \phi_{\theta_2}^2 dx\right)$$

$$\hat{J}_{22}^2 = \iint\limits_{L_{exp}} \phi_{\theta_2}(x_1) \cdot \phi_{\theta_2}(x_2) \cdot \hat{C}o_{ww}(\omega, \Delta x) dx_1 dx_2 \Bigg/ \left(\int_L \phi_{\theta_2}^2 dx\right)^2$$

Since $\phi_{z_1} = \phi_{\theta_2} = \sin \pi x / L$, and $\hat{C}o_{ww}(\omega, \Delta x) = \exp\left(-C_{wy} \cdot \omega \cdot \Delta x / V\right)$ the present situation is equivalent to that which was encountered in Example 6.2, and thus,

$$\left.\begin{array}{l}\hat{J}_{11}^2\\ \hat{J}_{21}^2\\ \hat{J}_{22}^2\end{array}\right\} = 4\cdot\psi(\omega) \qquad \text{where} \qquad \psi(\omega) = \frac{\hat{\omega}}{\hat{\omega}^2+\pi^2} + 2\pi^2\cdot\frac{1+\exp(-\hat{\omega})}{\left(\hat{\omega}^2+\pi^2\right)^2}$$

and where $\hat{\omega} = C_{ux}\,\omega L_{\exp}/V$. The normalised modal load matrix $\mathbf{S}_{\hat{Q}}$ is then given by

$$\mathbf{S}_{\hat{Q}}(\omega) = \begin{bmatrix} S_{\hat{Q}_1\hat{Q}_1} & S_{\hat{Q}_1\hat{Q}_2} \\ S_{\hat{Q}_2\hat{Q}_1} & S_{\hat{Q}_2\hat{Q}_2} \end{bmatrix} = \frac{(\rho VB)^2\cdot S_w(\omega)\cdot\psi(\omega)}{\left(\omega_1^2\tilde{m}_1\right)\cdot\left(\omega_2^2\tilde{m}_2\right)} \begin{bmatrix} C_L'^2\left(\dfrac{\omega_2}{\omega_1}\right)^2\dfrac{\tilde{m}_2}{\tilde{m}_1} & BC_L'C_M' \\ BC_L'C_M' & (BC_M')^2\left(\dfrac{\omega_1}{\omega_2}\right)^2\dfrac{\tilde{m}_1}{\tilde{m}_2} \end{bmatrix}$$

And thus, the spectral density response matrix at $x_r = L/2$ is given by (see Eqs. 4.81 and 4.82)

$$\mathbf{S}_{rr}(L/2,\omega) = \begin{bmatrix} S_{r_zr_z} & S_{r_zr_\theta} \\ S_{r_\theta r_z} & S_{r_\theta r_\theta} \end{bmatrix} = \mathbf{\Phi}_r(L/2)\cdot\mathbf{S}_\eta(\omega)\cdot\mathbf{\Phi}_r^T(L/2)$$

where: $\mathbf{S}_\eta(\omega) = \hat{\mathbf{H}}_\eta^*(\omega)\cdot\mathbf{S}_{\hat{Q}}(\omega)\cdot\hat{\mathbf{H}}_\eta^T(\omega)$ and $\mathbf{\Phi}_r(L/2) = \begin{bmatrix} 1 & 0 \\ 0 & 1 \end{bmatrix}$

Introducing the impedance matrix $\mathbf{E}(\omega) = \begin{bmatrix} E_{11} & E_{12} \\ E_{21} & E_{22} \end{bmatrix}$ where

$$E_{11} = 1-\left(\frac{\omega}{\omega_1}\right)^2 + 2i\frac{\omega}{\omega_1}\left(\zeta_1-\zeta_{ae_{11}}\right), \qquad E_{12} = -\kappa_{ae_{12}},$$

$$E_{21} = -2i\frac{\omega}{\omega_2}\zeta_{ae21} \qquad \text{and} \qquad E_{22} = 1-\kappa_{ae22}-\left(\frac{\omega}{\omega_2}\right)^2 + 2i\frac{\omega}{\omega_2}\left(\zeta_2-\zeta_{ae_{22}}\right)$$

Then
$$\hat{\mathbf{H}}_\eta(\omega) = \begin{bmatrix} \hat{H}_{11} & \hat{H}_{12} \\ \hat{H}_{21} & \hat{H}_{22} \end{bmatrix} = \mathbf{E}^{-1} = \frac{1}{\det\mathbf{E}}\begin{bmatrix} E_{22} & -E_{12} \\ -E_{21} & E_{11} \end{bmatrix}$$

rendering the following expression for the spectral density response matrix at $x_r = L/2$

$$\mathbf{S}_{rr}(L/2,\omega) = \frac{\rho B^2}{\tilde{m}_1}\cdot\frac{\rho B^4}{\tilde{m}_2}\cdot\left(\frac{V}{B\omega_1}\right)^2\cdot\left(\frac{V}{B\omega_2}\right)^2\cdot I_w^2\cdot\frac{S_w(\omega)}{\sigma_w^2}\cdot\psi(\omega)\cdot\begin{bmatrix} \hat{S}_{\eta 11} & \hat{S}_{\eta 12} \\ \hat{S}_{\eta 21} & \hat{S}_{\eta 22} \end{bmatrix}$$

where:

$$\hat{S}_{\eta 11}(\omega) = \gamma_{LL}\cdot\left(\hat{H}_{11}^*\hat{H}_{11}\right) + \gamma_{LM}\cdot\left(\hat{H}_{12}^*\hat{H}_{11} + \hat{H}_{11}^*\hat{H}_{12}\right) + \gamma_{MM}\cdot\left(\hat{H}_{12}^*\hat{H}_{12}\right)$$

$$\hat{S}_{\eta 12}(\omega) = \gamma_{LL}\cdot\left(\hat{H}_{11}^*\cdot\hat{H}_{21}\right) + \gamma_{LM}\cdot\left(\hat{H}_{12}^*\cdot\hat{H}_{21} + \hat{H}_{11}^*\cdot\hat{H}_{22}\right) + \gamma_{MM}\cdot\hat{H}_{12}^*\cdot\hat{H}_{22}$$

$$\hat{S}_{\eta 21}(\omega) = \gamma_{LL}\cdot\hat{H}_{11}^*\cdot\hat{H}_{21} + \gamma_{LM}\cdot\left(\hat{H}_{21}^*\cdot\hat{H}_{12} + \hat{H}_{22}^*\cdot\hat{H}_{11}\right) + \gamma_{MM}\cdot\hat{H}_{22}^*\cdot\hat{H}_{12}$$

$$\hat{S}_{\eta 22}(\omega) = \gamma_{LL} \cdot \hat{H}_{21}^* \hat{H}_{21} + \gamma_{LM} \cdot \left(\hat{H}_{21}^* \hat{H}_{22} + \hat{H}_{22}^* \hat{H}_{21} \right) + \gamma_{MM} \cdot \hat{H}_{22}^* \hat{H}_{22}$$

$$\gamma_{LL} = C_L'^2 \left(\frac{\omega_2}{\omega_1} \right)^2 \frac{\tilde{m}_2}{\tilde{m}_1} = 9375 \,, \quad \gamma_{LM} = B C_L' C_M' = 150 \text{ and } \gamma_{MM} = \left(B C_M' \right)^2 \left(\frac{\omega_1}{\omega_2} \right)^2 \frac{\tilde{m}_1}{\tilde{m}_2} = 2.4 \,.$$

Since we are mainly aiming at calculating the content of the covariance matrix

$$\mathbf{Cov}_{rr}(x_r = L/2) = \int_0^\infty \mathbf{S}_{rr}(L/2, \omega) d\omega = \begin{bmatrix} \sigma_{r_z r_z}^2 & Cov_{r_z r_\theta} \\ Cov_{r_\theta r_z} & \sigma_{r_\theta r_\theta}^2 \end{bmatrix}$$

it is only the absolute values that are of interest.

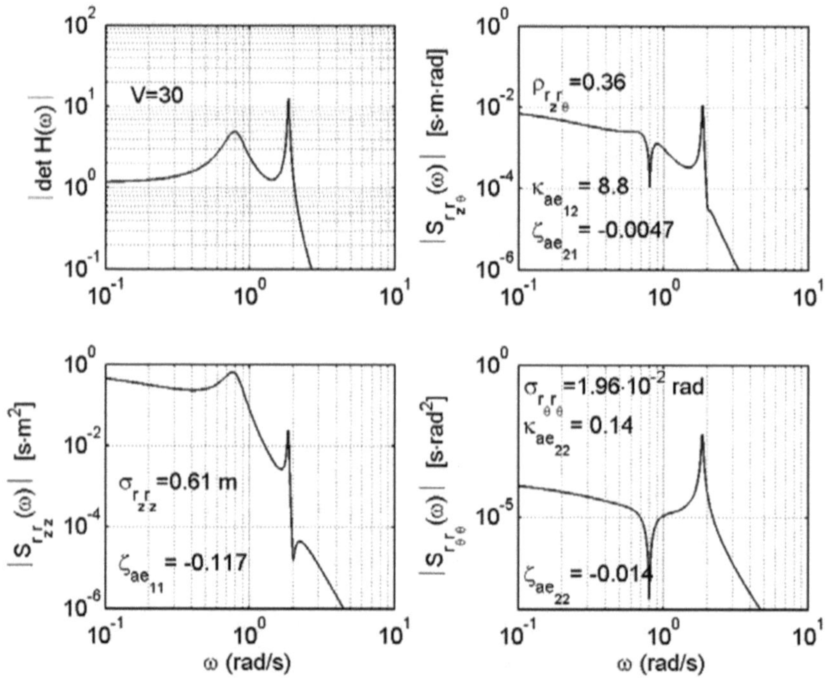

Fig. 6.10 *Top left: absolute value of frequency response function. Top right: cross spectrum between vertical and torsion response components. Lower left and right: spectra of components in vertical direction and torsion.* $V = 30 \ m/s$.

The absolute value of the determinant of the non-dimensional frequency at a mean wind velocity of $V = 30 \ m/s$ is shown in the top left hand side diagram in Fig. 6.10. The top right hand side diagram shows the amplitude of the cross spectrum between r_z and r_θ while the two lower diagrams show the spectral densities of r_z and r_θ, all at a mean wind velocity of $V = 30 \ m/s$. As can be seen, there are traces of modal coupling. In this case the coupling effects are exclusively motion induced. Comparing $|\det H(\omega)|$ shown in the top left hand side diagram of Fig. 6.10 to that which is shown in Fig. 6.9 it is seen that the resonance frequency associated with the second mode shape (in torsion) is no longer precisely at $2\ rad/s$, but slightly below. It is also seen that the resonance peaks are reduced, and particularly the peak associated with ϕ_{z_1} at $\omega_1 = 0.8$ rad/s. The standard deviation of the dynamic responses in the across wind direction (r_z) and in torsion (r_θ) at various mean wind velocities are shown on the two left hand side diagrams in Fig. 6.11. The circular points joined with a fully drawn line are based on the development shown above, i.e. they contain the effects of aerodynamic derivatives, while the broken line represents the situation that aerodynamic derivatives are ignored. As can be seen, the difference is considerable for the respone in the across wind vertical direction, but in torsion only at the highest mean wind velocity setting. It should be noted that the applicability of quasi static aerodynamic derivatives is in many cases questionable, and they should in general be replaced by values obtained from wind tunnel tests. The covariance coefficient between the dynamic responses r_z and r_θ is shown on the top right hand side diagram in Fig. 6.11, and again, circles and fully drawn line contain the effects of aerodynamic derivatives while for the broken line no motion induced effects have been included. The changes of the resonance frequency associated with the second mode shape (in torsion) at increasing mean wind velocities is shown on the lower right hand side diagram in Fig. 6.11. As can be seen, the reduction of the resonance frequency from $V = 0$ to $V = 40\ m/s$ is slightly less than 15 % (which without further iterations implies an overestimation of the torsion response).

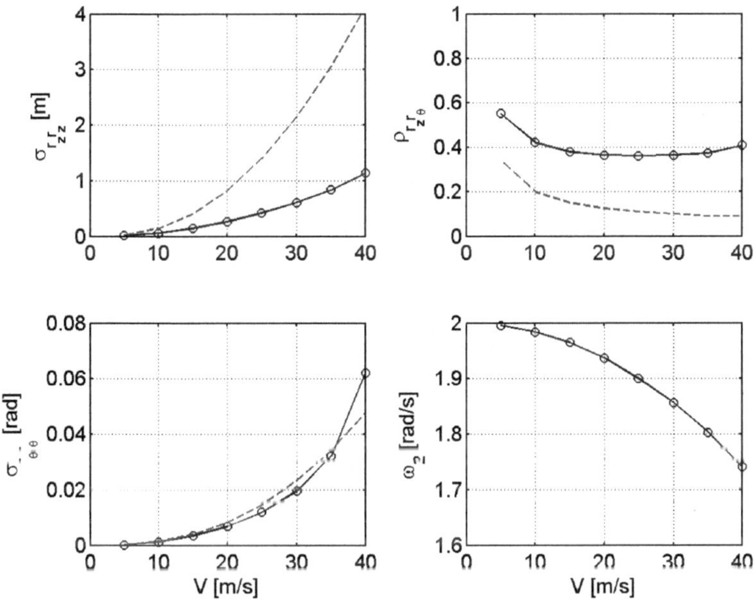

Fig. 6.11 *Top and lower left: dynamic response in vertical direction and torsion. Top right: covariance coefficient. Lower right: resonance frequency associated with 2nd mode. Full lines: including motion induced effects. Broken lines: without motion induced effects.*

6.4 *Vortex shedding*

As shown in chapter 5.3, the vortex shedding induced load effects at or in the vicinity of lock-in are dependent on the dynamic response of the structure, i.e. the total damping in each mode is unknown prior to any knowledge about the actual structural displacements. Thus, the calculation of vortex shedding induced dynamic response will inevitably involve iterations.

It should be acknowledged that the peak factor for vortex shedding response does not comply with the theory behind what may be obtained from Eq. 2.45. For an ultra-narrow-banded vortex shedding response the peak factor is close to 1.5 (theoretically $\sqrt{2}$, see Eq. 2.47). For broad-banded response Eq. 2.45 will most often render conservative results. Some time domain simulations of response spectra (see Appendix A) will give a good indication on what peak factor should be chosen.

Multi mode response calculations

The general solution of a multi mode approach to the problem of calculating vortex shedding induced dynamic response is identical to that which has been presented above for buffeting response calculations. I.e., the general solution to the calculation of the three by three cross spectra response matrix $\mathbf{S}_{rr}(x_r,\omega)$ is given in Eq. 4.80–4.82, while the corresponding covariance matrix is given in Eqs. 6.47 and 6.48. The N_{mod} by N_{mod} frequency response matrix $\hat{\mathbf{H}}_\eta(\omega)$ and the modal load matrix $\mathbf{S}_{\hat{Q}}(\omega)$ are given in Eqs. 4.69 and 4.75, except that for vortex shedding the motion induced load is assumed exclusively related to structural velocity, and its effect applies to the actual modal response and not to the individual Fourier components. As shown in Eq. 5.36, this implies that $\mathbf{K}_{ae} = 0$ and $\mathbf{C}_{ae} = \left(\rho B^2/2\right)\cdot\omega_i(V)\cdot diag\left[0 \quad H_1^* \quad B^2 A_2^*\right]$, and thus

$$\hat{\mathbf{H}}_\eta(\omega) = \left\{\mathbf{I} - \left(\omega\cdot diag\left[\frac{1}{\omega_i}\right]\right)^2 + 2i\omega\cdot diag\left[\frac{1}{\omega_i}\right]\cdot\left(\boldsymbol{\zeta} - \boldsymbol{\zeta}_{ae}\right)\right\}^{-1} \tag{6.65}$$

where $\boldsymbol{\zeta} = diag\left[\zeta_i\right]$ and the content of $\boldsymbol{\zeta}_{ae}$ is given by

$$\zeta_{ae_{ij}} = \frac{\omega_i}{2}\frac{\tilde{C}_{ae_{ij}}}{\omega_i^2 \tilde{M}_i} = \frac{\rho B^2}{4\tilde{m}_i}\cdot\frac{\int\limits_{L_{exp}}\left(\boldsymbol{\varphi}_i^T \cdot \hat{\mathbf{C}}_{ae}\cdot\boldsymbol{\varphi}_j\right)dx}{\int\limits_L\left(\boldsymbol{\varphi}_i^T\cdot\boldsymbol{\varphi}_i\right)dx}$$

$$= \frac{\rho B^2}{4\tilde{m}_i}\cdot\frac{\int\limits_{L_{exp}}\phi_{i_z}\phi_{j_z}H_1^* dx + B^2\int\phi_{i_\theta}\phi_{j_\theta}A_2^* dx}{\int\limits_L\left(\phi_{y_i}^2+\phi_{z_i}^2+\phi_{\theta_i}^2\right)dx} \tag{6.66}$$

where H_1^* and A_2^*, are given in Eq. 5.37 and where $\tilde{m}_i$ is defined in Eq. 6.41. If H_1^* and A_2^* are taken as modal constants and independent of span-wise position, then $\boldsymbol{\zeta}_{ae}$ becomes diagonal due to the orthogonal properties of the mode shapes, i.e.

$$\boldsymbol{\zeta}_{ae} = diag\left[\zeta_{ae_i}\right] \tag{6.67}$$

where

$$\zeta_{ae_i} = \frac{\rho B^2}{4\tilde{m}_i}\cdot\frac{H_1^*\int\limits_{L_{exp}}\phi_{i_z}^2 dx + B^2 A_2^*\int\phi_{i_\theta}^2 dx}{\int\limits_L\left(\phi_{i_y}^2+\phi_{i_z}^2+\phi_{i_\theta}^2\right)dx} \tag{6.68}$$

This implies that $\hat{\mathbf{H}}_\eta(\omega)$ is an N_{mod} by N_{mod} diagonal matrix. In vortex shedding induced vibration problems it is usually not essential to include the along wind load effects. The load vector may then be reduced to

$$\mathbf{q}(x,t) = \begin{bmatrix}0 & q_z & q_\theta\end{bmatrix}^T \tag{6.69}$$

and the corresponding Fourier transform is

$$\mathbf{a}_q(x,\omega) = \begin{bmatrix}0 & a_{q_z} & a_{q_\theta}\end{bmatrix}^T \tag{6.70}$$

The cross sectional load spectrum is defined by (see Eq. 4.78)

$$\mathbf{S}_{qq}(\Delta x,\omega) = \lim_{T\to\infty}\frac{1}{\pi T}\left(\mathbf{a}_q^*\mathbf{a}_q^T\right) = \lim_{T\to\infty}\frac{1}{\pi T}\begin{bmatrix}0 & 0 & 0\\ 0 & a_{q_z}^* a_{q_z} & a_{q_z}^* a_{q_\theta}\\ 0 & a_{q_\theta}^* a_{q_z} & a_{q_\theta}^* a_{q_\theta}\end{bmatrix} - \begin{bmatrix}0 & 0 & 0\\ 0 & S_{q_zq_z} & S_{q_zq_\theta}\\ 0 & S_{q_\theta q_z} & S_{q_\theta q_\theta}\end{bmatrix} \tag{6.71}$$

The problem is greatly simplified if the cross coupling between q_z and q_θ may be disregarded, in which case

$$\mathbf{S}_{qq}(\Delta x,\omega) \approx \begin{bmatrix} 0 & 0 & 0 \\ 0 & S_{q_z q_z} & 0 \\ 0 & 0 & S_{q_\theta q_\theta} \end{bmatrix} \tag{6.72}$$

where the cross spectra $S_{q_z q_z}$ and $S_{q_\theta q_\theta}$ are given by

$$\left.\begin{aligned} S_{q_z q_z} &= S_{q_z}(\omega)\cdot \hat{C}o_{q_z}(\Delta x) \\ S_{q_\theta q_\theta} &= S_{q_\theta}(\omega)\cdot \hat{C}o_{q_\theta}(\Delta x) \end{aligned}\right\} \tag{6.73}$$

The single point spectra S_{q_z} and S_{q_θ} are defined in Eq. 5.33, while the reduced co–spectra $\hat{C}o_{q_z}$ and $\hat{C}o_{q_\theta}$ are defined in Eq. 5.34. Thus, the elements of $\mathbf{S}_{\hat{Q}}$ (see Eqs. 4.75 – 4.78) are reduced to

$$\begin{aligned} S_{\hat{Q}_i\hat{Q}_j}(\omega) &= \frac{\iint\limits_{L_{\exp}} \boldsymbol{\varphi}_i^T(x_1)\cdot \mathbf{S}_{qq}(\Delta x,\omega)\cdot \boldsymbol{\varphi}_j(x_2)dx_1dx_2}{\left(\omega_i^2\tilde{M}_i\right)\cdot\left(\omega_j^2\tilde{M}_j\right)} \\ &= \frac{\iint\limits_{L_{\exp}} \left\{\phi_{i_z}(x_1)\phi_{j_z}(x_2)S_{q_z q_z} + \phi_{i\theta}(x_1)\phi_{j\theta}(x_2)S_{q_\theta q_\theta}\right\}dx_1dx_2}{\left(\omega_i^2\tilde{M}_i\right)\cdot\left(\omega_j^2\tilde{M}_j\right)} \\ &= \frac{S_{q_z}\iint\limits_{L_{\exp}} \phi_{i_z}(x_1)\phi_{j_z}(x_2)\hat{C}o_{q_z}dx_1dx_2 + S_{q_\theta}\iint\limits_{L_{\exp}} \phi_{i\theta}(x_1)\phi_{j\theta}(x_2)\hat{C}o_{q_\theta}dx_1dx_2}{\left(\omega_i^2\tilde{M}_i\right)\cdot\left(\omega_j^2\tilde{M}_j\right)} \end{aligned} \tag{6.74}$$

Furthermore, it is a reasonable assumption that the integral length–scale of the vortices λD is small as compared to the flow exposed length $L_{\exp}$ of the structure, and since q_z and q_θ are caused by the same vortices their coherence properties are likely to be identical, in which case [recalling that $\int_0^\infty \hat{C}o_{q_m}(\Delta x)d(\Delta x) \approx \lambda D$ (see Eq. 5.34) and adopting the integration procedure presented in example 6.1] the following is obtained:

$$S_{\hat{Q}_i\hat{Q}_j}(\omega) \approx \frac{2\lambda D\left[S_{q_z}\int\limits_{L_{\exp}}\phi_{i_z}(x)\phi_{j_z}(x)dx + S_{q_\theta}\int\limits_{L_{\exp}}\phi_{i_\theta}(x)\phi_{j_\theta}(x)dx\right]}{\left(\omega_i^2\tilde{M}_i\right)\cdot\left(\omega_j^2\tilde{M}_j\right)} \tag{6.75}$$

Again, due to the orthogonal properties of the mode shapes this implies that $\mathbf{S}_{\hat{Q}}$ becomes diagonal, i.e.

$$\mathbf{S}_{\hat{Q}} = diag\left[S_{\hat{Q}_i}\right] \tag{6.76}$$

where

$$S_{\hat{Q}_i}(\omega) = \frac{2\lambda D\left[S_{q_z}(\omega)\int\limits_{L_{\exp}}\phi_{i_z}^2 dx + S_{q_\theta}(\omega)\int\limits_{L_{\exp}}\phi_{i_\theta}^2 dx\right]}{\left(\omega_i^2\tilde{M}_i\right)^2} \tag{6.77}$$

The calculation of the spectral response matrix is given in Eqs. 4.80 – 4.82, though, it should be noted that if the simplifications above hold then both $\hat{\mathbf{H}}_\eta$ and $\mathbf{S}_{\hat{Q}}$ are diagonal, in which case

$$\begin{aligned}\mathbf{S}_{rr}(x_r,\omega) &= \mathbf{\Phi}_r(x_r)\cdot diag\left[S_{\eta_i}(\omega)\right]\cdot\mathbf{\Phi}_r^T(x_r) = \sum_{i=1}^{N_{\text{mod}}}\boldsymbol{\varphi}_i(x_r)\cdot\boldsymbol{\varphi}_i^T(x_r)\cdot S_{\eta_i}(\omega)\\ &= \sum_{i=1}^{N_{\text{mod}}}\begin{bmatrix}\phi_y^2(x_r) & \phi_y(x_r)\cdot\phi_z(x_r) & \phi_y(x_r)\cdot\phi_\theta(x_r)\\ & \phi_z^2(x_r) & \phi_z(x_r)\cdot\phi_\theta(x_r)\\ Sym. & & \phi_\theta^2(x_r)\end{bmatrix}_i\cdot S_{\eta_i}(\omega)\end{aligned} \tag{6.78}$$

where

$$S_{\eta_i}(\omega) = \left|\hat{H}_{\eta_i}(\omega)\right|^2\cdot S_{\hat{Q}_i}(\omega) \tag{6.79}$$

$\hat{H}_{\eta_i}$ is given by (see Eq. 6.65)

$$\hat{H}_{\eta_i}(\omega) = \left[1-\left(\frac{\omega}{\omega_i}\right)^2 + 2i\cdot\left(\zeta_i - \zeta_{ae_i}\right)\frac{\omega}{\omega_i}\right]^{-1} \tag{6.80}$$

and ζ_{ae_i} is given in Eq. 6.68 (see also 5.37).

The corresponding covariance response matrix $\mathbf{Cov}_{rr}(x_r)$ for the dynamic response at span-wise position x_r is then given by

$$\mathbf{Cov}_{rr}(x_r) = \int_0^\infty \mathbf{S}_{rr}(x_r,\omega)d\omega$$

$$= \begin{bmatrix} \sigma^2_{r_y r_y} & Cov_{r_y r_z} & Cov_{r_y r_\theta} \\ & \sigma^2_{r_z r_z} & Cov_{r_z r_\theta} \\ Sym. & & \sigma^2_{r_\theta r_\theta} \end{bmatrix} = \sum_{i=1}^{N_{\text{mod}}} \begin{bmatrix} \phi_y^2(x_r) & \phi_y(x_r)\cdot\phi_z(x_r) & \phi_y(x_r)\cdot\phi_\theta(x_r) \\ & \phi_z^2(x_r) & \phi_z(x_r)\cdot\phi_\theta(x_r) \\ Sym. & & \phi_\theta^2(x_r) \end{bmatrix}_i \sigma^2_{\eta_i} \tag{6.81}$$

where $$\sigma^2_{\eta_i} = \int_0^\infty S_{\eta_i} d\omega \tag{6.82}$$

is the variance contribution from an arbitrary mode i. Usually, vortex shedding induced dynamic response is largely resonant and narrow-banded. It will then usually suffice to only consider the resonant part of the frequency domain integration in Eq. 6.82, and discard the background part. Thus,

$$\sigma^2_{\eta_i} = \int_0^\infty S_{\eta_i} d\omega = \int_0^\infty \left|\hat{H}_{\eta_i}(\omega)\right|^2 \cdot S_{\hat{Q}_i}(\omega)d\omega \approx \int_0^\infty \left|\hat{H}_{\eta_i}(\omega)\right|^2 d\omega \cdot S_{\hat{Q}_i}(\omega_i) = \frac{\pi\omega_i \cdot S_{\hat{Q}_i}(\omega_i)}{4(\zeta_i - \zeta_{ae_i})} \tag{6.83}$$

where (see Eqs. 6.77 and 5.33)

$$S_{\hat{Q}_i}(\omega_i) = \frac{2\lambda D\left[S_{q_z}(\omega_i)\int_{L_{\exp}}\phi_{i_z}^2 dx + S_{q_\theta}(\omega_i)\int_{L_{\exp}}\phi_{i_\theta}^2 dx\right]}{\left(\omega_i^2\tilde{M}_i\right)^2}$$

$$= \frac{2\lambda D}{\left(\omega_i^2\tilde{M}_i\right)^2}\cdot\frac{\left(\rho V^2 B/2\right)^2}{\sqrt{\pi}\cdot\omega_s}\cdot\left\{\frac{\sigma^2_{q_z}}{b_z}\int_{L_{\exp}}\phi_{i_z}^2 dx\cdot\exp\left[-\left(\frac{1-\omega_i/\omega_s}{b_z}\right)^2\right]\right.$$
$$\left.+\frac{\left(B\sigma_{q_\theta}\right)^2}{b_\theta}\int_{L_{\exp}}\phi_{i_\theta}^2 dx\cdot\exp\left[-\left(\frac{1-\omega_i/\omega_s}{b_\theta}\right)^2\right]\right\} \tag{6.84}$$

and $\omega_s = 2\pi f_s$. As mentioned above, the calculations will inevitably demand iterations, because H_1^* and A_2^* are functions of $\sigma_{r_z r_z}$ and $\sigma_{r_\theta r_\theta}$. The iteration will take place on the difference between ζ_i and ζ_{ae_i}, which in general will be a small quantity.

Example 6.4

Let us consider a simply supported horizontal beam type of bridge with span $L = L_{\exp} = 500m$ and set out to calculate the vortex shedding induced dynamic response at $x_r = L/2$ which is associated with the three mode shapes

$$\boldsymbol{\varphi}_1 = \begin{bmatrix} 0 \\ \phi_{z_1} \\ 0 \end{bmatrix} = \begin{bmatrix} 0 \\ \sin\left(\dfrac{\pi x}{L}\right) \\ 0 \end{bmatrix} \quad \boldsymbol{\varphi}_2 = \begin{bmatrix} 0 \\ \phi_{z_2} \\ 0 \end{bmatrix} = \begin{bmatrix} 0 \\ \sin\left(\dfrac{3\pi x}{L}\right) \\ 0 \end{bmatrix} \quad \text{and} \quad \boldsymbol{\varphi}_3 = \begin{bmatrix} 0 \\ 0 \\ \phi_{\theta_3} \end{bmatrix} = \begin{bmatrix} 0 \\ 0 \\ \sin\left(\dfrac{\pi x}{L}\right) \end{bmatrix}$$

with corresponding eigen-frequencies 0.8, 1.6 and 2.5 rad/s. As can be seen, $\boldsymbol{\varphi}_1$ and $\boldsymbol{\varphi}_2$ contain only the displacement component in the across wind vertical direction while $\boldsymbol{\varphi}_3$ only contains torsion. Let us adopt the following structural properties:

ρ $\dfrac{kg}{m^3}$	B m	D m	m_z $\dfrac{kg}{m}$	m_θ $\dfrac{kgm^2}{m}$	$\omega_1 = \omega_{z_1}$ $\dfrac{rad}{s}$	$\omega_2 = \omega_{z_2}$ $\dfrac{rad}{s}$	$\omega_3 = \omega_{\theta_3}$ $\dfrac{rad}{s}$	$\zeta_1 = \zeta_3$ %	ζ_2 %
1.25	20	4	10^4	$6 \cdot 10^5$	0.8	1.6	2.5	0.5	0.75

and the following vortex induced wind load properties:

St	$\hat{\sigma}_{q_z}$	$\hat{\sigma}_{q\theta}$	b_z	b_θ	a_z	a_θ	$\lambda_z = \lambda_\theta$	K_{az0}	$K_{a\theta 0}$
0.1	0.9	0.3	0.15	0.1	0.4	0.1	1.2	0.2	0.02

where: $\hat{\sigma}_{q_z} = \sigma_{q_z} \Big/ \left(\frac{1}{2}\rho V^2 B\right)$ and $\hat{\sigma}_{q\theta} = \sigma_{q\theta} \Big/ \left(\frac{1}{2}\rho V^2 B^2\right)$

Since m_z and m_θ are constant along the span, then the modally equivalent and evenly distributed masses $\tilde{m}_1 = \tilde{m}_2 = m_z$ and $\tilde{m}_3 = m_\theta$.

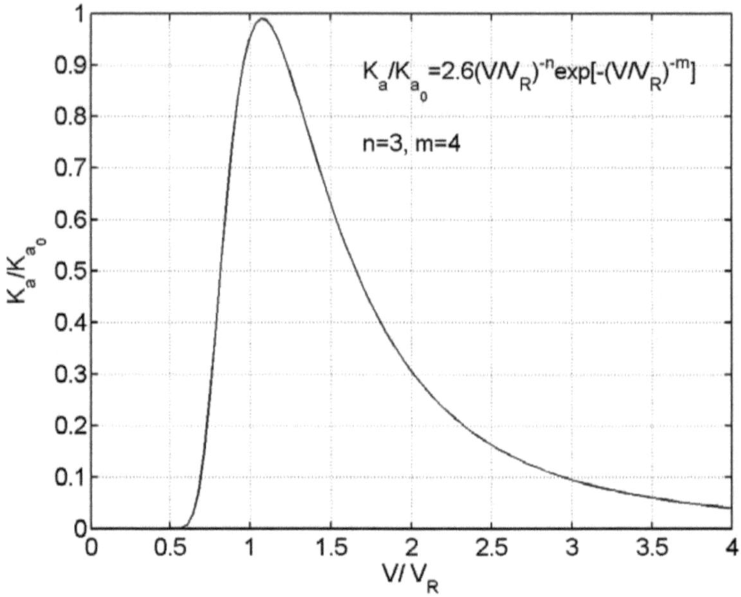

Fig. 6.12 *Aerodynamic damping coefficient*

Finally, let us adopt the following wind velocity variation of the relative aerodynamic damping coefficient (see Fig. 6.12)

$$K_a(V)/K_{a_0} = 2.6 \cdot \left(V/V_{R_i}\right)^{-n} \cdot \exp\left[-\left(V/V_{R_i}\right)^{-m}\right] \qquad \text{where} \qquad V_{R_i} = \frac{\omega_i}{2\pi} \cdot \frac{D}{St}$$

In this case (see Eq. 6.81)

$$\mathbf{Cov}_{rr}(x_r) = \int_0^\infty \mathbf{S}_{rr}(x_r,\omega)\,d\omega = \begin{bmatrix} \sigma^2_{r_y r_y} & Cov_{r_y r_z} & Cov_{r_y r_\theta} \\ & \sigma^2_{r_z r_z} & Cov_{r_z r_\theta} \\ Sym. & & \sigma^2_{r_\theta r_\theta} \end{bmatrix}$$

$$= \begin{bmatrix} 0 & 0 & 0 \\ 0 & \phi^2_{z1}(x_r) & 0 \\ 0 & 0 & 0 \end{bmatrix} \cdot \sigma^2_{\eta 1} + \begin{bmatrix} 0 & 0 & 0 \\ 0 & \phi^2_{z2}(x_r) & 0 \\ 0 & 0 & 0 \end{bmatrix} \cdot \sigma^2_{\eta 2} + \begin{bmatrix} 0 & 0 & 0 \\ 0 & 0 & 0 \\ 0 & 0 & \phi^2_{\theta 3}(x_r) \end{bmatrix} \cdot \sigma^2_{\eta 3}$$

and thus:

$$\mathbf{Cov}_{rr}(x_r) = \begin{bmatrix} 0 & 0 & 0 \\ 0 & \sigma^2_{r_z r_z} & 0 \\ 0 & 0 & \sigma^2_{r_\theta r_\theta} \end{bmatrix} = \begin{bmatrix} 0 & 0 & 0 \\ 0 & \phi^2_{z1}(x_r)\cdot\sigma^2_{\eta 1} + \phi^2_{z2}(x_r)\cdot\sigma^2_{\eta 2} & 0 \\ 0 & 0 & \phi^2_{\theta 3}(x_r)\cdot\sigma^2_{\eta 3} \end{bmatrix}$$

From Eqs. 6.83 and 6.84 (and taking it for granted that $\lambda_z = \lambda_\theta = \lambda$) the following variance contributions are obtained

$$\sigma_{\eta_1}^2 = \left(\frac{D}{2^{7/2}\pi^{7/4}} \cdot \frac{\rho BD}{\tilde{m}_1} \cdot \frac{\hat{\sigma}_{q_z}}{St^2} \right)^2 \cdot \frac{\lambda}{b_z} \cdot \frac{1}{\zeta_1 - \zeta_{ae_1}} \cdot \frac{D \int_{L_{exp}} \phi_{1_z}^2 dx}{\left(\int_L \phi_{1_z}^2 dx \right)^2} \cdot g_1^2\left(V_{R_1}, V\right)$$

$$\sigma_{\eta_2}^2 = \left(\frac{D}{2^{7/2}\pi^{7/4}} \cdot \frac{\rho BD}{\tilde{m}_2} \cdot \frac{\hat{\sigma}_{q_z}}{St^2} \right)^2 \cdot \frac{\lambda}{b_z} \cdot \frac{1}{\zeta_2 - \zeta_{ae_2}} \cdot \frac{D \int_{L_{exp}} \phi_{2_z}^2 dx}{\left(\int_L \phi_{2_z}^2 dx \right)^2} \cdot g_2^2\left(V_{R_2}, V\right)$$

$$\sigma_{\eta_3}^2 = \left(\frac{1}{2^{7/2}\pi^{7/4}} \cdot \frac{\rho (BD)^2}{\tilde{m}_3} \cdot \frac{\hat{\sigma}_{q_\theta}}{St^2} \right)^2 \cdot \frac{\lambda}{b_\theta} \cdot \frac{1}{\zeta_3 - \zeta_{ae_3}} \cdot \frac{D \int_{L_{exp}} \phi_{3_\theta}^2 dx}{\left(\int_L \phi_{3_\theta}^2 dx \right)^2} \cdot g_3^2\left(V_{R_3}, V\right)$$

where

$$\left.\begin{aligned} g_1\left(V_{R_1}, V\right) &= \left(\frac{V}{V_{R_1}} \right)^{3/2} \cdot \exp\left[-\frac{1}{2} \left(\frac{1 - V_{R_1}/V}{b_z} \right)^2 \right] \\ g_2\left(V_{R_2}, V\right) &= \left(\frac{V}{V_{R_2}} \right)^{3/2} \cdot \exp\left[-\frac{1}{2} \left(\frac{1 - V_{R_2}/V}{b_z} \right)^2 \right] \\ g_3\left(V_{R_3}, V\right) &= \left(\frac{V}{V_{R_3}} \right)^{3/2} \cdot \exp\left[-\frac{1}{2} \left(\frac{1 - V_{R_3}/V}{b_\theta} \right)^2 \right] \end{aligned}\right\} \text{ where } \quad V_{R_i} = \frac{\omega_i}{2\pi} \cdot \frac{D}{St} \qquad i = \begin{cases} 1 \\ 2 \\ 3 \end{cases}$$

What then remains are the aerodynamic damping contributions given in Eq. 6.68, from which the following is obtained:

$$\zeta_{ae_1}(V) = \frac{\rho B^2}{4\tilde{m}_1} \cdot H_1^* \cdot \frac{\int_{L_{exp}} \phi_{z_1}^2 dx}{\int_L \phi_{z_1}^2 dx} = \frac{\rho B^2}{4\tilde{m}_1} \cdot K_{a_z}\left(V_{R_1}, V\right) \cdot \left\{ 1 - \left[\frac{\sigma_{r_z r_z}(V)}{a_z D} \right]^2 \right\}$$

$$\zeta_{ae_2}(V) = \frac{\rho B^2}{4\tilde{m}_2} \cdot H_1^* \cdot \frac{\int_{L_{exp}} \phi_{z_2}^2 dx}{\int_L \phi_{z_2}^2 dx} = \frac{\rho B^2}{4\tilde{m}_2} \cdot K_{a_z}\left(V_{R_2}, V\right) \cdot \left\{ 1 - \left[\frac{\sigma_{r_z r_z}(V)}{a_z D} \right]^2 \right\}$$

$$\zeta_{ae_3}(V) = \frac{\rho B^4}{4\tilde{m}_3} \cdot A_2^* \cdot \frac{\int_{L_{exp}} \phi_{\theta_3}^2 dx}{\int_L \phi_{\theta_3}^2 dx} = \frac{\rho B^2}{4\tilde{m}_3} \cdot K_{a_\theta}\left(V_{R_3}, V\right) \cdot \left\{ 1 - \left[\frac{\sigma_{r_\theta r_\theta}(V)}{a_\theta} \right]^2 \right\}$$

The relevant response diagrams are shown in Figs. 6.13 and 6.14 below.

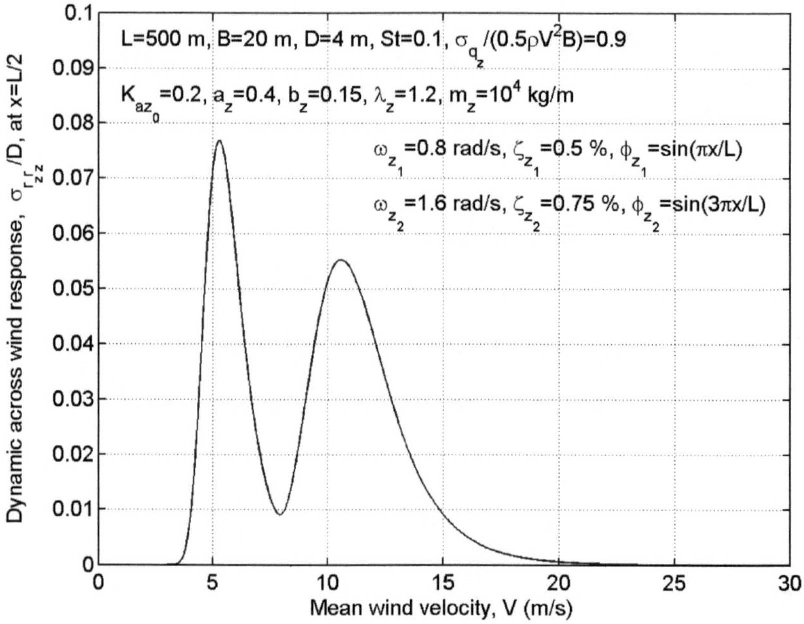

Fig. 6.13 *Vortex shedding induced across wind response*

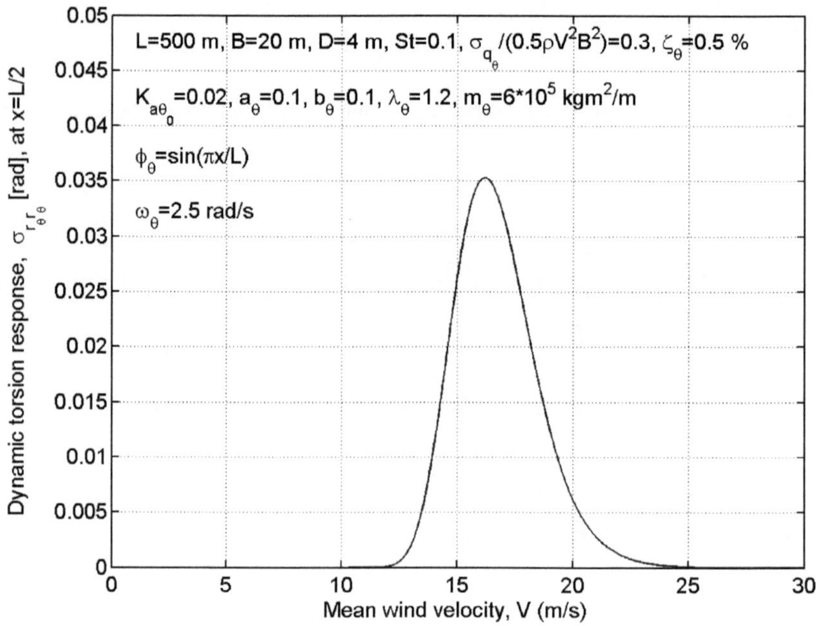

Fig. 6.14 *Vortex shedding induced torsion response*

Single mode single component response calculations

A single mode single component response calculation is in the following only considered relevant for displacements in the z direction and in torsion. Thus, it is only mode shapes that primarily contain either z or θ components that are relevant. I.e., it is taken for granted that any of the following two conditions apply

$$\left.\begin{aligned} \boldsymbol{\varphi}_i(x) &\approx \begin{bmatrix} 0 & \phi_z & 0 \end{bmatrix}_i^T \\ \boldsymbol{\varphi}_i(x) &\approx \begin{bmatrix} 0 & 0 & \phi_\theta \end{bmatrix}_i^T \end{aligned}\right\} \tag{6.85}$$

Off diagonal terms in Eq. 6.78 will then vanish, rendering all covariance quantities obsolete, and $\mathbf{S}_{rr}$ will simply contain the response variances of the excitation of each mode on its diagonal. Thus, the response spectrum and the displacement variance associated with the excitation of an arbitrary mode i are given by

$$\left.\begin{aligned} S_{r_n}(\omega) &= \phi_n^2(x_r)\cdot\left|\hat{H}_{\eta_n}(\omega)\right|^2\cdot S_{\hat{Q}_n}(\omega) \\ \sigma_{r_n}^2 &= \int_0^\infty S_{r_n}(\omega)d\omega \end{aligned}\right\} \qquad n = \begin{cases} z \\ \theta \end{cases} \tag{6.86}$$

where

$$\left.\begin{aligned} \hat{H}_{\eta_n}(\omega) &= \left[1-\left(\frac{\omega}{\omega_n}\right)^2 + 2i\cdot\left(\zeta_n - \zeta_{ae_n}\right)\cdot\frac{\omega}{\omega_n}\right]^{-1} \\ S_{\hat{Q}_n}(\omega) &= 2\lambda D\cdot\frac{S_{q_n}(\omega)\cdot\int\limits_{L_{\exp}}\phi_n^2(x)dx}{\left(\omega_n^2\tilde{M}_n\right)^2} \end{aligned}\right\} \tag{6.87}$$

and where aerodynamic damping properties may be extracted from Eq. 6.68, rendering

$$\left.\begin{aligned} \zeta_{ae_z} &= \frac{\tilde{C}_{ae_{zz}}}{2\omega_z\tilde{M}_z} = \frac{\rho B^2 H_1^*}{4\tilde{m}_z}\cdot\frac{\int\limits_{L_{\exp}}\phi_z^2dx}{\int\limits_L\phi_z^2dx} = \frac{\rho B^2}{4\tilde{m}_z}\cdot K_{a_z}\left[1-\left(\frac{\sigma_z}{a_z D}\right)^2\right]\cdot\frac{\int\limits_{L_{\exp}}\phi_z^2dx}{\int\limits_L\phi_z^2dx} \\ \zeta_{ae_\theta} &= \frac{\tilde{C}_{ae_{\theta\theta}}}{2\omega_\theta\tilde{M}_\theta} = \frac{\rho B^4 A_2^*}{4\tilde{m}_\theta}\cdot\frac{\int\limits_{L_{\exp}}\phi_\theta^2dx}{\int\limits_L\phi_\theta^2dx} = \frac{\rho B^4}{4\tilde{m}_\theta}\cdot K_{a_\theta}\left[1-\left(\frac{\sigma_\theta}{a_\theta}\right)^2\right]\cdot\frac{\int\limits_{L_{\exp}}\phi_\theta^2dx}{\int\limits_L\phi_\theta^2dx} \end{aligned}\right\} \tag{6.88}$$

Usually, vortex shedding induced dynamic response is largely resonant and narrow-banded. It will then suffice to only consider the resonant part of the frequency domain integration in Eq. 6.82, and discard the background part. Thus,

$$\sigma_{r_n}^2 = \int_0^\infty S_{r_n} d\omega \approx \phi_n^2(x_r) \cdot \int_0^\infty \left|\hat{H}_{\eta_n}(\omega)\right|^2 d\omega \cdot S_{\hat{Q}_n}(\omega_n)$$

$$\Rightarrow \sigma_{r_n}^2 = \phi_n^2(x_r) \cdot \frac{\pi \omega_n S_{\hat{Q}_n}(\omega_n)}{4\left(\zeta_n - \zeta_{ae_n}\right)} \qquad n = \begin{cases} z \\ \theta \end{cases} \tag{6.89}$$

As mentioned above, it is also a reasonable assumption that the integral length–scale λD for q_z and q_θ are identical. Adopting the convenient notation (see Eq. 6.41)

$$\tilde{M}_n = \tilde{m}_n \int_L \phi_n^2 dx \qquad n = \begin{cases} z \\ \theta \end{cases} \tag{6.90}$$

and introducing $S_{\hat{Q}_n}$ and S_{q_n} from Eqs. 6.87 and 5.33, then the following is obtained

$$\frac{\sigma_{r_z}}{D} = \frac{\left|\phi_z(x_r)\right|}{2^{7/2}\pi^{7/4}} \cdot \frac{\rho B D}{\tilde{m}_z} \cdot \frac{\hat{\sigma}_{q_z}}{St^2} \cdot \left[\frac{\lambda}{b_z \cdot \left(\zeta_z - \zeta_{ae_z}\right)}\right]^{1/2} \cdot \frac{\left(D \int_{L_{\exp}} \phi_z^2 dx\right)^{1/2}}{\int_L \phi_z^2 dx} \cdot g_z\left(V_{R_z}, V\right) \tag{6.91}$$

$$\sigma_{r_\theta} = \frac{\left|\phi_\theta(x_r)\right|}{2^{7/2}\pi^{7/4}} \cdot \frac{\rho (B D)^2}{\tilde{m}_\theta} \cdot \frac{\hat{\sigma}_{q_\theta}}{St^2} \cdot \left[\frac{\lambda}{b_\theta \cdot \left(\zeta_\theta - \zeta_{ae_\theta}\right)}\right]^{1/2} \cdot \frac{\left(D \int_{L_{\exp}} \phi_\theta^2 dx\right)^{1/2}}{\int_L \phi_\theta^2 dx} \cdot g_\theta\left(V_{R_\theta}, V\right) \tag{6.92}$$

where

$$g_n\left(V_{R_n}, V\right) = \left(\frac{V}{V_{R_n}}\right)^{3/2} \cdot \exp\left[-\frac{1}{2}\left(\frac{1 - V_{R_n}/V}{b_n}\right)^2\right] \qquad n = \begin{cases} z \\ \theta \end{cases} \tag{6.93}$$

and where $V_{R_n} = D\omega_n / (2\pi \cdot St)$.

Example 6.5

For a simple beam type of bridge let us set out to calculate the vortex shedding induced dynamic response at $x_r = L/2$ associated with the mode shape $\boldsymbol{\varphi} = \begin{bmatrix} 0 & \phi_z & 0 \end{bmatrix}^T$ with corresponding eigen-frequency $\omega_z = 0.8\ rad/s$. Let us again for simplicity assume that $\phi_z = \sin \pi x/L$. Thus, in this case it is only the across wind vertical direction that is of any interest. Typical variation of some basic data is illustrated in Fig. 6.15.

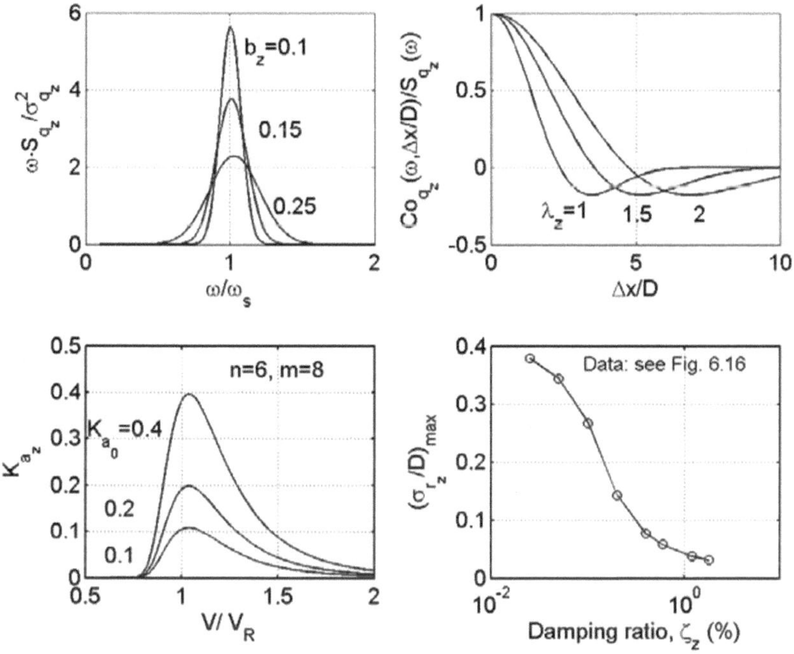

Fig. 6.15 *Top left and right: Non–dimensional cross sectional load spectrum and co–spectrum, lower left: aerodynamic damping coefficient, lower right: maximum vortex shedding induced dynamic response vs.* ζ_z

The top left hand side diagram shows the non–dimensional cross sectional load spectrum associated with vortex shedding in the across wind direction (see Eq. 5.33)

$$\frac{\omega S_{q_z}(\omega)}{\sigma_{q_z}^2} = \frac{1}{\sqrt{\pi} b_z} \cdot \frac{\omega}{\omega_s} \cdot \exp\left[-\left(\frac{1 - \omega/\omega_s}{b_z} \right)^2 \right]$$

where $\sigma_{q_z} = \frac{1}{2}\rho V^2 B \hat{\sigma}_{q_z}$. The load spectrum is shown for various relevant values of b_z, which is the parameter that controls the narrow-bandedness of the process. The reduced co-spectrum (see Eq. 5.34)

$$\hat{C}o_{q_z}(\Delta x) = \cos\left(\frac{2}{3}\frac{\Delta x}{\lambda_z D}\right)\cdot \exp\left[-\left(\frac{\Delta x}{3\lambda_z D}\right)^2\right]$$

at various values of λ_z is shown in the top right hand side diagram. It is this parameter that control the spanwise coherence (and thus, the length scale) of the vortices. The characteristic "lock-in" effect associated with vortex shedding induced dynamic response is controlled by the aerodynamic damping parameter K_{a_z}. Establishing data of the mean wind velocity variation of K_{a_z} will in general require wind tunnel experiments. As indicated in example 6.4 above, such data may often be fitted to an expression of the following type:

$$K_{a_z} = 2.6 \cdot K_{az0} \cdot \left(\frac{V}{V_{R_z}}\right)^{-n} \cdot \exp\left[-\left(\frac{V}{V_{R_z}}\right)^{-m}\right]$$

where $V_{R_z} = \omega_z D/(2\pi St)$ is the resonance velocity (see Eq. 5.32) and K_{a0} is the value at the apex of the K_{a_z} variation. See the lower left hand side diagram in Fig. 6.12, where $n = 6$ and $m = 8$.

Let us again consider a simply supported horizontal beam type of bridge with span $L = 500m$ that is elevated at a position $z_f = 50m$. Let us investigate the response variation with the mean wind velocity at various levels of structural eigen-damping. It is assumed that the entire span is flow exposed, i.e. $L_{\exp} = L$, and the expression for K_{a_z} given above is adopted. Let us allot the following values to the remaining constants that are necessary for a numerical calculation of $\sigma_{r_z}(x_r = L/2)$:

ρ (kg/m^3)	B (m)	D (m)	m_z (kg/m)	ω_z (rad/s)	St	$\hat{\sigma}_{q_z}$	b_z	a_z	λ_z	K_{a0}
1.25	20	4	10^4	0.8	0.1	0.9	0.15	0.4	1.2	0.2

Since m_z is constant along the span, then the modally equivalent and evenly distributed mass $\tilde{m}_z = m_z$. The dynamic response is given in Eq. 6.91, i.e.:

$$\frac{\sigma_{r_z}(x_r = L/2)}{D} = \frac{1}{2^{7/2}\pi^{7/4}} \cdot \frac{\rho BD}{\tilde{m}_z} \cdot \frac{\hat{\sigma}_{q_z}}{St^2} \cdot \left[\frac{2D\lambda}{b_z L(\zeta_z - \zeta_{ae_z})}\right]^{1/2} \cdot g_z(V_{R_z}, V)$$

where

$$g_z(V_{R_z}, V) = \left(\frac{V}{V_{R_z}}\right)^{3/2} \cdot \exp\left[-\frac{1}{2}\left(\frac{1 - V_{R_z}/V}{b_z}\right)^2\right]$$

The resonance mean wind velocity V_{R_z} is given by: $V_{R_z} = \dfrac{\omega_z D}{2\pi St} \approx 5.1\ m/s$.

Under these circumstances the equation above may be rewritten into the following fourth order polynomial

$$\hat{\sigma}_{r_z}^4 - \left(1-\hat{\zeta}\right)\hat{\sigma}_{r_z}^2 - \hat{\beta}^2 = 0$$

where

$$\hat{\zeta} = \frac{4\tilde{m}_z}{\rho B^2} \cdot \frac{\zeta_z}{K_{a_z}} \cdot \frac{\int_L \phi_z^2 dx}{\int_{L_{\exp}} \phi_z^2 dx} \quad \text{and} \quad \hat{\beta} = \frac{\left|\phi_z(x_r)\right|}{2^{5/2}\pi^{7/4}} \cdot \left(\frac{\rho D^3}{\tilde{m}_z \int_L \phi_z^2 dx} \cdot \frac{\lambda}{b_z K_{a_z}} \right)^{1/2} \cdot \frac{\hat{\sigma}_{q_z}}{St^2} \cdot \frac{g_z}{a_z}$$

and where $\hat{\sigma}_{r_z} = \sigma_{r_z} / (a_z D)$. Thus, the reduced standard deviation of the vortex shedding induced dynamic response is given by

$$\hat{\sigma}_{r_z} = \left\{ \frac{1-\hat{\zeta}}{2} + \left[\left(\frac{1-\hat{\zeta}}{2} \right)^2 + \hat{\beta}^2 \right]^{1/2} \right\}^{1/2}$$

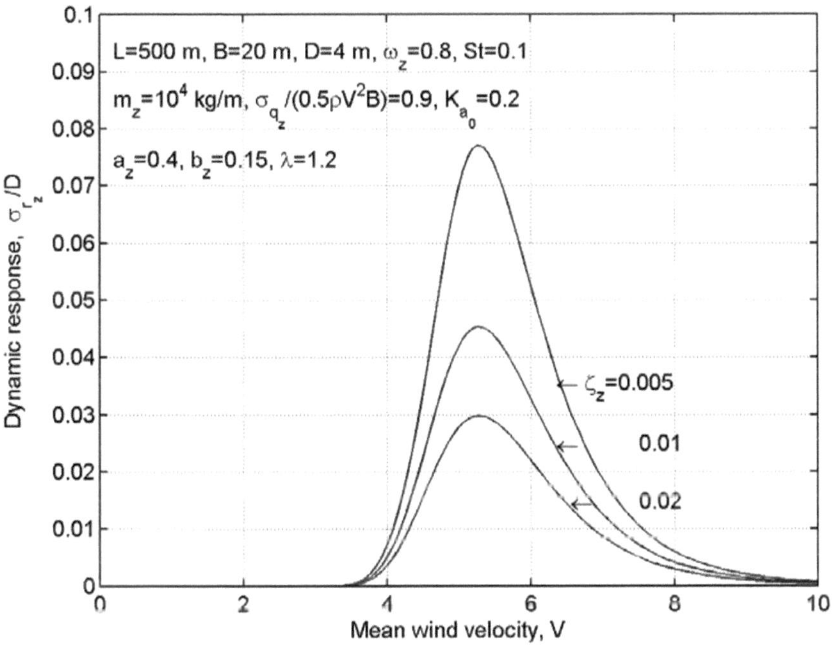

Fig. 6.16 *Vortex shedding induced dynamic response*

The variation of $\hat{\sigma}_{r_z}$ with the mean wind velocity at three levels of structural eigen–damping is shown in Fig. 6.16. As can be seen, the vortex shedding induced dynamic response is self-limiting and strongly damping dependant.

The maximum vortex shedding induced dynamic response will occur slightly above V_{R_z}, but for practical calculations the maximum value of σ_{r_z} may be obtained by setting $V = V_{R_z}$, in which case $g_z = 1$ and $K_{a_z} = K_{a0}$. As shown on the lower right hand side diagram in Fig. 6.15, the maximum value of σ_{r_z} is rapidly reduced with increased structural eigen-damping.

If $\boldsymbol{\varphi} = \begin{bmatrix} 0 & 0 & \phi_\theta \end{bmatrix}^T$ then

$$\hat{\sigma}_{r_\theta} = \left\{ \frac{1-\hat{\zeta}}{2} + \left[\left(\frac{1-\hat{\zeta}}{2} \right)^2 + \hat{\beta}^2 \right]^{1/2} \right\}^{1/2}$$

where $\hat{\sigma}_{r_\theta} = \sigma_{r_\theta} / a_\theta$ and

$$\hat{\zeta} = \frac{4\tilde{m}_\theta}{\rho B^4} \cdot \frac{\zeta_\theta}{K_{a_\theta}} \cdot \frac{\int_L \phi_\theta^2 dx}{\int_{L_{\exp}} \phi_\theta^2 dx} \quad \text{and} \quad \hat{\beta} = \frac{|\phi_\theta(x_r)|}{2^{5/2}\pi^{7/4}} \cdot \left(\frac{\rho D^5}{\tilde{m}_\theta \int_L \phi_\theta^2 dx} \cdot \frac{\lambda}{b_\theta K_{a_\theta}} \right)^{1/2} \cdot \frac{\hat{\sigma}_{q_\theta}}{St^2} \cdot \frac{g_\theta}{a_\theta}$$

Chapter 7

DETERMINATION OF CROSS SECTIONAL FORCES

7.1 Introduction

While we in chapter 6 focused exclusively on the determination of response displacements, we shall in this chapter deal with the determination of the corresponding cross sectional forces, i.e. the cross sectional stress resultants defined in chapter 1.3 (see Fig. 1.3.b). From a design point of view it is the maximum values of these quantities that decide the actual level of safety against structural failure. For a line like type of bridge structure the problem at hand is equivalent to that which is illustrated in Fig. 6.1, only that the response quantities we shall now set out to calculate are the cross sectional force components F (e.g. a bending moment, a torsion moment or a shear force) rather than the displacements which were in focus in chapter 6. The assumption of a Gaussian, stationary and homogeneous flow over the design period T (e.g. 10 min) is still valid, as well as the assumptions of linearity between load and load effects and a linear elastic structural behaviour. Thus, any cross sectional force component F may be described by the sum of its mean value and a fluctuating part that is Gaussian

$$F_{tot}(x,t) = \bar{F}(x) + F(x,t) \tag{7.1}$$

The time domain chain of events is illustrated in Fig. 7.1.a. Similar to that which was argued for the determination of displacements, it is in the following taken for granted that the fluctuating part of the cross sectional response forces are quantified by their standard deviation (σ_F), as illustrated in Fig. 7.1.b. The maximum value of a force component at spanwise position x_r is then given by

$$F_{\max}(x_r) = \bar{F}(x_r) + k_p \cdot \sigma_F(x_r) \tag{7.2}$$

where k_p is the peak factor (that depends on the type of response process). The chain of events for cross sectional forces is equivalent to that which is shown for structural displacements in Fig. 6.2 because the assumption of linear elastic structural behaviour implies that the relationship between structural displacements and cross sectional forces is also linear.

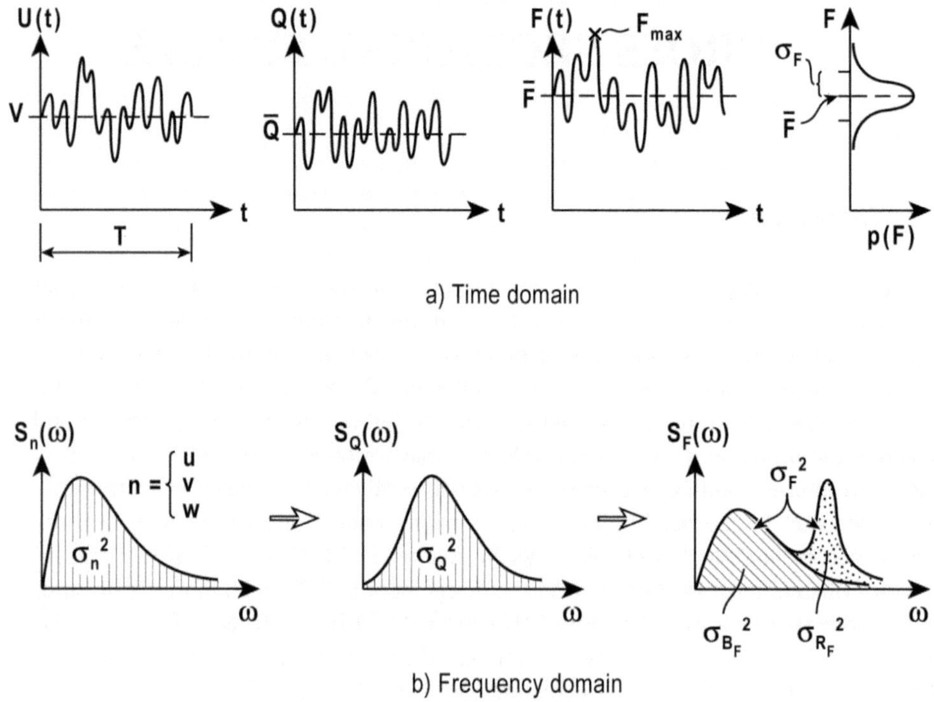

Fig. 7.1 *Time and frequency domain representations*

Thus, once the displacements have been determined, cross sectional forces may be obtained directly from the structural stiffness properties and the derivatives of the displacement functions according to usual structural mechanics procedures. While this is an appropriate strategy for the determination of the mean value $\overline{F}$, it is not an advisable strategy for the determination of σ_F . There are two reasons for this. First, dynamic response displacements are in general obtained from a modal solution in frequency domain that contains a chosen number of eigen modes which havebeen obtained from an eigen value solution that is based on the distributed stiffness and mass properties of the structure. The standard deviation of the total response displacements are then built up of the sum of contributions from each of these modes, either in a mode by mode approach (see Eqs. 4.15 and 4.49) or in a multi mode approach (see Eqs. 6.47 and 6.81). These eigen-modes are most often given as more or less ample vectors along the span of the structure, and their second and third order derivatives, which are required for the transfer from displacements to cross sectional forces, may in many cases be difficult to calculate with sufficient accuracy. It is therefore desirable (as indicated in Fig. 7.1.b), to split σ_F^2 into a background part $\sigma_{F_B}^2$ and a resonant part $\sigma_{F_R}^2$, such that

$$\sigma_F^2 \approx \sigma_{F_R}^2 + \sigma_{F_B}^2 \tag{7.3}$$

It is seen that this implies that the total response is sub-divided into a low frequency (background) part and a fluctuating (resonant) part that is centred on the eigen-frequency.

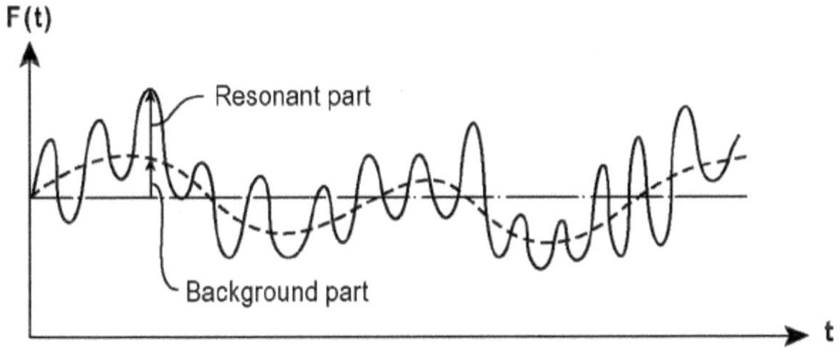

Fig. 7.2 *Background and resonant part in time domain*

This is further illustrated in Fig. 7.2. In time domain the background part is equivalent to a slowly varying process. Its contribution to inertia forces may therefore be disregarded, and thus, the load effects from this part may be regarded as quasi-static. Clearly, the quasi-static part of the load effects are more accurately determined from static shape functions or more directly from simple static equilibrium conditions, rather than a calculation based on the derivatives of eigen-modes.

The second motivation behind such a partition of σ_F^2 is the following. As previously described, when a structure is subject to a fluctuating wind field, the passing of the flow will generate fluctuating drag, lift and moment load components on the structure. These loads may cause the structure to oscillate. But in many cases the structure is stiff and its eigen-frequency is high (e.g. beyond 5 Hz), and then the displacements are small.

However, this does not imply that the fluctuations of cross sectional forces are insignificant. It only means that the resonant part of the force load effect is negligible (see chapter 2.10). For such a structure the total value of a cross sectional force component F at spanwise position x_r may be obtained from

$$F_{\max}(x_r) \approx \bar{F}(x_r) + k_p \cdot \sigma_{F_B}(x_r) \tag{7.4}$$

The entire solution, including σ_{F_B}, may then be obtained exclusively from static considerations, i.e. the determination of response spectra is obsolete. Since the solution contains the combined mean and fluctuating load effects, it represents the maximum value of the force load effect for a structure whose behaviour is defined as static.

The more general solution, covering static as well as dynamic structural behaviour is given in Eq. 7.2. Having split the fluctuating part of the response into a background and a resonant part, the maximum value of F at x_r may then be expressed by

$$F_{\max}(x_r) = \bar{F}(x_r) + k_p \cdot \sqrt{\sigma_{F_B}^2(x_r) + \sigma_{F_R}^2(x_r)} \tag{7.5}$$

where $\bar{F}$ and σ_{F_B} are obtained from static equilibrium conditions and σ_{F_R} is obtained from the resonant part of a modal frequency domain approach. For the determination of $\bar{F}$ the finite element type of approach that is shown below (chapter 7.2) is appropriate, unless the structural system is so simple that a direct analytical establishment of the equilibrium conditions is sufficient, in which case the solution is considered trivial. Similarly, for the determination of the background quasi-static part σ_{F_B} there are two alternatives. If the structural system is fairly complex a finite element approach is appropriate, but if the system is fairly simple a direct approach based on influence functions will suffice. Both methods are shown below (chapter 7.3).

For the determination of the resonant part σ_{F_R} there is the possibility of establishing an equivalent load based on the inertia forces, i.e. the product of response acceleration and the oscillating mass variation, but this option is only useful if the structural system is very simple because the equivalent load pattern must reproduce the actual structural displacements that are relevant for the mode shapes that have been excited. In chapter 7.4 a more general procedure is given, based on the linear relationship between cross sectional stress resultants and the corresponding spanwise derivatives of the resonant displacement response.

In a finite element formulation it is in the following assumed that the structural system has been modelled by nodes with six degrees of freedom as shown in Fig. 7.3 and by the use of beam or beam-column type of elements as shown in Fig. 7.4. At any level it is taken for granted that the load and load effect vectors can be split into a mean part and a fluctuating part, i.e. at a global system level

$$\mathbf{R}_{tot}(t)=\begin{bmatrix}R_1\\R_2\\R_3\\R_4\\R_5\\R_6\end{bmatrix}=\bar{\mathbf{R}}+\mathbf{R}(t)\quad\text{and}\quad\mathbf{r}_{tot}(t)=\begin{bmatrix}r_1\\r_2\\r_3\\r_4\\r_5\\r_6\end{bmatrix}=\bar{\mathbf{r}}+\mathbf{r}(t)\tag{7.6}$$

and at the local level for an arbitrary element m

$$\mathbf{F}_{tot_m}(t)=\begin{bmatrix}F_1\\F_2\\F_3\\F_4\\F_5\\F_6\end{bmatrix}_m=\bar{\mathbf{F}}_m+\mathbf{F}_m(t)\quad\text{and}\quad\mathbf{d}_{tot_m}(t)=\begin{bmatrix}d_1\\d_2\\d_3\\d_4\\d_5\\d_6\end{bmatrix}_m=\bar{\mathbf{d}}_m+\mathbf{d}_m(t)\tag{7.7}$$

The relationship between local forces and displacements is defined by the local stiffness matrix $\mathbf{k}_m$, i.e.

$$\mathbf{F}_{tot_m}=\mathbf{k}_m\cdot\mathbf{d}_{tot_m}\tag{7.8}$$

and the relationship between local and global degrees of freedom is defined by the matrix $\mathbf{A}_m$, i.e.

$$\mathbf{d}_{tot_m}=\mathbf{A}_m\cdot\mathbf{r}_{tot}\tag{7.9}$$

According to standard element method procedures the global stiffness matrix is then obtained by summation of contributions from all elements

$$\mathbf{K}=\sum_m\mathbf{A}_m^T\cdot\mathbf{k}_m\cdot\mathbf{A}_m\tag{7.10}$$

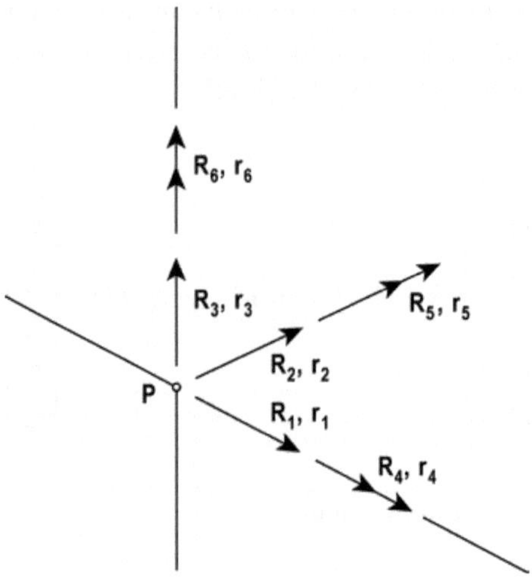

Fig. 7.3 *Definition of global load and displacement components*

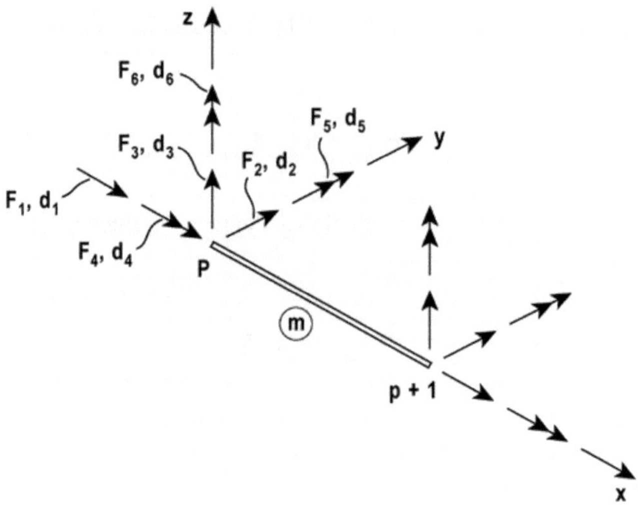

Fig. 7.4 *Definition of element forces and displacement components*

7.2 The mean value

For the calculation of the mean value of cross sectional forces all quantities are time invariants and thus, Eq. 6.3 still holds, implying that the global displacements are given by

$$\bar{\mathbf{r}} = \mathbf{K}^{-1} \cdot \bar{\mathbf{R}} \tag{7.11}$$

Similarly, the mean values of local forces and displacements (see Eqs. 7.8 and 7.9) are defined by

$$\left.\begin{aligned} \bar{\mathbf{F}}_m &= \mathbf{k}_m \cdot \bar{\mathbf{d}}_m \\ \bar{\mathbf{d}}_m &= \mathbf{A}_m \cdot \bar{\mathbf{r}} \end{aligned}\right\} \tag{7.12}$$

and thus, the mean value of cross sectional forces is given by

$$\mathbf{F}_m = \mathbf{k}_m \cdot \left(\mathbf{A}_m \cdot \bar{\mathbf{r}}\right) = \mathbf{k}_m \cdot \left[\mathbf{A}_m \cdot \left(\mathbf{K}^{-1} \cdot \bar{\mathbf{R}}\right)\right] \tag{7.13}$$

Eqs. 7.8 – 7.13 are identical to that which one will usually encounter in an ordinary finite element formulation. The establishment of $\mathbf{k}_m$ and $\mathbf{A}_m$ as well as the ensuing strategy for the calculation of global displacements and element force vectors may be found in many text books, see e.g. Hughes [25] or Cook et.al. [29]. Nonetheless, the brief summary presented above has been included for the sake of completeness. The only part that is special is the development of $\bar{\mathbf{R}}$, which has previously been shown in chapter 6.2.

7.3 The background quasi–static part

For the determination of the quasi-static part of the cross sectional response forces the mean part of the load as well as any motion induced contributions are obsolete. According to Eq. 5.8 the fluctuating part of the load on a line-like structure is given by

$$\mathbf{q}(x,t) = \begin{bmatrix} q_y(x,t) \\ q_z(x,t) \\ q_\theta(x,t) \end{bmatrix} = \mathbf{B}_q \cdot \mathbf{v} = \frac{\rho VB}{2} \cdot \hat{\mathbf{B}}_q \cdot \mathbf{v} \tag{7.14}$$

where $\mathbf{v}$ and $\mathbf{B}_q$ are defined in Eqs. 5.9 and 5.12, and recalling that this was developed for a horizontal type of structure. As mentioned above the quasi-static part may be determined by a formal finite element formulation, or alternatively, by the use of static influence functions based on a direct establishment of the equilibrium conditions.

Let us first pursue the more simple solution of a direct approach based on static influence functions. It is then taken for granted that the structure at hand is suitably uncomplicated, rendering straight forward equilibrium equations. Let us for the sake of simplicity consider the quasi-static load effect of the along wind component $q_y(x,t)$ on the horizontal simply supported beam shown in Fig. 7.5.

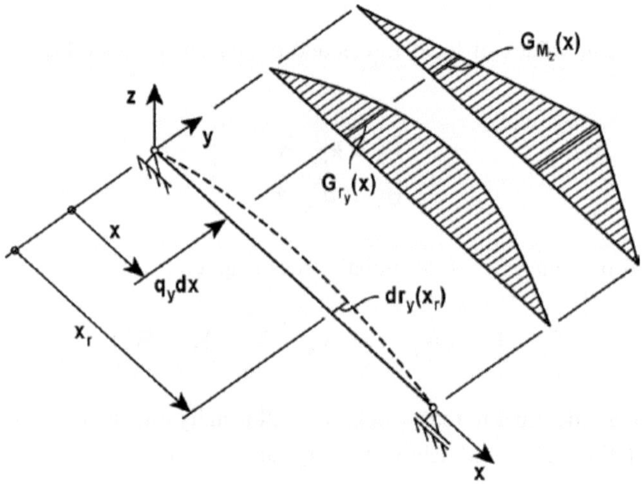

Fig. 7.5 *Along–wind load and two relevant response components*

As can be seen, the load effect is a horizontal displacement $r_y(x,t)$ associated with bending about the z-axis and shear in the direction of y. Let us focus on the background quasi-static part of the cross sectional bending moment $M_{zB}(x,t)$ at a chosen position x_r (e.g. at mid-span). It is seen from Eq. 7.14 (see also Eq. 5.12) that

$$q_y(x,t) = \frac{\rho VB}{2} \cdot \left[2\frac{D}{B}\bar{C}_D \cdot u(x,t) + \left(\frac{D}{B}C_D' - \bar{C}_L \right) \cdot w(x,t) \right] \tag{7.15}$$

As illustrated in Fig. 7.5, the bending moment M_{zB} at a chosen position x_r is given by

$$M_{zB}(x_r,t) = \int_{L_{\exp}} G_{M_z}(x) \cdot q_y(x,t) dx \tag{7.16}$$

where $L_{\exp}$ is the flow exposed part of the structure and G_{M_z} is the static influence function for M_z at x_r (defined as the function containing the values of M_z at x_r when the system is subject to a unit load $q_y = 1$ at arbitrary position x).

The variance of M_{zB} is then defined by

$$\sigma^2_{M_{zB}}(x_r) = E\left[\left\{M_{zB}(x_r,t)\right\}^2\right] = E\left[\left\{\int_{L_{\exp}} G_{M_z}(x)\cdot q_y(x,t)dx\right\}^2\right]$$
$$= \iint_{L_{\exp}} G_{M_z}(x_1)\cdot G_{M_z}(x_2)\cdot E\left[q_y(x_1,t)\cdot q_y(x_2,t)\right]dx_1 dx_2 \tag{7.17}$$

rendering a spatial and time domain averaging of the fluctuating cross sectional load. Introducing Eq. 7.15 then this space and time domain averaging is given by

$$E\left[q_y(x_1,t)\cdot q_y(x_2,t)\right] =$$
$$\left(\frac{\rho VB}{2}\right)^2 E\left[\left\{2\frac{D}{B}\bar{C}_D u_1 + \left(\frac{D}{B}C'_D - \bar{C}_L\right)w_1\right\}\cdot\left\{2\frac{D}{B}\bar{C}_D u_2 + \left(\frac{D}{B}C'_D - \bar{C}_L\right)w_2\right\}\right] \tag{7.18}$$

where $u_1 = u(x_1,t)$, $u_2 = u(x_2,t)$ and $w_1 = w(x_1,t)$, $w_2 = w(x_2,t)$. It is a usual assumption in wind engineering that cross-covariance between different velocity components is negligible, i.e. that

$$E\left[u(x_1,t)\cdot w(x_2,t)\right] = E\left[u(x_2,t)\cdot w(x_1,t)\right] \approx 0 \tag{7.19}$$

in which case

$$E\left[q_y(x_1,t)\cdot q_y(x_2,t)\right] =$$
$$\left(\frac{\rho VB}{2}\right)^2 \left\{\left(2\frac{D}{B}\bar{C}_D\right)^2 E\left[u(x_1,t)\cdot u(x_2,t)\right] + \left(\frac{D}{B}C'_D - \bar{C}_L\right)^2 E\left[w(x_1,t)\cdot w(x_2,t)\right]\right\} \tag{7.20}$$

Introducing (see chapters 2.2 and 3.3)

$$E\left[u(x_1,t)\cdot u(x_2,t)\right] = \sigma_u^2 \cdot \rho_{uu}(\Delta x)$$
$$E\left[w(x_1,t)\cdot w(x_2,t)\right] = \sigma_w^2 \cdot \rho_{ww}(\Delta x) \tag{7.21}$$

where ρ_{uu} and ρ_{ww} are the covariance coefficients of the *u*- and *w*-components, and where $\Delta x = |x_1 - x_2|$ is spanwise separation, then

$$E\left[q_y(x_1,t)\cdot q_y(x_2,t)\right]=$$
$$\left(\frac{\rho V^2 B}{2}\right)^2\cdot\left\{\left(2\frac{D}{B}\bar{C}_D I_u\right)^2\cdot\rho_{uu}(\Delta x)+\left[\left(\frac{D}{B}C_D'-\bar{C}_L\right)I_w\right]^2\cdot\rho_{ww}(\Delta x)\right\} \tag{7.22}$$

where $I_u=\sigma_u/V$ and $I_w=\sigma_w/V$ are the *u*– and *w*–component turbulence intensities. Thus, the variance of the background part is given by (see Eq. 7.17)

$$\sigma^2_{M_{zB}}(x_r)=\left(\frac{\rho V^2 B}{2}\right)^2\cdot\iint_{L_{\exp}} G_{M_z}(x_1)\cdot G_{M_z}(x_2)\cdot$$
$$\left\{\left(2\frac{D}{B}\bar{C}_D I_u\right)^2\cdot\rho_{uu}(\Delta x)+\left[\left(\frac{D}{B}C_D'-\bar{C}_L\right)I_w\right]^2\cdot\rho_{ww}(\Delta x)\right\}dx_1dx_2 \tag{7.23}$$

The volume integral in Eq. 7.23 represents a spatial averaging of the fluctuating load effect with respect to the bending component M_z at a certain spanwise position x_r. This is identical to that which has previously been dealt with in Chapter 2.10 (see Example 2.4).

While Eq. 7.23 provides the calculation procedure for the background part of the cross sectional force component M_z at x_r alone, it is convenient to establish more general procedures comprising the background response of several components, e.g. the bending moments M_y and M_z as well as the torsion moment M_x. These force components are in general given by

$$\mathbf{M}_B(x_r,t)=\begin{bmatrix}M_x\\M_y\\M_z\end{bmatrix}_B=\int_{L_{\exp}}\begin{bmatrix}G_{M_x}(x)\cdot q_\theta(x,t)\\G_{M_y}(x)\cdot q_y(x,t)\\G_{M_z}(x)\cdot q_z(x,t)\end{bmatrix}dx \tag{7.24}$$

where G_{M_n}, $n=x,y,z$, are the static influence functions for cross sectional force components M_x, M_y and M_z at x_r. By adopting the definition

$$\mathbf{G}_M(x)=\begin{bmatrix}0&0&G_{M_x}\\0&G_{M_y}&0\\G_{M_z}&0&0\end{bmatrix} \tag{7.25}$$

it follows from Eqs. 7.14 and 7.24 that

$$\mathbf{M}_B(x_r,t) = \int_{L_{exp}} \mathbf{G}_M(x)\cdot\mathbf{q}(x,t)dx = \frac{\rho VB}{2}\cdot\int_{L_{exp}} \mathbf{G}_M(x)\cdot\left\{\hat{\mathbf{B}}_q\cdot\mathbf{v}(x,t)\right\}dx \quad (7.26)$$

where $\hat{\mathbf{B}}_q$ is the load coefficient matrix defined in Eq. 5.12, i.e.

$$\hat{\mathbf{B}}_q(x) = \begin{bmatrix} 2(D/B)\bar{C}_D & \left((D/B)C'_D - \bar{C}_L\right) \\ 2\bar{C}_L & \left(C'_L + (D/B)\bar{C}_D\right) \\ 2B\bar{C}_M & BC'_M \end{bmatrix} \quad (7.27)$$

and where $\mathbf{v}(x,t) = \left[u(x,t) \quad w(x,t)\right]^T$ in the case of a horizontal bridge type of structure (see Eq. 5.9). The background covariance matrix

$$\mathbf{Cov}_{MM_B}(x_r) = \begin{bmatrix} \sigma^2_{M_xM_x} & Cov_{M_xM_y} & Cov_{M_xM_z} \\ Cov_{M_yM_x} & \sigma^2_{M_yM_y} & Cov_{M_yM_z} \\ Cov_{M_zM_x} & Cov_{M_zM_y} & \sigma^2_{M_zM_z} \end{bmatrix}_B \quad (7.28)$$

is then obtained from

$$\begin{aligned} \mathbf{Cov}_{MM_B}(x_r) &= E\left[\mathbf{M}_B(x_r,t)\cdot\mathbf{M}_B^T(x_r,t)\right] \\ &= \left(\frac{\rho VB}{2}\right)^2 E\left[\left\{\int_{L_{exp}} \mathbf{G}_M\cdot\left(\hat{\mathbf{B}}_q\cdot\mathbf{v}\right)dx\right\}\cdot\left\{\int_{L_{exp}} \mathbf{G}_M\cdot\left(\hat{\mathbf{B}}_q\cdot\mathbf{v}\right)dx\right\}^T\right] \\ &= \left(\frac{\rho VB}{2}\right)^2 \iint_{L_{exp}} \mathbf{G}_M(x_1)\cdot\left\{\hat{\mathbf{B}}_q\cdot E\left[\mathbf{v}(x_1,t)\cdot\mathbf{v}^T(x_2,t)\right]\hat{\mathbf{B}}_q^T\right\}\mathbf{G}_M^T(x_2)dx_1dx_2 \end{aligned} \quad (7.29)$$

Introducing Eq. 7.21 and adopting the assumptions in Eq. 7.19, then

$$E\left[\mathbf{v}(x_1,t)\cdot\mathbf{v}^T(x_2,t)\right] \approx \begin{bmatrix} \sigma_u^2\cdot\rho_{uu}(\Delta x) & 0 \\ 0 & \sigma_w^2\cdot\rho_{ww}(\Delta x) \end{bmatrix} = V^2\left\{\mathbf{I}_v^2 \;\; \boldsymbol{\rho}(\Delta x)\right\} \quad (7.30)$$

where

$$\mathbf{I}_v = \begin{bmatrix} I_u & 0 \\ 0 & I_w \end{bmatrix} \quad \text{and} \quad \boldsymbol{\rho}_v(\Delta x) = \begin{bmatrix} \rho_{uu} & 0 \\ 0 & \rho_{ww} \end{bmatrix} \tag{7.31}$$

and thus,

$$\mathbf{Cov}_{MM_B}(x_r) = \left(\frac{\rho V^2 B}{2}\right)^2 \iint_{L_{\exp}} \mathbf{G}_M(x_1) \cdot \left\{ \hat{\mathbf{B}}_q \cdot \left[\mathbf{I}_v^2 \cdot \boldsymbol{\rho}_v(\Delta x) \right] \cdot \hat{\mathbf{B}}_q^T \right\} \cdot \mathbf{G}_M^T(x_2) dx_1 dx_2 \tag{7.32}$$

The covariance matrix in Eq. 7.32 will be symmetric because x_1 and x_2 are interchangeable and ρ_{uu} and ρ_{ww} are only functions of the separation $\Delta x = |x_1 - x_2|$. In a fully expanded format the variance of the background response components are given by

$$\begin{bmatrix} \sigma_{M_x M_x}^2 \\ \sigma_{M_y M_y}^2 \\ \sigma_{M_z M_z}^2 \end{bmatrix}_B = \left(\frac{\rho V^2 B}{2}\right)^2 \iint_{L_{\exp}} \begin{bmatrix} g_{M_x M_x}(x_1, x_2) \\ g_{M_y M_y}(x_1, x_2) \\ g_{M_z M_z}(x_1, x_2) \end{bmatrix} dx_1 dx_2 \tag{7.33}$$

where

$$g_{M_x M_x} = B^2 G_{M_x}(x_1) G_{M_x}(x_2) \left[\left(2\bar{C}_M I_u\right)^2 \rho_{uu}(\Delta x) + \left(C_M' I_w\right)^2 \rho_{ww}(\Delta x) \right] \tag{7.34}$$

$$g_{M_y M_y} = G_{M_y}(x_1) G_{M_y}(x_2) \left\{ \left(2\bar{C}_L I_u\right)^2 \rho_{uu}(\Delta x) + \left[\left(C_L' + \frac{D}{B}\bar{C}_D \right) I_w \right]^2 \rho_{ww}(\Delta x) \right\} \tag{7.35}$$

$$g_{M_z M_z} = G_{M_z}(x_1) G_{M_z}(x_2) \cdot \left\{ \left(2\frac{D}{B}\bar{C}_D I_u \right)^2 \rho_{uu}(\Delta x) + \left[\left(\frac{D}{B} C_D' - \bar{C}_L \right) I_w \right]^2 \rho_{ww}(\Delta x) \right\} \tag{7.36}$$

Similarly, the corresponding covariance between background components may be expanded into

$$\begin{bmatrix} Cov_{M_x M_y} \\ Cov_{M_x M_z} \\ Cov_{M_y M_z} \end{bmatrix}_B = \left(\frac{\rho V^2 B}{2}\right)^2 \iint_{L_{\exp}} \begin{bmatrix} g_{M_x M_y}(x_1, x_2) \\ g_{M_x M_z}(x_1, x_2) \\ g_{M_y M_z}(x_1, x_2) \end{bmatrix} dx_1 dx_2 \tag{7.37}$$

where

$$g_{M_x M_y} = BG_{M_x}(x_1)G_{M_y}(x_2)\Big[4\bar{C}_L\bar{C}_M I_u^2 \rho_{uu}(\Delta x) + \left(C_L' + \frac{D}{B}\bar{C}_D\right)C_M' I_w^2 \rho_{ww}(\Delta x)\Big] \tag{7.38}$$

$$g_{M_x M_z} = BG_{M_x}(x_1)G_{M_z}(x_2)\Big[4\frac{D}{B}\bar{C}_D\bar{C}_M I_u^2 \rho_{uu}(\Delta x) + \left(\frac{D}{B}C_D' - \bar{C}_L\right)C_M' I_w^2 \rho_{ww}(\Delta x)\Big] \tag{7.39}$$

$$g_{M_y M_z} = G_{M_y}(x_1)G_{M_z}(x_2)\Big[4\frac{D}{B}\bar{C}_D\bar{C}_L I_u^2 \rho_{uu}(\Delta x) + \left(\frac{D}{B}C_D' - \bar{C}_L\right)\left(C_L' + \frac{D}{B}\bar{C}_D\right) I_w^2 \rho_{ww}(\Delta x)\Big] \tag{7.40}$$

Example 7.1

Let us set out to calculate the variances and covariance of the torsion and bending moments M_x, M_y and M_z at midspan of the simply supported beam type of bridge illustrated in Fig. 7.6. Let us for simplicity assume that it has a typical bridge type of cross section where C_D', $\bar{C}_L$ and $\bar{C}_M$ are negligible and $\bar{C}_D \cdot D/B \ll C_L'$. Then

$$\begin{bmatrix} \sigma_{M_x M_x}^2 \\ \sigma_{M_y M_y}^2 \\ \sigma_{M_z M_z}^2 \end{bmatrix} = \left(\frac{\rho V^2 B}{2}\right)^2 \cdot \int_0^L\int_0^L \begin{bmatrix} \left(BC_M' I_w\right)^2 \cdot G_{M_x}(x_1)\cdot G_{M_x}(x_2)\cdot \rho_{ww}(\Delta x) \\ \left(C_L' I_w\right)^2 \cdot G_{M_y}(x_1)\cdot G_{M_y}(x_2)\cdot \rho_{ww}(\Delta x) \\ \left(2\frac{D}{B}\bar{C}_D I_u\right)^2 \cdot G_{M_z}(x_1)\cdot G_{M_z}(x_2)\cdot \rho_{uu}(\Delta x) \end{bmatrix} dx_1 dx_2$$

$$\begin{bmatrix} Cov_{M_x M_y} \\ Cov_{M_x M_z} \\ Cov_{M_y M_z} \end{bmatrix} = \left(\frac{\rho V^2 B}{2}\right)^2 \cdot \int_0^L\int_0^L \begin{bmatrix} BC_L' C_M' I_w^2 \cdot G_{M_x}(x_1)\cdot G_{M_y}(x_2)\cdot \rho_{ww}(\Delta x) \\ 0 \\ 0 \end{bmatrix} dx_1 dx_2$$

where it has been taken for granted that $L_{\exp} = L$.

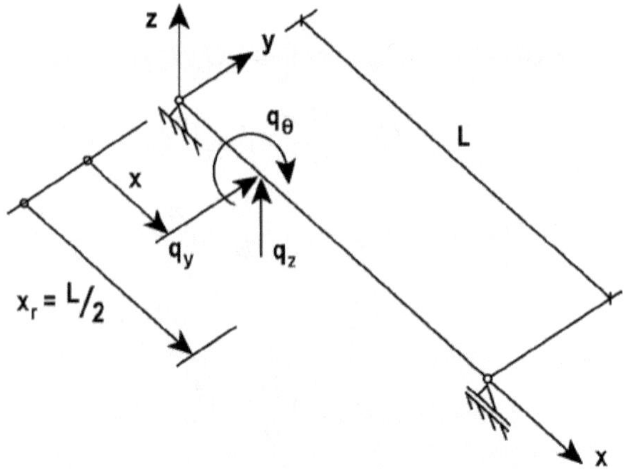

Fig. 7.6 *Simply supported beam type of bridge*

Introducing the non-dimensional spanwise coordinate $\hat{x} = x / L$ and the function

$$\psi(\hat{x}) = \begin{cases} \hat{x} & \text{when} \quad \hat{x} \le 1/2 \\ 1-\hat{x} & \text{when} \quad 1/2 < \hat{x} \le 1 \end{cases}$$

it is readily seen that the influence functions for M_x, M_y and M_z at midspan are given by

$$G_{M_x}(\hat{x}) = -\psi(\hat{x}) \qquad G_{M_y}(\hat{x}) = \frac{L}{2} \cdot \psi(\hat{x}) \qquad G_{M_x}(\hat{x}) = -\frac{L}{2} \cdot \psi(\hat{x})$$

Thus

$$\begin{bmatrix} \left(\dfrac{\sigma_{M_x M_x}}{\frac{1}{2}\rho V^2 B^2 L C_M' I_w} \right)^2 \\ \left(\dfrac{\sigma_{M_y M_y}}{\frac{1}{4}\rho V^2 B L^2 C_L' I_w} \right)^2 \\ \left(\dfrac{\sigma_{M_z M_z}}{\frac{1}{2}\rho V^2 D L^2 \bar{C}_D I_u} \right)^2 \end{bmatrix} = \int_0^1 \int_0^1 \psi(\hat{x}_1) \cdot \psi(\hat{x}_2) \cdot \begin{bmatrix} \rho_{ww}(\Delta\hat{x}) \\ \rho_{ww}(\Delta\hat{x}) \\ \rho_{uu}(\Delta\hat{x}) \end{bmatrix} d\hat{x}_1 d\hat{x}_2 = \begin{bmatrix} \hat{J}^2_{M_x M_x} \\ \hat{J}^2_{M_y M_y} \\ \hat{J}^2_{M_z M_z} \end{bmatrix}$$

and since $Cov_{M_x M_y} = -\sigma_{M_x M_x} \cdot \sigma_{M_y M_y}$ it will suffice to calculate the joint acceptance functions on the right hand side of the equation above. Let us divide the span into N segments and calculate the load effect at the midpoint of each of these segments. Then the volume integral is replaced by a double summation (and $d\hat{x} = 1/N$), i.e.

$$\begin{bmatrix} \hat{J}^2_{M_x M_x} \\ \hat{J}^2_{M_y M_y} \\ \hat{J}^2_{M_z M_z} \end{bmatrix} = \frac{1}{N^2} \sum_{n=1}^{N} \sum_{m=1}^{N} \psi(\hat{x}_n) \cdot \psi(\hat{x}_m) \cdot \begin{bmatrix} \rho_{ww}\left(\left|\hat{x}_n - \hat{x}_m\right|\right) \\ \rho_{ww}\left(\left|\hat{x}_n - \hat{x}_m\right|\right) \\ \rho_{uu}\left(\left|\hat{x}_n - \hat{x}_m\right|\right) \end{bmatrix}$$

and it is just a matter of choosing N sufficiently large. Let us assume that the position of the bridge is at $z_f = 50$ m and that the relevant length scales of the u and w components are given by (see Eq 3.36):

$$^{xf}L_u = 100\left(z_f/10\right)^{0.3} = 162 \text{ m}, \quad ^{yf}L_u = {}^{xf}L_u/3 = 54 \text{ m} \quad \text{and} \quad ^{yf}L_w = {}^{xf}L_u/16 = 10 \text{ m},$$

such that

$$\rho_{uu}(\Delta\hat{x}) = \exp(-c_u \cdot \Delta\hat{x}) \qquad \text{where} \qquad c_u = L/{}^{yf}L_u$$

$$\rho_{ww}(\Delta\hat{x}) = \exp(-c_w \cdot \Delta\hat{x}) \qquad \text{where} \qquad c_w = L/{}^{yf}L_w$$

Let us for simplicity set $L = {}^{xf}L_u$ and $N = 5$ (which in general will be far too small, as shown in Fig. 7.7). Thus, $c_u = 3$, $c_w = 16$.

The position vector and the influence function are given by

$$\hat{\mathbf{x}} = \begin{bmatrix} 0.1 & 0.3 & 0.5 & 0.7 & 0.9 \end{bmatrix}^T$$

$$\boldsymbol{\psi}(\hat{x}) = \begin{bmatrix} 0.1 & 0.3 & 0.5 & 0.3 & 0.1 \end{bmatrix}^T$$

The influence function multiplications $\psi(\hat{x}_n) \cdot \psi(\hat{x}_m)$ are then given by

	$\psi(\hat{x}_n)$: 0.1	0.3	0.5	0.3	0.1
$\psi(\hat{x}_m)$: 0.1	0.01	0.03	0.05	0.03	0.01
0.3	0.03	0.09	0.15	0.09	0.03
0.5	0.05	0.15	0.25	0.15	0.05
0.3	0.03	0.09	0.15	0.09	0.03
0.1	0.01	0.03	0.05	0.03	0.01

while the covariance coefficients associated with the u and w components are given by:

$\rho_{uu}(\Delta\hat{x})$:

	$\hat{x}_n$: 0.1	0.3	0.5	0.7	0.9
$\hat{x}_m$: 0.1	1	0.549	0.301	0.165	0.091
0.3	0.549	1	0.549	0.301	0.165
0.5	0.301	0.549	1	0.549	0.301
0.7	0.165	0.301	0.549	1	0.549
0.9	0.091	0.165	0.301	0.549	1

$\rho_{ww}(\Delta\hat{x})$:

$\hat{x}_m$ \ $\hat{x}_n$	0.1	0.3	0.5	0.7	0.9
0.1	1	0.0408	0.0017	0.0001	≈ 0
0.3	0.0408	1	0.0408	0.0017	0.0001
0.5	0.0017	0.0408	1	0.0408	0.0017
0.7	0.0001	0.0017	0.0408	1	0.0408
0.9	≈ 0	0.0001	0.0017	0.0408	1

The inner products $\psi(\hat{x}_n)\cdot\psi(\hat{x}_m)\cdot\rho_{uu}(\Delta\hat{x})$ and $\psi(\hat{x}_n)\cdot\psi(\hat{x}_m)\cdot\rho_{uu}(\Delta\hat{x})$ are then:

m \ n	1	2	3	4	5
	$10^2\cdot\psi(\hat{x}_n)\cdot\psi(\hat{x}_m)\cdot\rho_{uu}(\Delta\hat{x})$				
1	1	1.647	1.505	0.495	0.091
2	1.647	9	8.235	2.709	0.495
3	1.505	8.235	25	8.235	1.505
4	0.495	2.709	8.235	9	1.647
5	0.091	0.495	1.505	1.647	1

m \ n	1	2	3	4	5
	$10^2\cdot\psi(\hat{x}_n)\cdot\psi(\hat{x}_m)\cdot\rho_{ww}(\Delta\hat{x})$				
1	1	0.1225	0.0085	0.0003	≈ 0
2	0.1225	9	0.612	0.0153	0.0003
3	0.0085	0.612	25	0.612	0.0085
4	0.0003	0.0153	0.612	9	0.1224
5	≈ 0	0.0003	0.0085	0.1224	1

The normalised joint acceptance functions are given by

$$\hat{J}_{M_zM_z} = \left[\frac{1}{5^2}\cdot\sum_{n=1}^{5}\sum_{m=1}^{5}\psi(\hat{x}_n)\cdot\psi(\hat{x}_m)\cdot\rho_{uu}(\Delta\hat{x})\right]^{\frac{1}{2}} \approx 0.2$$

$$\hat{J}_{M_xM_x} = \hat{J}_{M_yM_y} = \left[\frac{1}{5^2}\cdot\sum_{n=1}^{5}\sum_{m=1}^{5}\psi(\hat{x}_n)\cdot\psi(\hat{x}_m)\cdot\rho_{ww}(\Delta\hat{x})\right]^{\frac{1}{2}} \approx 0.14$$

and thus

$$\sigma_{M_xM_x} \approx 0.14\cdot\left(\frac{1}{2}\rho V^2B^2LC'_MI_w\right)$$

$$\sigma_{M_yM_y} \approx 0.07\cdot\left(\frac{1}{2}\rho V^2BL^2C'_LI_w\right)$$

$$\sigma_{M_zM_z} \approx 0.2\cdot\left(\frac{1}{2}\rho V^2DL^2\bar{C}_DI_u\right)$$

This solution has been based on $L_{\exp} = L = {}^{y_f}L_u$ and $N = 5$. If $N = 40$ then the integration coefficient 0.14 is reduced to 0.11, the coefficient 0.07 is reduced to 0.055 while the coefficient 0.2 is reduced to 0.188. The problem of choosing a sufficiently large number of integration points is illustrated in Fig. 7.7, where the normalised joint acceptance function $\hat{J}_{nn}$ for an arbitrary force component whose influence function is linear and with a maximum of 0.5 at midspan is plotted versus N for three different values of the ratio between $L_{\exp}$ and the relevant length scale ${}^xL_j$, $j = u$ or w (and $|x| = |y_f|$). It is seen that the necessary number of integration points is in general considerable. The reason for this is that ρ_{jj} ($j = u$ or w) is a rapidly decaying function. Similarly, in Fig. 7.8 the joint acceptance function has been plotted versus the ratio between the length of the span ($L_{\exp} = L$) and the relevant integral length scale ${}^xL_j$, $j = u$ or w. The case $\lim\left(L / {}^xL_j\right) \to 0$ is identical to the situation with an evenly distributed load along the entire span. As can be seen, $\hat{J}_{nn}$ is a rapidly decreasing function with increasing values of $L/{}^xL_j$. At a large value of $L/{}^xL_j$ it is close to 0.05.

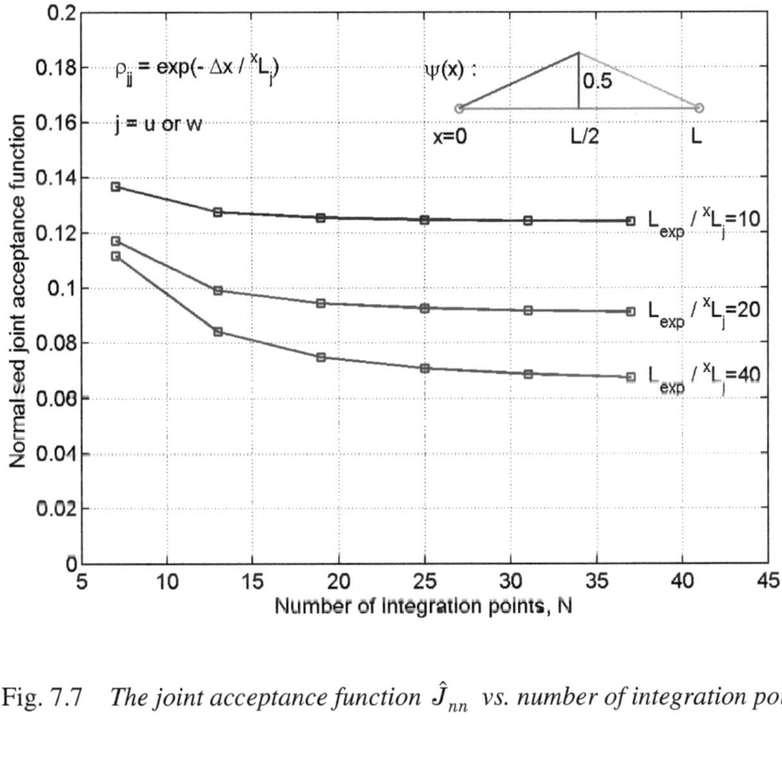

Fig. 7.7 *The joint acceptance function* $\hat{J}_{nn}$ *vs. number of integration points*

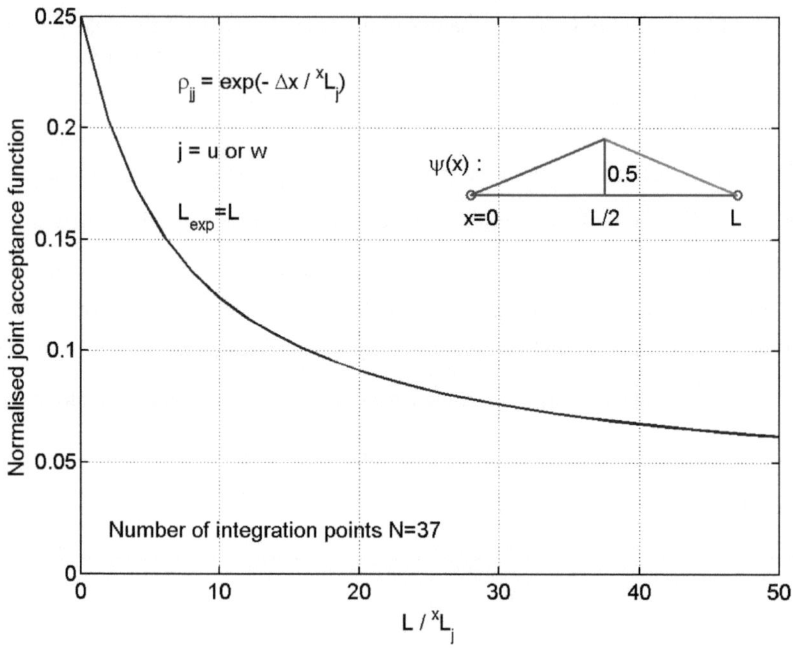

Fig. 7.8 *The joint acceptance function* $\hat{J}_{nn}$ *at various span length (* $L_{\exp} = L$ *)*

While the solution strategy based on influence functions shown above is suitable for many cases of fairly simple structural systems, a formulation within the finite element method is more suitable for a general approach. Recalling that Eq. 5.8 was developed for a horizontal type of structure, it is seen from Fig. 7.9 that in general $\mathbf{v} = [u \quad w]^T$ for a horizontal element and $\mathbf{v} = [u \quad v]^T$ for a vertical element. Defining the non-dimensional instantaneous fluctuating wind velocity vector at an arbitrary node p

$$\hat{\mathbf{v}}_p = \frac{1}{V}\begin{bmatrix} u(t) \\ v(t) \\ w(t) \end{bmatrix} \tag{7.41}$$

and the matrix

$$\boldsymbol{\psi}_m = \begin{cases} \begin{bmatrix} 1 & 0 & 0 \\ 0 & 0 & 1 \end{bmatrix} & \text{for a horisontal element} \\ \begin{bmatrix} 1 & 0 & 0 \\ 0 & 1 & 0 \end{bmatrix} & \text{for a vertical element} \end{cases} \tag{7.42}$$

which is associated with the direction of an adjoining element m, it is seen from Fig. 7.9 that the distributed load vector acting on this element is given by

$$\mathbf{q}_m(x,t) = \frac{\rho V^2 B}{2} \cdot \hat{\mathbf{B}}_q \cdot \left(\boldsymbol{\psi}_m \cdot \hat{\mathbf{v}}_p \right) \tag{7.43}$$

where $\hat{\mathbf{B}}_q$ is given in Eq. 7.27 and where all cross sectional quantities are those that are applicable to element m.

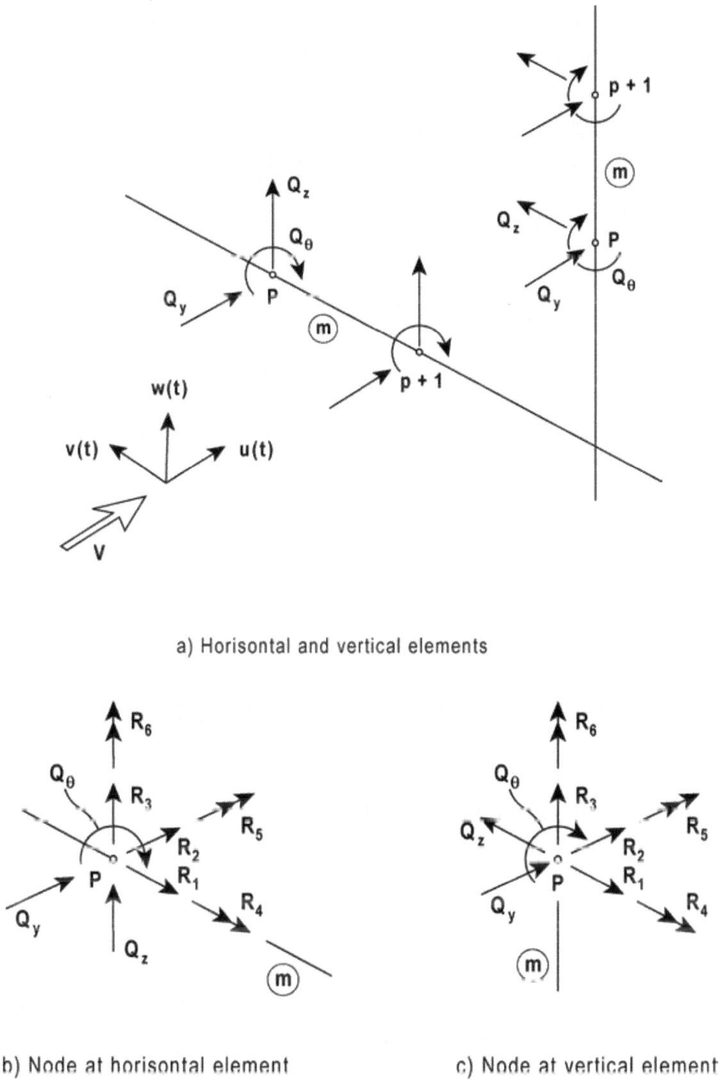

Fig. 7.9 *Wind induced load components*

The contribution from element m to concentrated loads in node p is then given by (see Fig. 7.9.a)

$$\mathbf{Q}_{p_m}(t) = \begin{bmatrix} Q_y \\ Q_z \\ Q_\theta \end{bmatrix}_{p_m} = \mathbf{q}_m(x,t) \cdot \frac{L_m}{2} = \left(\frac{\rho V^2}{2}\right)_p \cdot \left(\frac{BL}{2}\right)_m \cdot \hat{\mathbf{B}}_q \cdot \left(\boldsymbol{\psi}_m \cdot \hat{\mathbf{v}}_p\right) \tag{7.44}$$

where L_m is the element length, and where it has been assumed that the nodal discretisation is such that all wind velocity properties with sufficient accuracy may be allotted to node p, and that they are constants along the span of the element. (This is a simplification that is not mandatory, but otherwise, an element integration scheme has to be adopted.) Comparing the definition of nodal loads shown in Fig. 7.3 to the element load components shown in Fig. 7.9, it is seen that the contribution from element m to the load vector in node p is defined by (see Fig. 7.9.b and c)

$$\mathbf{R}_{p_m} = \begin{bmatrix} R_1 \\ R_2 \\ R_3 \\ R_4 \\ R_5 \\ R_6 \end{bmatrix}_{p_m} = \begin{cases} \begin{bmatrix} 0 & Q_y & Q_z & -Q_\theta & 0 & 0 \end{bmatrix}^{\mathrm{T}} & \text{for a horizontal element} \\[2ex] \begin{bmatrix} -Q_z & Q_y & 0 & 0 & 0 & -Q_\theta \end{bmatrix}^{\mathrm{T}} & \text{for a vertical element} \end{cases} \tag{7.45}$$

and thus,

$$\mathbf{R}_{p_m}(t) = \boldsymbol{\theta}_m \cdot \mathbf{Q}_{p_m} = \left(\frac{\rho V^2}{2}\right)_p \cdot \left(\frac{BL}{2}\right)_m \boldsymbol{\theta}_m \cdot \left[\hat{\mathbf{B}}_{q_m} \cdot \left(\boldsymbol{\psi}_m \cdot \hat{\mathbf{v}}_p\right)\right] \tag{7.46}$$

where indices m and p indicate quantities associated with element m at node p, and where

$$\boldsymbol{\theta}_m = \begin{cases} \begin{bmatrix} 0 & 1 & 0 & 0 & 0 & 0 \\ 0 & 0 & 1 & 0 & 0 & 0 \\ 0 & 0 & 0 & -1 & 0 & 0 \end{bmatrix}^T & \text{for a horizontal element} \\[2ex] \begin{bmatrix} 0 & 1 & 0 & 0 & 0 & 0 \\ -1 & 0 & 0 & 0 & 0 & 0 \\ 0 & 0 & 0 & 0 & 0 & -1 \end{bmatrix}^T & \text{for a vertical element} \end{cases} \tag{7.47}$$

Thus, the load vector in node p is obtained by adding up the contributions from all adjoining elements. i.e.

$$\mathbf{R}_p(t) = \sum_m \mathbf{R}_{p_m} = \left(\frac{\rho V^2}{2}\right)_p \left[\sum_m \left(\frac{BL}{2}\right)_m \left(\boldsymbol{\theta} \cdot \hat{\mathbf{B}}_q \cdot \boldsymbol{\psi}\right)_m\right] \hat{\mathbf{v}}_p = \left(\frac{\rho V^2}{2} \cdot \hat{\mathbf{Q}} \cdot \hat{\mathbf{v}}\right)_p \tag{7.48}$$

where

$$\hat{\mathbf{Q}}_p = \sum_m \left(\frac{BL}{2} \cdot \boldsymbol{\theta} \cdot \hat{\mathbf{B}}_q \cdot \boldsymbol{\psi}\right)_m \tag{7.49}$$

The total system load vector is then given by

$$\mathbf{R}(t) = \begin{bmatrix} \mathbf{R}_1 & \cdots & \mathbf{R}_p & \cdots & \mathbf{R}_N \end{bmatrix}^T \tag{7.50}$$

where N is the total number of nodes. Since the content of this load vector is considered quasi-static the relationship $\mathbf{K} \cdot \mathbf{r}(t) = \mathbf{R}(t)$ holds, and because $u(t)$ and $w(t)$ are both zero mean variables then $\mathbf{R}(t)$ as well as $\mathbf{r}(t)$ are also zero mean variables. Thus, it is seen from to Eqs. 7.6 – 7.9 that the fluctuating background quasi-static part of the element force vector $\mathbf{F}_m(t)$ is given by

$$\mathbf{F}_m(t) = \begin{bmatrix} F_1 \\ F_2 \\ F_3 \\ F_4 \\ F_5 \\ F_6 \end{bmatrix}_m = \mathbf{k}_m \cdot \mathbf{d}_m(t) = \mathbf{k}_m \cdot \left[\mathbf{A}_m \cdot \mathbf{r}(t)\right] = \mathbf{k}_m \cdot \left\{\mathbf{A}_m \cdot \left[\mathbf{K}^{-1} \cdot \mathbf{R}(t)\right]\right\} \tag{7.51}$$

The covariance matrix between cross sectional force components

$$\mathbf{Cov}_{F_m F_m} = \begin{bmatrix} \sigma_{F_1}^2 & Cov_{F_1F_2} & Cov_{F_1F_3} & Cov_{F_1F_4} & Cov_{F_1F_5} & Cov_{F_1F_6} \\ & \sigma_{F_2}^2 & Cov_{F_2F_3} & Cov_{F_2F_4} & Cov_{F_2F_5} & Cov_{F_2F_6} \\ & & \sigma_{F_3}^2 & Cov_{F_3F_4} & Cov_{F_3F_5} & Cov_{F_3F_6} \\ & & & \sigma_{F_4}^2 & Cov_{F_4F_5} & Cov_{F_4F_6} \\ & Sym. & & & \sigma_{F_5}^2 & Cov_{F_5F_6} \\ & & & & & \sigma_{F_6}^2 \end{bmatrix} \tag{7.52}$$

is then defined by

$$\begin{aligned}\mathbf{Cov}_{F_mF_m} &= E\left[\mathbf{F}_m \cdot \mathbf{F}_m^T\right] = E\left[\left\{\mathbf{k}_m\left[\mathbf{A}_m\left(\mathbf{K}^{-1}\mathbf{R}\right)\right]\right\}\cdot\left\{\mathbf{k}_m\left[\mathbf{A}_m\left(\mathbf{K}^{-1}\mathbf{R}\right)\right]\right\}^T\right]\\ &= \mathbf{k}_m \cdot \left\{\mathbf{A}_m \cdot \left(\mathbf{K}^{-1}\cdot E\left[\mathbf{R}\cdot\mathbf{R}^T\right]\cdot\left(\mathbf{K}^{-1}\right)^T\right)\cdot\mathbf{A}_m^T\right\}\cdot\mathbf{k}_m^T\\ &= \mathbf{k}_m \cdot \left\{\mathbf{A}_m \cdot \left[\mathbf{K}^{-1}\cdot \mathbf{Cov}_{RR}\cdot\left(\mathbf{K}^{-1}\right)^T\right]\cdot\mathbf{A}_m^T\right\}\cdot\mathbf{k}_m^T\end{aligned} \tag{7.53}$$

where the $6N$ by $6N$ nodal load covariance matrix

$$\mathbf{Cov}_{RR} = E\left[\mathbf{R}\cdot\mathbf{R}^T\right] = \begin{bmatrix} \ddots & \vdots & ⋰ \\ \cdots & \mathbf{Cov}_{R_pR_k} & \cdots \\ ⋰ & \vdots & \ddots \end{bmatrix} \quad \text{where} \quad \left.\begin{matrix} p \\ k \end{matrix}\right\} = 1,2,3,\ldots,N \tag{7.54}$$

Its content is N nubers of 6 by 6 covariance matrices between force components associated with nodes p and k, each is given by

$$\begin{aligned}\mathbf{Cov}_{R_pR_k} &= E\left[\mathbf{R}_p\cdot\mathbf{R}_k^T\right] = E\left[\left(\frac{\rho V^2}{2}\cdot\hat{\mathbf{Q}}\cdot\hat{\mathbf{v}}\right)_p\cdot\left(\frac{\rho V^2}{2}\cdot\hat{\mathbf{Q}}\cdot\hat{\mathbf{v}}\right)_k^T\right]\\ &= \left(\frac{\rho V^2}{2}\right)_p\cdot\left(\frac{\rho V^2}{2}\right)_k\cdot\hat{\mathbf{Q}}_p\cdot E\left[\hat{\mathbf{v}}_p\cdot\hat{\mathbf{v}}_k^T\right]\cdot\hat{\mathbf{Q}}_k^T\\ &= \left(\frac{\rho V^2}{2}\right)_p\cdot\left(\frac{\rho V^2}{2}\right)_k\cdot\hat{\mathbf{Q}}_p\cdot\hat{\mathbf{Cov}}_{v_pv_k}\cdot\hat{\mathbf{Q}}_k^T\end{aligned} \tag{7.55}$$

where

$$\hat{\mathbf{Cov}}_{v_pv_k} = E\left[\hat{\mathbf{v}}_p\cdot\hat{\mathbf{v}}_k^T\right] = \frac{1}{V^2}\cdot E\begin{bmatrix} u_pu_k & u_pv_k & u_pw_k \\ v_pu_k & v_pv_k & v_pw_k \\ w_pu_k & w_pv_k & w_pw_k \end{bmatrix} \tag{7.56}$$

As previously mentioned (see Eq. 7.19), it is a usual assumption in wind engineering that cross-covariance between different velocity components is insignificant, i.e. that all off diagonal terms in Eq. 7.56 may be neglected, in which case

$$\hat{\mathbf{Cov}}_{v_pv_k} \approx \mathbf{I}_p\cdot\mathbf{I}_k\cdot\boldsymbol{\rho}_{pk} \tag{7.57}$$

where

$$\mathbf{I}_j = diag\left[I_u \quad I_v \quad I_w\right]_j \quad j = \begin{cases} p \\ k \end{cases} \tag{7.58}$$

$$\boldsymbol{\rho}_{pk}\left(\Delta s_{pk}\right) = diag\left[\rho_{uu} \quad \rho_{vv} \quad \rho_{ww}\right] \tag{7.59}$$

The auto covariance functions of the fluctuating flow components u, v and w are defined by (see Eq. 3.35)

$$\rho_{nn}\left(\Delta s_{pk}\right) = \exp\left\{-\left[\left(\frac{\Delta x_{pk}}{{}^xL_n}\right)^2 + \left(\frac{\Delta z_{pk}}{{}^zL_n}\right)^2\right]^{1/2}\right\} \quad \text{where} \quad n = \begin{cases} u \\ v \\ w \end{cases} \tag{7.60}$$

and where

$$\begin{aligned} \Delta x_{pk} &= \left|x_p - x_k\right| \\ \Delta z_{pk} &= \left|z_p - z_k\right| \end{aligned} \tag{7.61}$$

are defined in Fig. 7.10.

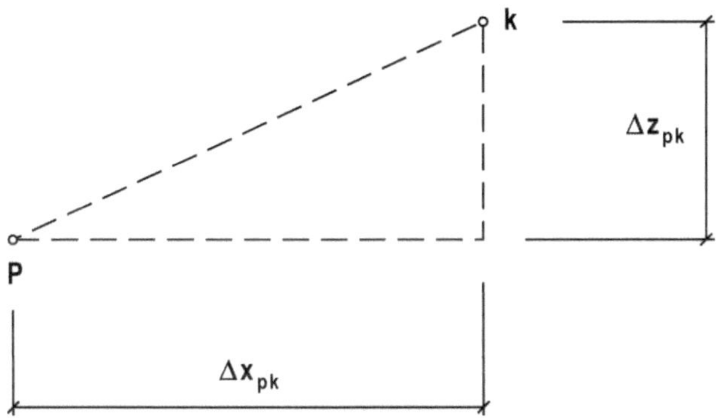

Fig. 7.10 *Spatial separation between nodes p and k*

Example 7.2

Let us consider the simply supported beam shown in Fig. 7.11, and assume that it has been subdivided into N elements with equal length L/N. Let us for simplicity assume that all cross sectional quantities are constants along the span of the bridge, and that it has a typical type of cross section where C_D', C_L and $\bar{C}_M$ are negligible and $\bar{C}_D \cdot D/B \ll C_L'$. The reduced wind load vector associated with an arbitrary node p is then given by

$$\hat{\mathbf{Q}}_p = 2 \cdot \left(\frac{B \cdot L/N}{2} \cdot \boldsymbol{\theta}_m \cdot \hat{\mathbf{B}}_{q_m} \cdot \boldsymbol{\psi}_m\right)$$

where $\hat{\mathbf{B}}_{q_m} = \begin{bmatrix} 2\frac{D}{B}\bar{C}_D & 0 \\ 0 & C'_L \\ 0 & BC'_M \end{bmatrix}$, $\boldsymbol{\theta}_m = \begin{bmatrix} 0 & 0 & 0 \\ 1 & 0 & 0 \\ 0 & 1 & 0 \\ 0 & 0 & -1 \\ 0 & 0 & 0 \\ 0 & 0 & 0 \end{bmatrix}$ and $\boldsymbol{\psi}_m = \begin{bmatrix} 1 & 0 & 0 \\ 0 & 0 & 1 \end{bmatrix}$

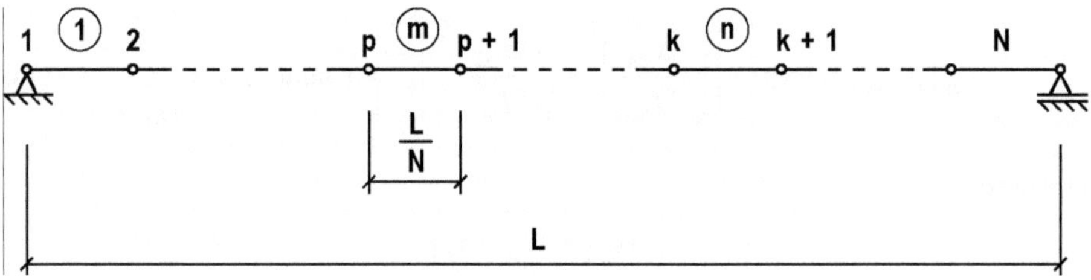

Fig. 7.11 *Simply supported beam type of bridge*

The wind load covariance matrix associated with the cross product between external load components in arbitrary nodes p and k is given by (see Eqs. 7.55 -7.59)

$$\mathbf{Cov}_{R_pR_k} = \left(\frac{\rho V^2 B}{2}\right)^2 \cdot \hat{\mathbf{Q}}_p \cdot \mathbf{Cov}_{v_pv_k} \cdot \hat{\mathbf{Q}}_k^T \qquad \text{where} \qquad \mathbf{Cov}_{v_pv_k} = \mathbf{I}_p \cdot \mathbf{I}_k \cdot \boldsymbol{\rho}_{pk}\left(\Delta x_{pk}\right)$$

and $\mathbf{I}_p = \mathbf{I}_k = diag\left[I_u \quad I_v \quad I_w\right]$, $\rho_{pk}\left(\Delta x_{pk}\right) = diag\left[\rho_{uu} \quad \rho_{vv} \quad \rho_{ww}\right]$, and where Δx_{pk} is the absolute value of the distance between nodes p and k. Thus,

$$\mathbf{Cov}_{R_pR_k} = \left(\frac{\rho V^2 BL}{2N}\right)^2 \cdot \boldsymbol{\theta}_m \cdot \hat{\mathbf{B}}_{q_m} \cdot \boldsymbol{\psi}_m \cdot \left(\mathbf{I}_p \cdot \mathbf{I}_k \cdot \boldsymbol{\rho}_{pk}\right) \cdot \boldsymbol{\theta}_m^T \cdot \hat{\mathbf{B}}_{q_m}^T \cdot \boldsymbol{\psi}_m^T$$

$$\Rightarrow \mathbf{Cov}_{R_pR_k} = \left(\frac{\rho V^2 BL}{2N}\right)^2 \cdot \hat{\mathbf{Cov}}_{R_pR_k}$$

where the non-dimensional cross covariance matrix $\hat{\mathbf{Cov}}_{R_pR_k}$ is given by

$$\hat{\mathbf{Cov}}_{R_pR_k} =$$

$$\begin{bmatrix} 0 & 0 & 0 & 0 & 0 & 0 \\ 0 & \left(2\frac{D}{B}\bar{C}_D I_u\right)^2 \cdot \rho_{uu}\left(\Delta x_{pk}\right) & 0 & 0 & 0 & 0 \\ 0 & 0 & \left(C'_L I_w\right)^2 \cdot \rho_{ww}\left(\Delta x_{pk}\right) & -C'_L BC'_M I_w^2 \cdot \rho_{ww}\left(\Delta x_{pk}\right) & 0 & 0 \\ 0 & 0 & -C'_L BC'_M I_w^2 \cdot \rho_{ww}\left(\Delta x_{pk}\right) & \left(BC'_M I_w\right)^2 \cdot \rho_{ww}\left(\Delta x_{pk}\right) & 0 & 0 \\ 0 & 0 & 0 & 0 & 0 & 0 \\ 0 & 0 & 0 & 0 & 0 & 0 \end{bmatrix}$$

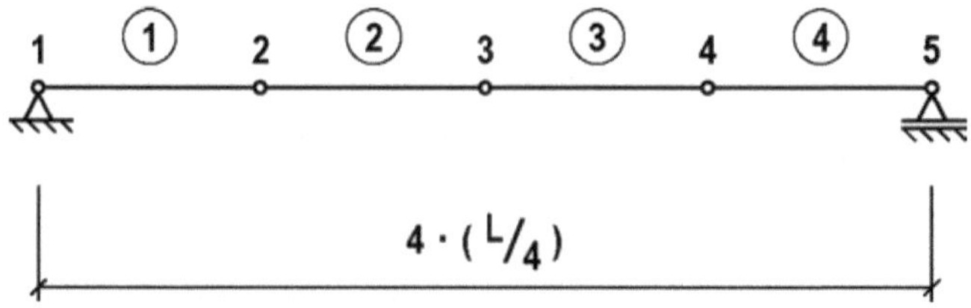

Fig. 7.12 *Four element simply supported beam*

Let us consider the simple case of $N = 4$ as illustrated in Fig. 7.12, and adopt the data given in the table below.

B (m)	D (m)	$\bar{C}_D$	C'_L	C'_M	I_u	I_w	$\frac{L}{{}^{xf}L_u}$	$\frac{{}^{yf}L_u}{{}^{xf}L_u}$	$\frac{{}^{yf}L_w}{{}^{xf}L_u}$
20	4	0.7	5	1.5	0.2	0.1	1/4	1/3	1/16

$$\Rightarrow \rho_{uu} = \exp\left(-\frac{\Delta x}{{}^{yf}L_u}\right) = \exp\left(-\frac{3}{4}\Delta\hat{x}\right) \quad \text{and} \quad \rho_{ww} = \exp\left(-\frac{\Delta x}{{}^{yf}L_w}\right) = \exp(-4\Delta\hat{x})$$

where $\Delta x = \left|\hat{x}_p - \hat{x}_k\right|$ and $\hat{x} = x/L$. In this case there are only two separations:

$$\Delta\hat{x} = \frac{1}{4} \Rightarrow \begin{cases} \rho_{uu}(L/4) = 0.82 \\ \rho_{ww}(L/4) = 0.36 \end{cases} \quad \text{and} \quad \Delta\hat{x} = \frac{1}{2} \Rightarrow \begin{cases} \rho_{uu}(L/4) = 0.69 \\ \rho_{ww}(L/4) = 0.14 \end{cases}$$

Then $$\mathbf{Cov}_{RR} = \left(\frac{\rho V^2 BL}{2 \cdot 4}\right)^2 \begin{bmatrix} \hat{\mathbf{Cov}}_{R_2R_2} & \hat{\mathbf{Cov}}_{R_2R_3} & \hat{\mathbf{Cov}}_{R_2R_4} \\ & \hat{\mathbf{Cov}}_{R_3R_3} & \hat{\mathbf{Cov}}_{R_3R_4} \\ Sym. & & \hat{\mathbf{Cov}}_{R_4R_4} \end{bmatrix}$$

where

$$\hat{\mathbf{Cov}}_{R_2R_2} = \hat{\mathbf{Cov}}_{R_3R_3} = \hat{\mathbf{Cov}}_{R_4R4} = \begin{bmatrix} 0 & 0 & 0 & 0 & 0 & 0 \\ 0 & 0.003136 & 0 & 0 & 0 & 0 \\ 0 & 0 & 0.25 & -1.5 & 0 & 0 \\ 0 & 0 & 1.5 & 9 & 0 & 0 \\ 0 & 0 & 0 & 0 & 0 & 0 \\ 0 & 0 & 0 & 0 & 0 & 0 \end{bmatrix}$$

$$\hat{\mathbf{Cov}}_{R_2R_3} = \hat{\mathbf{Cov}}_{R_3R_4} = \begin{bmatrix} 0 & 0 & 0 & 0 & 0 & 0 \\ 0 & 0.002572 & 0 & 0 & 0 & 0 \\ 0 & 0 & 0.09 & -0.54 & 0 & 0 \\ 0 & 0 & -0.54 & 3.24 & 0 & 0 \\ 0 & 0 & 0 & 0 & 0 & 0 \\ 0 & 0 & 0 & 0 & 0 & 0 \end{bmatrix}$$

$$\hat{\mathbf{Cov}}_{R_2R_4} = \begin{bmatrix} 0 & 0 & 0 & 0 & 0 & 0 \\ 0 & 0.002164 & 0 & 0 & 0 & 0 \\ 0 & 0 & 0.035 & -0.21 & 0 & 0 \\ 0 & 0 & -0.21 & 1.26 & 0 & 0 \\ 0 & 0 & 0 & 0 & 0 & 0 \\ 0 & 0 & 0 & 0 & 0 & 0 \end{bmatrix}$$

The covariance matrix of a chosen set of cross sectional forces is given in Eq. 7.53.

7.4 The resonant part

Fluctuating cross sectional forces at an arbitrary spanwise position x_r that are exclusively ascribed to the resonant part of the response may in general be extracted from the derivatives of predetermined modal displacements

$$\mathbf{r}(x_r,t) = \mathbf{\Phi}_r(x_r) \cdot \mathbf{\eta}(t) \tag{7.62}$$

as defined in Eqs 4.7 and 4.8 (see also Eq. 4.79). The direct transition from the variance of the fluctuating displacement response quantities to the variances of corresponding dynamic cross sectional forces is therefore presented below. The procedures for the calculation of response displacements are in a general format shown in Chapter 4. For the special cases of buffeting or vortex shedding induced dynamic response the procedure is shown in Chapter 6. For simplicity it is in the following as usual assumed that we are dealing with a line-like horizontal (bridge) type of structure where axial forces may be disregarded, in which case the force component F_1 in Eq. 7.7 may be omitted. (Axial forces may in general be determined by the product of the axial stiffness of a beam type of element and the difference between the axial displacements at its end nodes.)

It follows from the definition of cross sectional forces in Fig. 1.3 that the connection between the fluctuating force vector at an arbitrary spanwise position x_r and the corresponding cross sectional stress resultant is

$$\mathbf{F}(x_r,t) = \begin{bmatrix} F_2 & F_3 & F_4 & F_5 & F_6 \end{bmatrix}^T = \begin{bmatrix} V_y & V_z & M_x & M_y & M_z \end{bmatrix}^T \tag{7.63}$$

where V_y, V_z are the shear forces in the direction of the y and z axes, M_y, M_z are the bending moments about the same axes, and where M_x is the torsion moment.

It is taken for granted that the material behaviour is linear elastic and that displacements are small (i.e. there is no geometric non-linearity). The relationship between cross sectional stress resultants and the derivatives of the corresponding displacements are then given by the following differential equations (see e.g. Chen & Atsuta [27]):

$$\left.\begin{aligned} M_x(x_r,t) &= GI_t \cdot r'_\theta(x_r,t) - EI_w \cdot r'''_\theta(x_r,t) \\ M_y(x_r,t) &= -EI_y \cdot r''_z(x_r,t) \\ M_z(x_r,t) &= EI_z \cdot r''_y(x_r,t) \\ V_y(x_r,t) &= -M'_z(x_r,t) = -EI_z \cdot r'''_y(x_r,t) \\ V_z(x_r,t) &= M'_y(x_r,t) = -EI_y \cdot r'''_z(x_r,t) \end{aligned}\right\} \tag{7.64}$$

where the prime behind symbols indicate derivation with respect to x. Defining the cross sectional property matrix $\mathbf{T}(x_r)$

$$\mathbf{T}(x_r) = \begin{bmatrix} 0 & -EI_z & 0 & 0 & 0 & 0 \\ 0 & 0 & 0 & -EI_y & 0 & 0 \\ 0 & 0 & 0 & 0 & GI_t & -EI_w \\ 0 & 0 & -EI_y & 0 & 0 & 0 \\ EI_z & 0 & 0 & 0 & 0 & 0 \end{bmatrix} \tag{7.65}$$

then $\mathbf{F}(x_r,t)$ as defined in Eq. 7.63 is given by

$$\mathbf{F}(x_r,t) = \mathbf{T} \cdot \begin{bmatrix} r''_y & r'''_y & r''_z & r'''_z & r'_\theta & r'''_\theta \end{bmatrix}^T \tag{7.66}$$

<u>*Multi mode approach*</u>

Introducing the six by N_{mod} mode shape derivative matrix

$$\boldsymbol{\beta}(x_r) = \begin{bmatrix} \phi''_{y1} & \cdots & \phi''_{yi} & \cdots & \phi''_{yN_{\text{mod}}} \\ \phi'''_{y1} & \cdots & \phi'''_{yi} & \cdots & \phi'''_{yN_{\text{mod}}} \\ \phi''_{z1} & \cdots & \phi''_{zi} & \cdots & \phi''_{zN_{\text{mod}}} \\ \phi'''_{z1} & \cdots & \phi'''_{zi} & \cdots & \phi'''_{zN_{\text{mod}}} \\ \phi'_{\theta 1} & \cdots & \phi'_{\theta i} & \cdots & \phi'_{\theta N_{\text{mod}}} \\ \phi'''_{\theta 1} & \cdots & \phi'''_{\theta i} & \cdots & \phi'''_{\theta N_{\text{mod}}} \end{bmatrix} \tag{7.67}$$

where i is an arbitrary mode number and N_{mod} is the total number of modes, it then follows from Eq. 7.62 that

$$\mathbf{F}(x_r,t) = \mathbf{T}(x_r)\cdot\left[\boldsymbol{\beta}(x_r)\cdot\boldsymbol{\eta}(t)\right] \tag{7.68}$$

Taking the Fourier transform on either side

$$\mathbf{a}_F(x_r,\omega) = \begin{bmatrix} a_{V_y} & a_{V_z} & a_{M_x} & a_{M_y} & a_{M_z} \end{bmatrix}^T = \mathbf{T}(x_r)\cdot\left[\boldsymbol{\beta}(x_r)\cdot\mathbf{a}_\eta(\omega)\right] \tag{7.69}$$

where $\mathbf{a}_\eta(\omega) = \begin{bmatrix} a_{\eta_1} & \cdots & a_{\eta_i} & \cdots & a_{\eta N_{\text{mod}}} \end{bmatrix}^T$, and defining the matrix

$$\mathbf{S}_F(x_r,\omega) = \begin{bmatrix} S_{V_yV_y} & S_{V_yV_z} & S_{V_yM_x} & S_{V_yM_y} & S_{V_yM_z} \\ & S_{V_zV_z} & S_{V_zM_x} & S_{V_zM_y} & S_{V_zM_z} \\ & & S_{M_xM_x} & S_{M_xM_y} & S_{M_xM_z} \\ & Sym. & & S_{M_yM_y} & S_{M_yM_z} \\ & & & & S_{M_zM_z} \end{bmatrix} \tag{7.70}$$

containing auto spectral densities and cross spectral densities of all force components, then the following is obtained:

$$\mathbf{S}_F(x_r,\omega) = \lim_{T\to\infty}\frac{1}{\pi T}\mathbf{a}_F^*\cdot\mathbf{a}_F^T = \lim_{T\to\infty}\frac{1}{\pi T}\left[\mathbf{T}\cdot\left(\boldsymbol{\beta}\cdot\mathbf{a}_\eta^*\right)\right]\cdot\left[\mathbf{T}\cdot\left(\boldsymbol{\beta}\cdot\mathbf{a}_\eta\right)\right]^T$$

$$\Rightarrow \mathbf{S}_F(x_r,\omega) = \mathbf{T}\cdot\boldsymbol{\beta}\cdot\left[\lim_{T\to\infty}\frac{1}{\pi T}\left(\mathbf{a}_\eta^*\cdot\mathbf{a}_\eta^T\right)\right]\cdot\boldsymbol{\beta}^T\cdot\mathbf{T}^T = \mathbf{T}\cdot\boldsymbol{\beta}\cdot\mathbf{S}_\eta\cdot\boldsymbol{\beta}^T\cdot\mathbf{T}^T \tag{7.71}$$

where $\mathbf{S}_\eta$ is given in Eq. 4.74. However, because $\mathbf{S}_\eta$ contains the entire dynamic response, i.e. background as well as resonant, it requires reduction to include only the resonant part. The extraction of the resonant part is equivalent to a white noise type of load assumption, and thus

$$\mathbf{S}_{\eta R}(\omega) = \hat{\mathbf{H}}_\eta^*(\omega)\cdot\mathbf{S}_{\hat{Q}_R}\cdot\hat{\mathbf{H}}_\eta^T(\omega) \tag{7.72}$$

where $\hat{\mathbf{H}}_\eta$ is given in Eqs. 4.69 and where (see Eq. 4.75)

$$\mathbf{S}_{\hat{Q}_R} = \begin{bmatrix} \ddots & & \cdot^{\cdot^{\cdot}} \\ & S_{\hat{Q}_i\hat{Q}_j} & \\ \cdot^{\cdot^{\cdot}} & & \ddots \end{bmatrix} \tag{7.73}$$

whose elements on row i column j are given by

$$S_{\hat{Q}_i\hat{Q}_j} = \frac{\iint\limits_{L_{\exp}} \boldsymbol{\varphi}_i^T(x_1)\cdot \mathbf{S}_{qq}(\Delta x,\omega_i)\cdot \boldsymbol{\varphi}_j(x_2)dx_1dx_2}{\left(\omega_i^2\tilde{M}_i\right)\cdot\left(\omega_j^2\tilde{M}_j\right)} \tag{7.74}$$

where $\Delta x = |x_1 - x_2|$, and where $\mathbf{S}_{qq}(\Delta x,\omega_i)$ is the spectral density matrix of cross sectional loads at the eigen-frequency ω_i (see Eq. 4.78), i.e.

$$\mathbf{S}_{qq}(\Delta x,\omega_i) = \begin{bmatrix} S_{q_yq_y} & S_{q_yq_z} & S_{q_yq_\theta} \\ S_{q_zq_y} & S_{q_zq_z} & S_{q_zq_\theta} \\ S_{q_\theta q_y} & S_{q_\theta q_z} & S_{q_\theta q_\theta} \end{bmatrix} \tag{7.75}$$

It is seen that $\mathbf{S}_{\hat{Q}_R}$ is frequency independent. The resonant part of the auto and cross spectral density matrix of all force components is then given by

$$\mathbf{S}_{F_R}(x_r,\omega) = \mathbf{T}\cdot\boldsymbol{\beta}\cdot\mathbf{S}_{\eta_R}(\omega)\cdot\boldsymbol{\beta}^T\cdot\mathbf{T}^T = \mathbf{T}\cdot\boldsymbol{\beta}\cdot\left[\hat{\mathbf{H}}_\eta^*(\omega)\cdot\mathbf{S}_{\hat{Q}_R}\cdot\hat{\mathbf{H}}_\eta^T(\omega)\right]\cdot\boldsymbol{\beta}^T\cdot\mathbf{T}^T \tag{7.76}$$

The corresponding matrix containing the resonant part of the variance and covariance of cross sectional stress resultants is obtained by frequency domain integration, i.e.

$$\mathbf{Cov}_{FF_R}(x_r) = \int_0^\infty \mathbf{S}_{F_R}(x_r,\omega)d\omega$$
$$= \begin{bmatrix} \sigma_{V_yV_y}^2 & Cov_{V_yV_z} & Cov_{V_yM_x} & Cov_{V_yM_y} & Cov_{V_yM_z} \\ & \sigma_{V_zV_z}^2 & Cov_{V_zM_x} & Cov_{V_zM_y} & Cov_{V_zM_z} \\ & & \sigma_{M_xM_x}^2 & Cov_{M_xM_y} & Cov_{M_xM_z} \\ & Sym. & & \sigma_{M_yM_y}^2 & Cov_{M_yM_z} \\ & & & & \sigma_{M_zM_z}^2 \end{bmatrix}_R \tag{7.77}$$

If the load case is wind buffeting as described in chapter 6.3, and the assumption of negligible cross spectra between fluctuating flow components is adopted, then $\mathbf{S}_{qq}(\Delta x,\omega)$ is given in Eq. 6.57 and $\mathbf{S}_v(\Delta x,\omega) = V^2 \mathbf{I}_v \hat{\mathbf{S}}_v(\Delta x,\omega)$, see Eq. 6.40. Thus,

$$\mathbf{S}_{qq}(\Delta x,\omega_i) = \left(\frac{\rho V^2 B}{2}\right)^2 \cdot \hat{\mathbf{B}}_q \cdot \left[\mathbf{I}_v^2 \cdot \hat{\mathbf{S}}_v(\Delta x,\omega_i)\right] \cdot \hat{\mathbf{B}}_q^T \tag{7.78}$$

where $\mathbf{I}_v = diag\begin{bmatrix} I_u & I_w \end{bmatrix}$ and

$$\hat{\mathbf{S}}_v(\Delta x,\omega_i) = \begin{bmatrix} \hat{S}_{uu} & 0 \\ 0 & \hat{S}_{ww} \end{bmatrix} = \begin{bmatrix} \dfrac{S_u(\omega_i)}{\sigma_u^2} \cdot \hat{C}_{uu}(\Delta x,\omega_i) & 0 \\ 0 & \dfrac{S_w(\omega_i)}{\sigma_w^2} \cdot \hat{C}_{ww}(\Delta x,\omega_i) \end{bmatrix} \tag{7.79}$$

and where $\hat{C}_{uu}(\Delta x,\omega_i)$ and $\hat{C}_{ww}(\Delta x,\omega_i)$ are the reduced co-spectra defined in Eq.6.64. Introducing the evenly distributed and modally equivalent masses $\tilde{m}_i = \tilde{M}_i / \int_L \left(\boldsymbol{\varphi}_i^T \cdot \boldsymbol{\varphi}_i\right) dx$ and $\tilde{m}_j = \tilde{M}_j / \int_L \left(\boldsymbol{\varphi}_j^T \cdot \boldsymbol{\varphi}_j\right) dx$ (see Eq. 6.41), then the content of $\mathbf{S}_{\hat{Q}_R}$ on row i column j is given by

$$S_{\hat{Q}_i \hat{Q}_j}(\omega_i) = \frac{\rho B^3}{2\tilde{m}_i} \cdot \frac{\rho B^3}{2\tilde{m}_j} \cdot \left(\frac{V}{B\omega_i}\right)^2 \cdot \left(\frac{V}{B\omega_j}\right)^2 \cdot \hat{J}_{ij}^2(\omega_i) \tag{7.80}$$

where

$$\hat{J}_{ij}^2(\omega_i) = \frac{\displaystyle\iint_{L_{\exp}} \boldsymbol{\varphi}_i^T(x_1) \cdot \left\{\hat{\mathbf{B}}_q \cdot \left[\mathbf{I}_v^2 \cdot \hat{\mathbf{S}}_v(\Delta x,\omega_i)\right] \cdot \hat{\mathbf{B}}_q^T\right\} \cdot \boldsymbol{\varphi}_j(x_2) dx_1 dx_2}{\left(\displaystyle\int_L \boldsymbol{\varphi}_i^T \cdot \boldsymbol{\varphi}_i dx\right) \cdot \left(\displaystyle\int_L \boldsymbol{\varphi}_j^T \cdot \boldsymbol{\varphi}_j dx\right)} \tag{7.81}$$

Single mode three component approach

In many cases where eigen-frequencies are well separated and flow induced coupling effects are negligible a multi mode procedure as presented above may with sufficient accuracy be replaced by a mode by mode approach. Then all modes are uncoupled, and therefore, the covariance contributions between force components from different modes are zero, and thus, the total covariance of cross sectional forces may be obtained as the sum of the covariance contributions from each mode, i.e.

$$\mathbf{Cov}_{FF_R}(x_r) = \sum_{i=1}^{N_{\text{mod}}} \mathbf{Cov}_{FF_{R_i}} \tag{7.82}$$

For an arbitrary mode i Eq. 7.76 is then reduced to

$$\mathbf{S}_{F_{R_i}}(x_r,\omega) = \mathbf{T}\cdot\boldsymbol{\beta}_i\cdot\left[\left|\hat{H}_i(\omega)\right|^2\cdot S_{\hat{Q}_i\hat{Q}_i}(\omega_i)\right]\cdot\boldsymbol{\beta}_i^T\cdot\mathbf{T}^T \tag{7.83}$$

where $\boldsymbol{\beta}_i = \begin{bmatrix}\phi_y'' & \phi_y''' & \phi_z'' & \phi_z''' & \phi_\theta' & \phi_\theta'''\end{bmatrix}^T$, and thus

$$\begin{aligned}\mathbf{Cov}_{FF_{R_i}}(x_r) &= \int_0^\infty \mathbf{S}_{F_{R_i}}(x_r,\omega)d\omega = \mathbf{T}\cdot\boldsymbol{\beta}_i\cdot\boldsymbol{\beta}_i^T\cdot\mathbf{T}^T\cdot S_{\hat{Q}_i\hat{Q}_i}\cdot\int_0^\infty\left|\hat{H}_i(\omega)\right|d\omega \\ &= \frac{\pi\omega_i\cdot S_{\hat{Q}_i\hat{Q}_i}}{4\left(1-\kappa_{ae_i}\right)\cdot\left(\zeta_i-\zeta_{ae_i}\right)}\cdot\mathbf{T}\cdot\boldsymbol{\beta}_i\cdot\boldsymbol{\beta}_i^T\cdot\mathbf{T}^T\end{aligned} \tag{7.84}$$

where κ_{ae_i} and ζ_{ae_i} are given in Eqs. 6.45 and 6.46, and where

$$S_{\hat{Q}_i\hat{Q}_i} = \left[\frac{\rho B^3}{2\tilde{m}_i}\cdot\left(\frac{V}{B\omega_i}\right)^2\cdot\hat{J}_{ii}\right]^2 \tag{7.85}$$

$$\hat{J}_{ii}^2 = \frac{\iint_{L_{\exp}} \boldsymbol{\varphi}_i^T(x_1)\cdot\left\{\hat{\mathbf{B}}_q\cdot\left[\mathbf{I}_v^2\cdot\hat{\mathbf{S}}_v(\Delta x,\omega_i)\right]\cdot\hat{\mathbf{B}}_q^T\right\}\cdot\boldsymbol{\varphi}_i(x_2)dx_1dx_2}{\left(\int_L \boldsymbol{\varphi}_i^T\cdot\boldsymbol{\varphi}_i dx\right)^2} \tag{7.86}$$

Performing the multiplication $\mathbf{T} \cdot \boldsymbol{\beta}_i \cdot \boldsymbol{\beta}_i^T \cdot \mathbf{T}^T$ and then the following is obtained:

$$\mathbf{T} \cdot \boldsymbol{\beta}_i \cdot \boldsymbol{\beta}_i^T \cdot \mathbf{T}^T =$$

$$\begin{bmatrix} \left(\phi_y'''EI_z\right)^2 & \left(\phi_y'''EI_z\right)\left(\phi_z'''EI_y\right) & \begin{matrix} -\left\{\left(\phi_y'''EI_z\right)\cdot \right. \\ \left. \left(\phi_\theta' GI_t - \phi_\theta''' EI_w\right)\right\} \end{matrix} & \left(\phi_y'''EI_z\right)\left(\phi_z''EI_y\right) & -\left(\phi_y'''EI_z\right)\left(\phi_y''EI_z\right) \\ & \left(\phi_z'''EI_y\right)^2 & \begin{matrix} -\left\{\left(\phi_z'''EI_y\right)\cdot \right. \\ \left. \left(\phi_\theta' GI_t - \phi_\theta''' EI_w\right)\right\} \end{matrix} & \left(\phi_z'''EI_y\right)\left(\phi_z''EI_y\right) & -\left(\phi_z'''EI_y\right)\left(\phi_y''EI_z\right) \\ & & \left(\phi_\theta' GI_t - \phi_\theta''' EI_w\right)^2 & \begin{matrix} \left\{\left(\phi_\theta' GI_t - \phi_\theta''' EI_w\right) \right. \\ \left. \cdot\left(\phi_z''EI_y\right)\right\} \end{matrix} & \begin{matrix} \left\{\left(\phi_\theta' GI_t - \phi_\theta''' EI_w\right) \right. \\ \left. \cdot\left(\phi_y''EI_z\right)\right\} \end{matrix} \\ & Sym. & & \left(\phi_z''EI_y\right)^2 & -\left(\phi_z''EI_y\right)\left(\phi_y''EI_z\right) \\ & & & & \left(\phi_y''EI_z\right)^2 \end{bmatrix}_i \tag{7.87}$$

It is readily seen from Eq. 7.87 that the covariance matrix associated with an arbitrary mode i has the properties

$$\mathbf{Cov}_{FFR_i}\left(x_r\right) =$$

$$\begin{bmatrix} \sigma_{V_yV_y}^2 & \sigma_{V_yV_y} \cdot \sigma_{V_zV_z} & -\sigma_{V_yV_y} \cdot \sigma_{M_xM_x} & \sigma_{V_yV_y} \cdot \sigma_{M_yM_y} & -\sigma_{V_yV_y} \cdot \sigma_{M_zM_z} \\ & \sigma_{V_zV_z}^2 & -\sigma_{V_zV_z} \cdot \sigma_{M_xM_x} & \sigma_{V_zV_z} \cdot \sigma_{M_yM_y} & -\sigma_{V_zV_z} \cdot \sigma_{M_zM_z} \\ & & \sigma_{M_xM_x}^2 & \sigma_{M_xM_x} \cdot \sigma_{M_yM_y} & \sigma_{M_xM_x} \cdot \sigma_{M_zM_z} \\ & Sym. & & \sigma_{M_yM_y}^2 & -\sigma_{M_yM_y} \cdot \sigma_{M_zM_z} \\ & & & & \sigma_{M_zM_z}^2 \end{bmatrix}_i \tag{7.88}$$

I.e., for an arbitrary mode i

$$\rho_{mn_i} = \frac{Cov_{mn_i}}{\sigma_{mm_i} \cdot \sigma_{nn_i}} = \begin{cases} +1 \text{ for } mn = V_yV_z,\, V_yM_y,\, V_zM_y,\, M_xM_y,\, M_xM_z \\ -1 \text{ for } mn = V_yM_x,\, V_yM_z,\, V_zM_x,\, V_zM_z,\, M_yM_z \end{cases} \tag{7.89}$$

which could be expected, because within a single mode all coupling between cross sectional force components are caused by the structural properties already contained in the relevant mode shape, and thus, all covariance coefficients will either be plus or minus unity (depending on the chosen sign conventions). Thus, the problem is reduced to the calculation of variance contributions from each of the modes that have be considered necessary for a sufficiently accurate solution. It follows from Eqs. 7.84 – 7.87 that

$$\begin{bmatrix} \sigma_{V_yV_y} \\ \sigma_{V_zV_z} \\ \sigma_{M_xM_x} \\ \sigma_{M_yM_y} \\ \sigma_{M_zM_z} \end{bmatrix}_i = \frac{\rho B^3}{4\tilde{m}_i}\left(\frac{V}{B\omega_i}\right)^2 \hat{J}_{ii} \left[\frac{\pi\omega_i}{\left(1-\kappa_{ae_i}\right)\cdot\left(\zeta_i-\zeta_{ae_i}\right)}\right]^{\frac{1}{2}} \begin{bmatrix} \left|\phi_y'''EI_z\right| \\ \left|\phi_z'''EI_y\right| \\ \left|\phi_\theta' GI_t - \phi_\theta'''EI_w\right| \\ \left|\phi_z''EI_y\right| \\ \left|\phi_y''EI_z\right| \end{bmatrix}_i \qquad (7.90)$$

where $\hat{J}_{ii}$ is given in Eq. 7.86. The total variances and covariance coefficients are then given by Eqs. 7.81 and 7.88.

Single mode single component approach

In some cases a single mode single component approach will suffice. The necessary calculations are then further reduced. Let us first consider a single mode that only contains an along wind y component, i.e. $\phi = \left[\phi_y \quad 0 \quad 0\right]^T$, and whose eigen-frequency is ω_y. Then the necessary calculations are reduced to

$$\begin{bmatrix} \sigma_{V_yV_y} \\ \sigma_{M_zM_z} \end{bmatrix} = \frac{\rho B^3}{4\tilde{m}_y}\cdot\left(\frac{V}{B\omega_y}\right)^2 \cdot \hat{J}_y\left(\omega_y\right)\cdot\left[\frac{\pi\omega_y}{\left(1-\kappa_{ae_y}\right)\cdot\left(\zeta_y-\zeta_{ae_y}\right)}\right]^{\frac{1}{2}} \cdot \begin{bmatrix} \left|\phi_y'''EI_z\right| \\ \left|\phi_y''EI_z\right| \end{bmatrix} \qquad (7.91)$$

where $\tilde{m}_y$ is defined in Eq. 6.20, κ_{ae_y} and ζ_{ae_y} are defined in Eq. 6.24, and where $\hat{J}_y$ is given in Eq. 6.22 (see also Eq. 6.19).

Similarly, if the relevant mode only contains an across wind z component, i.e. $\phi = \left[0 \quad \phi_z \quad 0\right]^T$, whose eigen-frequency is ω_z, then

$$\begin{bmatrix} \sigma_{V_zV_z} \\ \sigma_{M_yM_y} \end{bmatrix} = \frac{\rho B^3}{4\tilde{m}_z}\cdot\left(\frac{V}{B\omega_z}\right)^2 \cdot \hat{J}_z\left(\omega_z\right)\cdot\left[\frac{\pi\omega_z}{\left(1-\kappa_{ae_z}\right)\cdot\left(\zeta_z-\zeta_{ae_z}\right)}\right]^{\frac{1}{2}} \cdot \begin{bmatrix} \left|\phi_z'''EI_y\right| \\ \left|\phi_z''EI_y\right| \end{bmatrix} \qquad (7.92)$$

where $\tilde{m}_z$ is defined in Eq. 6.27, κ_{ae_z} and ζ_{ae_z} are defined in Eq. 6.31, and where $\hat{J}_z$ is given in Eq. 6.28.

Finally, if the relevant mode only contains a cross sectional rotation component θ, i.e. $\phi = [0 \quad 0 \quad \phi_\theta]^T$, whose eigen-frequency is ω_θ, then

$$\sigma_{M_x M_x} = \frac{\rho B^4}{4\tilde{m}_\theta}\left(\frac{V}{B\omega_\theta}\right)^2 \hat{J}_\theta(\omega_\theta)\left[\frac{\pi\omega_\theta}{\left(1-\kappa_{ae_\theta}\right)\cdot\left(\zeta_\theta-\zeta_{ae_\theta}\right)}\right]^{\frac{1}{2}} \left|\phi_\theta' GI_t - \phi_\theta''' EI_w\right| \quad (7.93)$$

where $\tilde{m}_\theta$ is defined in Eq. 6.27, κ_{ae_θ} and ζ_{ae_θ} are defined in Eq. 6.31, and where $\hat{J}_\theta$ is given in Eq. 6.29.

Example 7.3

Let us again consider the simply supported beam shown in Fig. 7.6, and as usual, let us for simplicity assume that all cross sectional quantities are constants along the span of the bridge, and that it has a typical bridge type of cross section where C_D', $\bar{C}_L$ and $\bar{C}_M$ are negligible and $\bar{C}_D \cdot D/B \ll C_L'$. Let us set out to determine the covariance matrix associated with cross sectional forces at spanwise positions $x_r = 0$ and $x_r = L/2$ that is caused by resonant oscillations in a chosen mode

$$\boldsymbol{\varphi}_i = \begin{bmatrix} \phi_y \\ \phi_z \\ \phi_\theta \end{bmatrix}_i = \begin{bmatrix} a_y \\ a_z \\ a_\theta \end{bmatrix} \cdot \sin\frac{\pi}{L}x$$

whose eigen-frequency, eigen-damping-ratio and modally equivalent and evenly distributed mass are ω_i, ζ_i and $\tilde{m}_i$. The necessary calculations are given in Eq. 7.83, i.e.

$$\mathbf{Cov}_{FF_{R_i}}(x_r) = \frac{\pi\omega_i \cdot S_{\hat{Q}_i\hat{Q}_i}}{4\left(1-\kappa_{ae_i}\right)\cdot\left(\zeta_i-\zeta_{ae_i}\right)} \cdot \mathbf{T}\cdot\boldsymbol{\beta}_i\cdot\boldsymbol{\beta}_i^T\cdot\mathbf{T}^T$$

where $S_{\hat{Q}_i\hat{Q}_i}$ and $\mathbf{T}\cdot\boldsymbol{\beta}_i\cdot\boldsymbol{\beta}_i^T\cdot\mathbf{T}^T$ are given in Eqs. 7.84 and 7.86. Since (see Ex. 7.2)

$$\hat{\mathbf{B}}_q = \begin{bmatrix} 2\frac{D}{B}\bar{C}_D & 0 \\ 0 & C_L' \\ 0 & BC_M' \end{bmatrix} \quad \text{and} \quad \mathbf{I}_v^2 \cdot \hat{\mathbf{S}}_v(\Delta x, \omega_i) = \begin{bmatrix} I_u^2 \hat{S}_{uu} & 0 \\ 0 & I_w^2 \hat{S}_{ww} \end{bmatrix}$$

then

$$J_{ii}^2 = \iint\limits_{L_{\exp}} \left\{ \left[2\frac{D}{B}\bar{C}_D I_u \right]^2 \cdot \phi_{y_i}(x_1)\cdot\phi_{y_i}(x_2)\cdot\hat{S}_{uu}(\Delta x,\omega_i) + \left[(C_L' I_w)^2 \cdot \phi_{z_i}(x_1)\cdot\phi_{z_i}(x_2) + (BC_M' I_w)^2 \cdot \phi_{\theta_i}(x_1)\cdot\phi_{\theta_i}(x_2) \right]\cdot\hat{S}_{ww}(\Delta x,\omega_i) + C_L' BC_M' I_w^2 \left[\phi_{z_i}(x_1)\cdot\phi_{\theta_i}(x_2) + \phi_{\theta_i}(x_1)\cdot\phi_{z_i}(x_2) \right]\cdot\hat{S}_{ww}(\Delta x,\omega_i) \right\} dx_1 dx_2$$

and

$$\hat{J}_{ii} = \frac{J_{ii}}{\int\limits_L \left(\phi_{y_i}^2 + \phi_{z_i}^2 + \phi_{\theta_i}^2 \right) dx}$$

Introducing the sinusoidal mode shapes, and

$$\hat{S}_{uu}(\Delta x,\omega_i) = \frac{S_u(\omega_i)}{\sigma_u^2}\cdot\hat{C}o_{uu}(\Delta x,\omega_i) \qquad \text{where} \qquad \hat{C}o_{uu}(\Delta x,\omega_i) = \exp\left(-C_{uy}\cdot\frac{\omega_i\cdot\Delta x}{V} \right)$$

$$\hat{S}_{ww}(\Delta x,\omega_i) = \frac{S_w(\omega_i)}{\sigma_w^2}\cdot\hat{C}o_{ww}(\Delta x,\omega_i) \qquad \text{where} \qquad \hat{C}o_{ww}(\Delta x,\omega_i) = \exp\left(-C_{wy}\cdot\frac{\omega_i\cdot\Delta x}{V} \right)$$

then

$$\hat{J}_{ii}^2 = \frac{\left(2\frac{D}{B}\bar{C}_D a_y I_u \right)^2 \cdot \frac{S_u(\omega_i)}{\sigma_u^2}\cdot\psi_u(\omega_i) + \left[(C_L' a_z + BC_M' a_\theta) I_w \right]^2 \cdot \frac{S_w(\omega_i)}{\sigma_w^2}\cdot\psi_w(\omega_i)}{\left(a_y^2 + a_z^2 + a_\theta^2 \right)^2}$$

where

$$\begin{bmatrix} \psi_u(\omega) \\ \psi_w(\omega) \end{bmatrix} = \frac{1}{\int\limits_L \left(\sin\frac{\pi}{L}x \right)^2 dx} \cdot \iint\limits_{L_{\exp}} \sin\frac{\pi}{L}x_1 \cdot \sin\frac{\pi}{L}x_2 \cdot \begin{bmatrix} \hat{C}o_{uu}(\Delta x,\omega) \\ \hat{C}o_{ww}(\Delta x,\omega) \end{bmatrix} dx_1 dx_2$$

This integral has previously been solved in Example 6.1, and thus

$$\psi_u(\omega_i) = 4\cdot\left[\frac{\hat{\omega}_u}{\hat{\omega}_u^2 + \pi^2} + 2\pi^2 \cdot \frac{1+\exp(-\hat{\omega}_u)}{\left(\hat{\omega}_u^2 + \pi^2 \right)^2} \right] \qquad \text{where} \qquad \hat{\omega}_u = C_{uy_f}\cdot\frac{\omega_i L_{\exp}}{V}$$

$$\psi_w(\omega_i) = 4\cdot\left[\frac{\hat{\omega}_w}{\hat{\omega}_w^2 + \pi^2} + 2\pi^2 \cdot \frac{1+\exp(-\hat{\omega}_w)}{\left(\hat{\omega}_w^2 + \pi^2 \right)^2} \right] \qquad \text{where} \qquad \hat{\omega}_w = C_{wy_f}\cdot\frac{\omega_i L_{\exp}}{V}$$

Thus, $S_{\hat{Q}_i\hat{Q}_i} = \left[\frac{\rho B^3}{2\tilde{m}_i}\cdot\left(\frac{V}{B\omega_i} \right)^2 \cdot \hat{J}_{ii} \right]^2$ is defined. At $x_r = 0$

$$\mathbf{Cov}_{FF_{R_i}}(x_r = 0) = \frac{\pi\omega_i \cdot S_{\hat{Q}_i\hat{Q}_i}}{4\left(1-\kappa_{ae_i} \right)\cdot\left(\zeta_i - \zeta_{ae_i} \right)}\cdot\mathbf{T}\cdot\boldsymbol{\beta}_i(x_r = 0)\cdot\boldsymbol{\beta}_i^T(x_r = 0)\cdot\mathbf{T}^T$$

where

$$\mathbf{T}\cdot\boldsymbol{\beta}_i(x_r=0)\cdot\boldsymbol{\beta}_i^T(x_r=0)\cdot\mathbf{T}^T=$$

$$\begin{bmatrix} a_y^2\left(\frac{\pi}{L}\right)^6(EI_z)^2 & a_ya_z\left(\frac{\pi}{L}\right)^6 EI_zEI_y & -a_ya_\theta\left(\frac{\pi}{L}\right)^4 EI_z\left(GI_t+\frac{\pi^2EI_w}{L^2}\right) & 0 & 0 \\ & a_z^2\left(\frac{\pi}{L}\right)^6(EI_y)^2 & -a_za_\theta\left(\frac{\pi}{L}\right)^4 EI_y\left(GI_t+\frac{\pi^2EI_w}{L^2}\right) & 0 & 0 \\ & & a_\theta^2\left(\frac{\pi}{L}\right)^2\left(GI_t+\frac{\pi^2EI_w}{L^2}\right) & 0 & 0 \\ Sym. & & & 0 & 0 \\ & & & & 0 \end{bmatrix}$$

As could be expected, at $x_r=0$ it is only V_y, V_z and M_x that applies, and thus

$$\begin{bmatrix}\sigma_{V_yV_y}\\ \sigma_{V_zV_z}\\ \sigma_{M_xM_x}\end{bmatrix}=\frac{\rho B^3}{4\tilde{m}_i}\cdot\left(\frac{V}{B\omega_i}\right)^2\cdot\hat{J}_{ii}\cdot\left[\frac{\pi\omega_i}{\left(1-\kappa_{ae_i}\right)\cdot\left(\zeta_i-\zeta_{ae_i}\right)}\right]^{\frac{1}{2}}\cdot\begin{bmatrix} a_y\left(\frac{\pi}{L}\right)^3 EI_z \\ a_z\left(\frac{\pi}{L}\right)^3 EI_y \\ a_\theta\left(\frac{\pi}{L}\right)^2\left(GI_t+\frac{\pi^2EI_w}{L^2}\right)\end{bmatrix}$$

and $\rho_{V_yV_z}=1$, $\rho_{V_yM_x}=\rho_{V_zM_x}=-1$. At $x_r=L/2$

$$\mathbf{Cov}_{FFR_i}\left(x_r=\frac{L}{2}\right)=\frac{\pi\omega_i\cdot S_{\hat{Q}_i\hat{Q}_i}}{4\left(1-\kappa_{ae_i}\right)\cdot\left(\zeta_i-\zeta_{ae_i}\right)}\cdot\mathbf{T}\cdot\boldsymbol{\beta}_i\left(x_r=\frac{L}{2}\right)\cdot\boldsymbol{\beta}_i^T\left(x_r=\frac{L}{2}\right)\cdot\mathbf{T}^T$$

where

$$\mathbf{T}\cdot\boldsymbol{\beta}_i\left(x_r=\frac{L}{2}\right)\cdot\boldsymbol{\beta}_i^T\left(x_r=\frac{L}{2}\right)\cdot\mathbf{T}^T=\begin{bmatrix} 0 & 0 & 0 & 0 & 0 \\ & 0 & 0 & 0 & 0 \\ & & 0 & 0 & 0 \\ Sym. & & & a_z^2\left(\frac{\pi}{L}\right)^4(EI_y)^2 & -a_za_y\left(\frac{\pi}{L}\right)^6 EI_yEI_z \\ & & & & a_y^2\left(\frac{\pi}{L}\right)^4(EI_z)^2 \end{bmatrix}$$

As could be expected, at $x_r=L/2$ it is only M_y and M_z that applies, and thus

$$\begin{bmatrix}\sigma_{M_yM_y}\\ \sigma_{M_zM_z}\end{bmatrix}=\frac{\rho B^3}{4\tilde{m}_i}\cdot\left(\frac{V}{B\omega_i}\right)^2\cdot\hat{J}_{ii}\cdot\left[\frac{\pi\omega_i}{\left(1-\kappa_{ae_i}\right)\cdot\left(\zeta_i-\zeta_{ae_i}\right)}\right]^{\frac{1}{2}}\cdot\begin{bmatrix} a_z\left(\frac{\pi}{L}\right)^2 EI_y \\ a_y\left(\frac{\pi}{L}\right)^2 EI_z\end{bmatrix}$$

Let us consider the case that $L_{\exp} = L = 500m$, $V = 40\,m/s$ and $\begin{bmatrix} a_y \\ a_z \\ a_\theta \end{bmatrix} = \begin{bmatrix} 1 \\ 0.5 \\ 0.1 \end{bmatrix}$ and adopt the numerical values given in the following tables:

B m	D m	$\tilde{m}_i$ Kg/m	ω_i Rad/s	ζ_i %	EI_y Nm^2	EI_z Nm^2	GI_t Nm^2	EI_w Nm^4
20	4	10^5	0.8	0.5	10^{11}	$2 \cdot 10^{12}$	$5 \cdot 10^{10}$	$2 \cdot 10^{12}$

ρ kg/m^3	$\bar{C}_D$	C'_L	C'_M	C_{uyf}	C_{wyf}	I_u	I_w	$^{xf}L_u$ m	$^{xf}L_w$ m
1.25	0.7	5	1.5	1.5	1.0	0.2	0.1	162	13.5

Thus,

$$\hat{\omega}_u = C_{uyf} \cdot \frac{\omega_i L_{\exp}}{V} = 16 \qquad \Rightarrow \psi_u(\omega_i) = 4 \cdot \left[\frac{\hat{\omega}_u}{\hat{\omega}_u^2 + \pi^2} + 2\pi^2 \cdot \frac{1 + \exp(-\hat{\omega}_u)}{\left(\hat{\omega}_u^2 + \pi^2\right)^2} \right] = 0.24$$

$$\hat{\omega}_w = C_{wyf} \cdot \frac{\omega_i L_{\exp}}{V} = 10 \qquad \Rightarrow \psi_w(\omega_i) = 4 \cdot \left[\frac{\hat{\omega}_w}{\hat{\omega}_w^2 + \pi^2} + 2\pi^2 \cdot \frac{1 + \exp(-\hat{\omega}_w)}{\left(\hat{\omega}_w^2 + \pi^2\right)^2} \right] = 0.37$$

Let us adopt the typical Kaimal type of turbulence spectra

$$\frac{S_u(\omega)}{\sigma_u^2} = \frac{1.08\,^{xf}L_u/V}{\left(1 + 1.62\omega\,^{xf}L_u/V\right)^{5/3}} \qquad \text{and} \qquad \frac{S_w(\omega)}{\sigma_w^2} = \frac{1.5\,^{xf}L_w/V}{\left(1 + 2.25\omega\,^{xf}L_w/V\right)^{5/3}}$$

$$\Rightarrow \qquad \frac{S_u(\omega_i)}{\sigma_u^2} = 0.206 \qquad \text{and} \qquad \frac{S_w(\omega_i)}{\sigma_w^2} = 0.229$$

The joint acceptance is then

$$\hat{J}_{ii}^2 = \frac{\left(2\frac{D}{B}\bar{C}_D a_y I_u\right)^2 \frac{S_u(\omega_i)}{\sigma_u^2}\psi_u(\omega_i) + \left[(C'_L a_z + BC'_M a_\theta) I_w\right]^2 \frac{S_w(\omega_i)}{\sigma_w^2}\psi_w(\omega_i)}{\left(a_y^2 + a_z^2 + a_\theta^2\right)^2} = 0.127$$

Let us for simplicity also adopt quasi-steady values to the aerodynamic derivatives:

$$\Rightarrow \begin{bmatrix} P_1^* \\ H_1^* \\ A_1^* \end{bmatrix} = -\frac{V}{B\omega_i}\begin{bmatrix} 2(D/B)\bar{C}_D \\ C_L' \\ C_M' \end{bmatrix} = -\begin{bmatrix} 0.7 \\ 12.5 \\ 3.75 \end{bmatrix} \quad \text{and} \quad \begin{bmatrix} H_3^* \\ A_3^* \end{bmatrix} = \left(\frac{V}{B\omega_i}\right)^2 \begin{bmatrix} C_L' \\ C_M' \end{bmatrix} = \begin{bmatrix} 31.25 \\ 9.375 \end{bmatrix}$$

Thus,

$$\kappa_{ae_i} = \frac{\rho B^2}{2\tilde{m}_i}\cdot\frac{\int_{L_{\exp}}\left(\phi_z\phi_\theta BH_3^* + \phi_\theta^2 B^2 A_3^*\right)dx}{\int_L\left(\phi_y^2+\phi_z^2+\phi_\theta^2\right)dx} = \frac{\rho B^2}{2\tilde{m}_i}\cdot\frac{a_z a_\theta BH_3^* + a_\theta^2 B^2 A_3^*}{a_y^2+a_z^2+a_\theta^2} = 0.136$$

$$\zeta_{ae_i} = \frac{\rho B^2}{4\tilde{m}_i}\cdot\frac{\int_{L_{\exp}}\left(\phi_y^2 P_1^* + \phi_z^2 H_1^* + \phi_\theta\phi_z BA_1^*\right)dx}{\int_L\left(\phi_y^2+\phi_z^2+\phi_\theta^2\right)dx} = \frac{\rho B^2}{4\tilde{m}_i}\cdot\frac{a_y^2 P_1^* + a_z^2 H_1^* + a_\theta a_z BA_1^*}{a_y^2+a_z^2+a_\theta^2} = -0.0069$$

The following is then obtained at $x_r = 0$

$$\begin{bmatrix} \sigma_{V_yV_y} \\ \sigma_{V_zV_z} \\ \sigma_{M_xM_x} \end{bmatrix} = \frac{\rho B^3}{4\tilde{m}_i}\cdot\left(\frac{V}{B\omega_i}\right)^2\cdot\hat{J}_{ii}\cdot\left[\frac{\pi\omega_i}{\left(1-\kappa_{ae_i}\right)\cdot\left(\zeta_i-\zeta_{ae_i}\right)}\right]^{\frac{1}{2}}\cdot\begin{bmatrix} a_y\left(\frac{\pi}{L}\right)^3 EI_z \\ a_z\left(\frac{\pi}{L}\right)^3 EI_y \\ a_\theta\left(\frac{\pi}{L}\right)^2\left(GI_t+\frac{\pi^2 EI_w}{L^2}\right) \end{bmatrix}$$

$$\Rightarrow \begin{bmatrix} \sigma_{V_yV_y}(x_r=0) \\ \sigma_{V_zV_z}(x_r=0) \\ \sigma_{M_xM_x}(x_r=0) \end{bmatrix} = \begin{bmatrix} 154\text{ kN} \\ 3.8\text{ kN} \\ 61\text{ kNm} \end{bmatrix}$$

The following is obtained at $x_r = L/2$:

$$\begin{bmatrix} \sigma_{M_yM_y} \\ \sigma_{M_zM_z} \end{bmatrix} = \frac{\rho B^3}{4\tilde{m}_i}\cdot\left(\frac{V}{B\omega_i}\right)^2\cdot\hat{J}_{ii}\cdot\left[\frac{\pi\omega_i}{\left(1-\kappa_{ae_i}\right)\cdot\left(\zeta_i-\zeta_{ae_i}\right)}\right]^{\frac{1}{2}}\cdot\begin{bmatrix} a_z\left(\frac{\pi}{L}\right)^2 EI_y \\ a_y\left(\frac{\pi}{L}\right)^2 EI_z \end{bmatrix}$$

$$\Rightarrow \begin{bmatrix} \sigma_{M_yM_y}(x_r=L/2) \\ \sigma_{M_zM_z}(x_r=L/2) \end{bmatrix} = \begin{bmatrix} 612\text{ kNm} \\ 24480\text{ kNm} \end{bmatrix}$$

Chapter 8

MOTION INDUCED INSTABILITIES

8.1 *Introduction*

Static as well as dynamic structural response will in general increase with increasing mean wind velocity. In some cases the response may develop towards what is perceived as unstable behaviour, i.e. the response is rapidly increasing for even a small increase of the mean wind velocity, as indicated in Fig. 6.3. It is seen from Eqs. 6.48 and 6.49 (see also Eqs. 4.69 and 4.82) that in the limit the structural displacement response will become infinitely large if the absolute value of the determinant to the non-dimensional N_{mod} by N_{mod} impedance matrix

$$\hat{\mathbf{E}}_{\eta}(\omega,V)=\left\{\mathbf{I}-\boldsymbol{\kappa}_{ae}-\left(\omega\cdot diag\left[\frac{1}{\omega_i}\right]\right)^2+2i\omega\cdot diag\left[\frac{1}{\omega_i}\right]\cdot(\boldsymbol{\zeta}-\boldsymbol{\zeta}_{ae})\right\} \tag{8.1}$$

is zero. Thus, any stability limit may be revealed by studying the properties of the impedance matrix. Obviously, unstable behaviour is caused by the effects of $\boldsymbol{\kappa}_{ae}$ and $\boldsymbol{\zeta}_{ae}$. The effects of $\boldsymbol{\zeta}_{ae}$ is to change the damping properties of the combined structure and flow system, while the effects of $\boldsymbol{\kappa}_{ae}$ is to change the stiffness properties. While we in the entire chapter 6 ignored any motion induced changes to resonance frequencies (defined as the frequency positions of the apexes of the modal frequency response function) this can not be accepted in the search for any relevant instability limit. The reason is explained in chapter 5.2, and as shown in Eq. 5.24, it involves taking into account that the aerodynamic derivatives are modal quantities that have been normalised by and are functions of the mean wind velocity dependent resonance frequencies. Thus (see Eqs. 5.24, 6.51 and 6.52) the content of

$$\boldsymbol{\kappa}_{ae}=\begin{bmatrix}\ddots & & \cdot^{\cdot^{\cdot}}\\ & \kappa_{ae_{ij}} & \\ \cdot^{\cdot^{\cdot}} & & \ddots\end{bmatrix}\quad\text{and}\quad\boldsymbol{\zeta}_{ae}=\begin{bmatrix}\ddots & & \cdot^{\cdot^{\cdot}}\\ & \zeta_{ae_{ij}} & \\ \cdot^{\cdot^{\cdot}} & & \ddots\end{bmatrix} \tag{8.2}$$

are now given by

$$\kappa_{ae_{ij}} = \frac{\tilde{K}_{ae_{ij}}}{\omega_i^2 \tilde{M}_i} = \frac{\rho B^2}{2\tilde{m}_i} \cdot \left[\frac{\omega_i(V)}{\omega_i} \right]^2 \cdot \frac{\int_{L_{\exp}} \left(\boldsymbol{\varphi}_i^T \cdot \hat{\mathbf{K}}_{ae} \cdot \boldsymbol{\varphi}_j \right) dx}{\int_L \left(\boldsymbol{\varphi}_i^T \cdot \boldsymbol{\varphi}_i \right) dx} \tag{8.3}$$

$$\zeta_{ae_{ij}} = \frac{\omega_i}{2} \frac{\tilde{C}_{ae_{ij}}}{\omega_i^2 \tilde{M}_i} = \frac{\rho B^2}{4\tilde{m}_i} \cdot \frac{\omega_i(V)}{\omega_i} \cdot \frac{\int_{L_{\exp}} \left(\boldsymbol{\varphi}_i^T \cdot \hat{\mathbf{C}}_{ae} \cdot \boldsymbol{\varphi}_j \right) dx}{\int_L \left(\boldsymbol{\varphi}_i^T \cdot \boldsymbol{\varphi}_i \right) dx} \tag{8.4}$$

where $\omega_i(V)$ is the mean wind velocity dependent resonance frequency associated with mode i and $\omega_i = \omega_i(V=0)$ (or as calculated in vacuum). The solution to Eq. 8.1 is an eigen-value problem with N_{mod} roots. Each of these eigen-values represents a limiting behaviour in which the structural response is nominally infinitely large (or irrelevant). I.e., the condition

$$\left| \det\left(\hat{\mathbf{E}}_\eta(\omega, V) \right) \right| = 0 \tag{8.5}$$

will formally reveal N_{mod} stability limits associated with all the relevant mode shapes contained in $\hat{\mathbf{E}}_\eta$, static or dynamic. In general $\det\left(\hat{\mathbf{E}}_\eta\right)$ will contain complex quantities, and therefore $\left|\det\left(\hat{\mathbf{E}}_\eta\right) = 0\right|$ implies the simultaneous conditions that

$$\text{Re}\left(\det\left(\hat{\mathbf{E}}_\eta \right) \right) = 0 \qquad \text{and} \qquad \text{Im}\left(\det\left(\hat{\mathbf{E}}_\eta \right) \right) = 0 \tag{8.6}$$

As shown above, $\hat{\mathbf{E}}_\eta$ is a function of the frequency and of the mean wind velocity, and thus, each root will contain a pair of ω and V values which may be used to identify the relevant stability problem. For a static stability limit $\omega = 0$, and thus, such a limit may simply be identified by a critical wind velocity V_{cr}. For a dynamic stability limit the response is narrow-banded and centred on an in-wind preference or resonance frequency associated with a certain mode or combination of modes. Thus, the outcome of the eigen-value solution to Eq. 8.5 will identify a dynamic stability limit by a critical velocity V_{cr} and the corresponding in-wind preference or resonance frequency ω_r.

Of all the eigen-values that may be extracted from Eq. 8.5 the main focus is on the one that represents the stability limit at the lowest mean wind velocity, i.e. it is the lowest V_{cr} (and corresponding ω_r) that has priority.

Cases of structural behaviour close to a stability limit may in general be classified according to the response type of displacement that develops. The problem of identification is greatly simplified if the impedance is taken directly from the characteristic behaviour of each stability problem as known from full scale or experimental observations. For a bridge section there are four types of such behaviour. First, there is the possibility of a static type of unstable behaviour in torsion, called static divergence. Second, there is the possibility of a dynamic type of unstable behaviour in the across wind vertical (z) direction, called galloping. Third, there is a possible unstable type of dynamic response in pure torsion, and finally, there is the possibility of an unstable type of dynamic response in a combined motion of vertical displacements and torsion, called flutter. Thus, it is always either r_z, r_θ or both that are the critical response quantities. It is then only necessary to search for the instability limits associated with the two most onerous modes, $\boldsymbol{\varphi}_1$ and $\boldsymbol{\varphi}_2$ with corresponding eigen-frequencies ω_1 and ω_2, of which one contain a predominant ϕ_z component and the other contain a predominant ϕ_θ component. Therefore, the impedance matrix may be reduced to

$$\hat{\mathbf{E}}_\eta(\omega_r, V_{cr}) = \left\{ \begin{bmatrix} 1 & 0 \\ 0 & 1 \end{bmatrix} - \begin{bmatrix} \kappa_{ae_{11}} & \kappa_{ae_{12}} \\ \kappa_{ae_{21}} & \kappa_{ae_{22}} \end{bmatrix} - \begin{bmatrix} (\omega_r/\omega_1)^2 & 0 \\ 0 & (\omega_r/\omega_2)^2 \end{bmatrix} \right.$$
$$\left. + 2i \begin{bmatrix} \omega_r/\omega_1 & 0 \\ 0 & \omega_r/\omega_2 \end{bmatrix} \cdot \begin{bmatrix} \zeta_1 - \zeta_{ae_{11}} & -\zeta_{ae_{12}} \\ -\zeta_{ae_{21}} & \zeta_2 - \zeta_{ae_{22}} \end{bmatrix} \right\} \tag{8.7}$$

where (see Eqs. 8.3 and 8.4)

$$\frac{\kappa_{ae_{ij}}}{\dfrac{\rho B^2}{2\tilde{m}_i}} = \left(\frac{\omega_i(V)}{\omega_i} \right)^2 \left[\int_{L_{\exp}} \left(\phi_{y_i}\phi_{y_j} P_4^* + \phi_{z_i}\phi_{y_j} H_6^* + \phi_{\theta_i}\phi_{y_j} B A_6^* + \phi_{y_i}\phi_{z_j} P_6^* + \phi_{z_i}\phi_{z_j} H_4^* \right.\right.$$
$$\left.\left. + \phi_{\theta_i}\phi_{z_j} B A_4^* + \phi_{y_i}\phi_{\theta_j} B P_3^* + \phi_{z_i}\phi_{\theta_j} B H_3^* + \phi_{\theta_i}\phi_{\theta_j} B^2 A_3^* \right) dx \right] \Bigg/ \left[\int_L \left(\phi_{y_i}^2 + \phi_{z_i}^2 + \phi_{\theta_i}^2 \right) dx \right] \tag{8.8}$$

$$\frac{\zeta_{ae_{ij}}}{\dfrac{\rho B^2}{4\tilde{m}_i}} = \frac{\omega_i(V)}{\omega_i} \cdot \left[\int_{L_{\exp}} \left(\phi_{y_i}\phi_{y_j} P_1^* + \phi_{z_i}\phi_{y_j} H_5^* + \phi_{\theta_i}\phi_{y_j} BA_5^* + \phi_{y_i}\phi_{z_j} P_5^* + \phi_{z_i}\phi_{z_j} H_1^* \right.\right.$$

$$\left.\left. + \phi_{\theta_i}\phi_{z_j} BA_1^* + \phi_{y_i}\phi_{\theta_j} BP_2^* + \phi_{z_i}\phi_{\theta_j} BH_2^* + \phi_{\theta_i}\phi_{\theta_j} B^2 A_2^* \right) dx \right] \Bigg/ \left[\int_L \left(\phi_{y_i}^2 + \phi_{z_i}^2 + \phi_{\theta_i}^2 \right) dx \right] \tag{8.9}$$

and $\left.\begin{matrix} i \\ j \end{matrix}\right\} = 1,2$. The problem is further simplified if

$$\left.\begin{aligned} \boldsymbol{\varphi}_1(x) &\approx [0 \quad \phi_z \quad 0]_1^T \\ \boldsymbol{\varphi}_2(x) &\approx [0 \quad 0 \quad \phi_\theta]_2^T \end{aligned}\right\} \tag{8.10}$$

with corresponding eigen-frequencies $\omega_1 = \omega_z$ and $\omega_2 = \omega_\theta$, modal eigen-damping ratios $\zeta_1 = \zeta_z$ and $\zeta_2 = \zeta_\theta$, and with modal mass properties $\tilde{m}_1 = \tilde{m}_z$ and $\tilde{m}_2 = \tilde{m}_\theta$. In that particular case

$$\hat{\mathbf{E}}_\eta(\omega_r, V_{cr}) = \left\{ \begin{bmatrix} 1 & 0 \\ 0 & 1 \end{bmatrix} - \begin{bmatrix} \kappa_{ae_{zz}} & \kappa_{ae_{z\theta}} \\ \kappa_{ae_{\theta z}} & \kappa_{ae_{\theta\theta}} \end{bmatrix} - \begin{bmatrix} (\omega_r/\omega_z)^2 & 0 \\ 0 & (\omega_r/\omega_\theta)^2 \end{bmatrix} \right.$$

$$\left. + 2i \begin{bmatrix} \omega_r/\omega_z & 0 \\ 0 & \omega_r/\omega_\theta \end{bmatrix} \cdot \begin{bmatrix} \zeta_z - \zeta_{ae_{zz}} & -\zeta_{ae_{z\theta}} \\ -\zeta_{ae_{\theta z}} & \zeta_\theta - \zeta_{ae_{\theta\theta}} \end{bmatrix} \right\} \tag{8.11}$$

and

$$\kappa_{ae_{zz}} = \frac{\rho B^2}{2\tilde{m}_z} \left(\frac{\omega_z(V)}{\omega_z} \right)^2 H_4^* \frac{\int_{L_{\exp}} \phi_z^2 dx}{\int_L \phi_z^2 dx} \qquad \kappa_{ae_{z\theta}} = \frac{\rho B^3}{2\tilde{m}_z} \left(\frac{\omega_z(V)}{\omega_z} \right)^2 H_3^* \frac{\int_{L_{\exp}} \phi_z \phi_\theta dx}{\int_L \phi_z^2 dx} \tag{8.12}$$

$$\kappa_{ae_{\theta\theta}} = \frac{\rho B^4}{2\tilde{m}_\theta} \left(\frac{\omega_\theta(V)}{\omega_\theta} \right)^2 A_3^* \frac{\int_{L_{\exp}} \phi_\theta^2 dx}{\int_L \phi_\theta^2 dx} \qquad \kappa_{ae_{\theta z}} = \frac{\rho B^3}{2\tilde{m}_\theta} \left(\frac{\omega_\theta(V)}{\omega_\theta} \right)^2 A_4^* \frac{\int_{L_{\exp}} \phi_\theta \phi_z dx}{\int_L \phi_\theta^2 dx} \tag{8.13}$$

$$\zeta_{ae_{zz}} = \frac{\rho B^2}{4\tilde{m}_z}\frac{\omega_z(V)}{\omega_z}H_1^* \frac{\int\limits_{L_{\exp}} \phi_z^2 dx}{\int\limits_L \phi_z^2 dx} \qquad \zeta_{ae_{z\theta}} = \frac{\rho B^3}{4\tilde{m}_z}\frac{\omega_z(V)}{\omega_z}H_2^* \frac{\int\limits_{L_{\exp}} \phi_z\phi_\theta dx}{\int\limits_L \phi_z^2 dx} \tag{8.14}$$

$$\zeta_{ae_{\theta\theta}} = \frac{\rho B^4}{4\tilde{m}_\theta}\frac{\omega_\theta(V)}{\omega_\theta}A_2^* \frac{\int\limits_{L_{\exp}} \phi_\theta^2 dx}{\int\limits_L \phi_\theta^2 dx} \qquad \zeta_{ae_{\theta z}} = \frac{\rho B^3}{4\tilde{m}_\theta}\frac{\omega_\theta(V)}{\omega_\theta}A_1^* \frac{\int\limits_{L_{\exp}} \phi_\theta\phi_z dx}{\int\limits_L \phi_\theta^2 dx} \tag{8.15}$$

where $\omega_z(V)$ and $\omega_\theta(V)$ are the mean wind velocity dependent resonance frequencies associated with $\boldsymbol{\varphi}_1(x) \approx \begin{bmatrix} 0 & \phi_z & 0 \end{bmatrix}_1^T$ and $\boldsymbol{\varphi}_2(x) \approx \begin{bmatrix} 0 & 0 & \phi_\theta \end{bmatrix}_2^T$. A purely single mode unstable behaviour contains motion either in the vertical direction (i.e. galloping) or in torsion. Such an instability limit may then be identified from the first or the second row of the matrices in Eq. 8.11 alone, in which case $\omega_r = \omega_z(V_{cr})$ or $\omega_r = \omega_\theta(V_{cr})$. Otherwise, the unstable behaviour contains a combined motion in the vertical direction and torsion (i.e. flutter), in which case the instability limit may be identified from Eq. 8.11, and $\omega_r = \omega_z(V_{cr}) = \omega_\theta(V_{cr})$. Motion induced coupling effects between r_z and r_θ (i.e. flutter) will only occur if the off–diagonal terms in Eq. 8.11 are unequal to zero, i.e. if $\int\limits_{L_{\exp}} \phi_z\phi_\theta dx \neq 0$ (see Eqs. 8.12 – 8.15).

8.2 *Static divergence*

Let $\boldsymbol{\varphi}_2$ be the mode shape in predominantly torsion that has the lowest eigen–frequency. Let us for simplicity assume that

$$\boldsymbol{\varphi}_2 \approx \begin{bmatrix} 0 & 0 & \phi_\theta \end{bmatrix}^T \tag{8.16}$$

At $\omega_r = 0$, the instability effect is static and not dynamic. It is simply a problem of loosing torsion stiffness due to interaction effects with the air flow. Thus, the impedance in Eq. 8.11 is reduced to

$$\hat{E}_\eta(\omega_r = 0, V_{cr}) = 1 - \kappa_{ae_{\theta\theta}} \tag{8.17}$$

where: $$\kappa_{ae_{\theta\theta}} = \frac{\rho B^4}{2\tilde{m}_\theta}\left(\frac{\omega_\theta(V_{cr})}{\omega_\theta}\right)^2 A_3^* \frac{\int_{L_{\exp}} \phi_\theta^2 dx}{\int_L \phi_\theta^2 dx}$$

It is seen that $\hat{E}_\eta(\omega_r = 0, V_{cr}) = 0$ when $\kappa_{ae_{\theta\theta}} = 1$. Thus, a static divergence type of instability limit may be identified under the condition that

$$\frac{\rho B^4}{2\tilde{m}_\theta}\left(\frac{\omega_\theta(V_{cr})}{\omega_\theta}\right)^2 A_3^* \frac{\int_{L_{\exp}} \phi_\theta^2 dx}{\int_L \phi_\theta^2 dx} = 1 \tag{8.18}$$

Since this is a purely static type of unstable behaviour the quasi-static version of A_3^* from Eq. 5.26 applies, and thus, the following critical mean wind velocity for static divergence is obtained

$$V_{cr} = B \cdot \omega_\theta \cdot \left(\frac{2\tilde{m}_\theta}{\rho B^4 C_M'} \cdot \frac{\int_L \phi_\theta^2 dx}{\int_{L_{\exp}} \phi_\theta^2 dx}\right)^{1/2} \tag{8.19}$$

8.3 Galloping

Let $\boldsymbol{\varphi}_1$ be the mode shape with the lowest eigen-frequency $\omega_1 = \omega_z$ whose main component is ϕ_z, i.e.

$$\boldsymbol{\varphi}_1 \approx [0 \quad \phi_z \quad 0]^T \tag{8.20}$$

Since the resonance frequency associated with this mode is $\omega_z(V)$, then

$$\omega_r = \omega_z(V_{cr}) \tag{8.21}$$

and the impedance in Eq. 8.11 is reduced to

$$\hat{E}_\eta(\omega_r, V_{cr}) = 1 - \kappa_{ae_{zz}} - (\omega_r/\omega_z)^2 + 2i\left(\zeta_z - \zeta_{ae_{zz}}\right)\omega_r/\omega_z \tag{8.22}$$

where

$$\kappa_{ae_{zz}} = \frac{\rho B^2}{2\tilde{m}_z}\left(\frac{\omega_r}{\omega_z}\right)^2 H_4^* \frac{\int\limits_{L_{exp}} \phi_z^2 dx}{\int\limits_L \phi_z^2 dx} \quad \text{and} \quad \zeta_{ae_{zz}} = \frac{\rho B^2}{4\tilde{m}_z}\frac{\omega_r}{\omega_z} H_1^* \frac{\int\limits_{L_{exp}} \phi_z^2 dx}{\int\limits_L \phi_z^2 dx}$$

Setting the real and imaginary parts of Eq. 8.22 equal to zero, a dynamic stability limit may then be identified at an in-wind resonance frequency

$$\omega_r = \omega_z \left(1 + \frac{\rho B^2}{2\tilde{m}_z} H_4^* \frac{\int\limits_{L_{exp}} \phi_z^2 dx}{\int\limits_L \phi_z^2 dx}\right)^{-1/2} \tag{8.23}$$

when the damping properties are such that

$$\zeta_z = \zeta_{ae_{zz}} = \frac{\rho B^2}{4\tilde{m}_z}\frac{\omega_r}{\omega_z} H_1^* \frac{\int\limits_{L_{exp}} \phi_z^2 dx}{\int\limits_L \phi_z^2 dx} \tag{8.24}$$

This type of stability problem is called galloping. It is seen that a galloping instability can only occur if H_1^* attains positive values. (For a flat plate H_1^* is consistently negative, see Fig. 5.3, but this is a property that vanishes for cross sections with increasing bluffness.)

Adopting the quasi-static versions of the aerodynamic derivatives given in Eq. 5.26, then the stability limit is defined by the following mean wind velocity

$$V_{cr} = B\omega_z \cdot \frac{\zeta_z}{-\left(C_L' + \bar{C}_D \cdot D/B\right)} \cdot \frac{4\tilde{m}_z}{\rho B^2} \cdot \frac{\int\limits_L \phi_z^2 dx}{\int\limits_{L_{exp}} \phi_z^2 dx} \tag{8.25}$$

An analytical solution to the problem of galloping was first presented by den Hartog [29], showing that galloping can only occur if $C_L' < -\bar{C}_D \cdot D/B$.

8.4 Dynamic stability limit in torsion

A stability problem in torsion is related to galloping in the sense that it involves a single mode type of motion. Let $\boldsymbol{\varphi}_2$ be the mode shape with the lowest eigen-frequency $\omega_2 = \omega_\theta$ whose main component is ϕ_θ, i.e.

$$\boldsymbol{\varphi}_2 \approx \begin{bmatrix} 0 & 0 & \phi_\theta \end{bmatrix}^T \tag{8.26}$$

Since the resonance frequency associated with this mode is $\omega_\theta(V)$, then

$$\omega_r = \omega_\theta(V_{cr}) \tag{8.27}$$

and the impedance in Eq. 8.11 is reduced to

$$\hat{E}_\eta(\omega_r, V_{cr}) = 1 - \kappa_{ae_{\theta\theta}} - (\omega_r/\omega_\theta)^2 + 2i\left(\zeta_\theta - \zeta_{ae_{\theta\theta}}\right)\omega_r/\omega_\theta \tag{8.28}$$

where

$$\kappa_{ae_{\theta\theta}} = \frac{\rho B^4}{2\tilde{m}_\theta}\left(\frac{\omega_r}{\omega_\theta}\right)^2 A_3^* \frac{\int_{L_{\exp}} \phi_\theta^2 dx}{\int_L \phi_\theta^2 dx} \quad \text{and} \quad \zeta_{ae_{\theta\theta}} = \frac{\rho B^4}{4\tilde{m}_\theta}\frac{\omega_r}{\omega_\theta} A_2^* \frac{\int_{L_{\exp}} \phi_\theta^2 dx}{\int_L \phi_\theta^2 dx}$$

Setting the real and imaginary parts of Eq. 8.28 equal to zero, a dynamic stability limit may then be identified at an in-wind resonance frequency

$$\omega_r = \omega_\theta \left(1 + \frac{\rho B^4}{2\tilde{m}_\theta} \cdot A_3^* \cdot \frac{\int_{L_{\exp}} \phi_\theta^2 dx}{\int_L \phi_\theta^2 dx}\right)^{-1/2} \tag{8.29}$$

when the damping properties are such that

$$\zeta_\theta = \zeta_{ae_{\theta\theta}} = \frac{\rho B^4}{4\tilde{m}_\theta}\frac{\omega_r}{\omega_\theta} A_2^* \frac{\int_{L_{\exp}} \phi_\theta^2 dx}{\int_L \phi_\theta^2 dx} \tag{8.30}$$

It is seen that an instability in pure torsion can only occur if A_2^* attains positive values. (For a flat plate A_2^* is consistently negative, see Fig. 5.3.) Since the quasi-static value of A_2^* is zero, it is futile to define a stability limit based on the quasi-static theory.

8.5 Flutter

As mentioned above, flutter is a dynamic stability problem where r_z couples with r_θ. Such coupling occurs via the off-diagonal terms $\kappa_{ae_{z\theta}}$ and $\kappa_{ae_{\theta z}}$ in Eq. 8.11 above, and therefore, it is most prone to occur between modes $\boldsymbol{\varphi}_1$ and $\boldsymbol{\varphi}_2$ that are shape-wise similar and whose main components are ϕ_z and ϕ_θ. Experimental observations show that it is usually the aerodynamic forces associated with the motion in torsion that are the driving forces in the coupling process.

Let $\boldsymbol{\varphi}_2$ be the mode shape with the lowest eigen-frequency $\omega_2 = \omega_\theta$ whose main component is ϕ_θ, i.e.

$$\boldsymbol{\varphi}_2 \approx \begin{bmatrix} 0 & 0 & \phi_\theta \end{bmatrix}^T \tag{8.31}$$

Let $\boldsymbol{\varphi}_1$ be another mode that shape-wise is similar to $\boldsymbol{\varphi}_2$ and whose main component is ϕ_z, i.e.

$$\boldsymbol{\varphi}_1 \approx \begin{bmatrix} 0 & \phi_z & 0 \end{bmatrix}^T \tag{8.32}$$

and whose eigen–frequency is $\omega_1 = \omega_z$. A flutter stability limit is then identified by $\left|\det\left(\hat{\mathbf{E}}_\eta\left(\omega_r, V_{cr}\right)\right)\right| = 0$ where $\hat{\mathbf{E}}_\eta\left(\omega_r, V_{cr}\right)$ is given in Eq. 8.11. Since r_z couples with r_θ into a joint resonant motion, then

$$\omega_r = \omega_z\left(V_{cr}\right) = \omega_\theta\left(V_{cr}\right) \tag{8.33}$$

From a computational point of view it is convenient to split $\hat{\mathbf{E}}_\eta$ into four parts, i.e.

$$\hat{\mathbf{E}}_\eta = \hat{\mathbf{E}}_1 + \hat{\mathbf{E}}_2 + 2i\left(\hat{\mathbf{E}}_3 + \hat{\mathbf{E}}_4\right) \tag{8.34}$$

where

$$\left.\begin{array}{ll}
\hat{\mathbf{E}}_1 = \begin{bmatrix} 1-\kappa_{ae_{zz}}-\left(\omega_r/\omega_z\right)^2 & 0 \\ -\kappa_{ae_{\theta z}} & 0 \end{bmatrix} & \hat{\mathbf{E}}_2 = \begin{bmatrix} 0 & -\kappa_{ae_{z\theta}} \\ 0 & 1-\kappa_{ae_{\theta\theta}}-\left(\omega_r/\omega_\theta\right)^2 \end{bmatrix} \\
\hat{\mathbf{E}}_3 = \begin{bmatrix} \left(\zeta_z-\zeta_{ae_{zz}}\right)\cdot\omega_r/\omega_z & 0 \\ -\zeta_{ae_{\theta z}}\cdot\omega_r/\omega_\theta & 0 \end{bmatrix} & \hat{\mathbf{E}}_4 = \begin{bmatrix} 0 & -\zeta_{ae_{z\theta}}\cdot\omega_r/\omega_z \\ 0 & \left(\zeta_\theta-\zeta_{ae_{\theta\theta}}\right)\cdot\omega_r/\omega_\theta \end{bmatrix}
\end{array}\right\} \tag{8.35}$$

The stability limit is then defined by the following two conditions

$$\mathrm{Re}\left(\det\left(\hat{\mathbf{E}}_\eta\right)\right)=\det\left(\hat{\mathbf{E}}_1+\hat{\mathbf{E}}_2\right)-4\cdot\det\left(\hat{\mathbf{E}}_3+\hat{\mathbf{E}}_4\right)=0 \tag{8.36}$$

$$\mathrm{Im}\left(\det\left(\hat{\mathbf{E}}_\eta\right)\right)=2\cdot\left[\det\left(\hat{\mathbf{E}}_1+\hat{\mathbf{E}}_4\right)+\det\left(\hat{\mathbf{E}}_2+\hat{\mathbf{E}}_3\right)\right]=0 \tag{8.37}$$

Fully expanded these equations become

$$\begin{aligned}
\mathrm{Re}\left(\det\left(\hat{\mathbf{E}}_\eta\right)\right) = {} & 1-\kappa_{ae_{zz}}-\kappa_{ae_{\theta\theta}}+\kappa_{ae_{zz}}\cdot\kappa_{ae_{\theta\theta}}-\kappa_{ae_{z\theta}}\cdot\kappa_{ae_{\theta z}}- \\
& 4\cdot\left[\left(\zeta_z-\zeta_{ae_{zz}}\right)\cdot\left(\zeta_\theta-\zeta_{ae_{\theta\theta}}\right)-\zeta_{ae_{z\theta}}\cdot\zeta_{ae_{\theta z}}\right]\cdot\left(\omega_r/\omega_z\right)\cdot\left(\omega_r/\omega_\theta\right)- \\
& \left(1-\kappa_{ae_{\theta\theta}}\right)\cdot\left(\omega_r/\omega_z\right)^2-\left(1-\kappa_{ae_{zz}}\right)\cdot\left(\omega_r/\omega_\theta\right)^2+\left(\omega_r/\omega_z\right)^2\cdot\left(\omega_r/\omega_\theta\right)^2 \\
= {} & 0
\end{aligned} \tag{8.38}$$

$$\begin{aligned}
\mathrm{Im}\left(\det\left(\hat{\mathbf{E}}_\eta\right)\right) = 2\cdot\Big\{ & \left[\left(1-\kappa_{ae_{\theta\theta}}\right)\cdot\left(\zeta_z-\zeta_{zz}\right)-\kappa_{ae_{\theta z}}\cdot\zeta_{ae_{z\theta}}\right]\cdot\omega_r/\omega_z+ \\
& \left[\left(1-\kappa_{ae_{zz}}\right)\cdot\left(\zeta_\theta-\zeta_{ae_{\theta\theta}}\right)-\kappa_{ae_{z\theta}}\cdot\zeta_{ae_{\theta z}}\right]\cdot\omega_r/\omega_\theta- \\
& \left(\zeta_\theta-\zeta_{ae_{\theta\theta}}\right)\cdot\left(\omega_r/\omega_\theta\right)\cdot\left(\omega_r/\omega_z\right)^2-\left(\zeta_z-\zeta_{ae_{zz}}\right)\cdot\left(\omega_r/\omega_z\right)\cdot\left(\omega_r/\omega_\theta\right)^2\Big\} \\
= 0 &
\end{aligned} \tag{8.39}$$

where (see Eqs. 8.12 – 8.15)

$$\kappa_{ae_{zz}}=\frac{\rho B^2}{2\tilde{m}_z}\left(\frac{\omega_r}{\omega_z}\right)^2 H_4^*\frac{\int_{L_{\exp}}\phi_z^2 dx}{\int_L \phi_z^2 dx} \qquad \kappa_{ae_{z\theta}}=\frac{\rho B^3}{2\tilde{m}_z}\left(\frac{\omega_r}{\omega_z}\right)^2 H_3^*\frac{\int_{L_{\exp}}\phi_z\phi_\theta dx}{\int_L \phi_z^2 dx}$$

$$\kappa_{ae_{\theta\theta}}=\frac{\rho B^4}{2\tilde{m}_\theta}\left(\frac{\omega_r}{\omega_\theta}\right)^2 A_3^*\frac{\int_{L_{\exp}}\phi_\theta^2 dx}{\int_L \phi_\theta^2 dx} \qquad \kappa_{ae_{\theta z}}=\frac{\rho B^3}{2\tilde{m}_\theta}\left(\frac{\omega_r}{\omega_\theta}\right)^2 A_4^*\frac{\int_{L_{\exp}}\phi_\theta\phi_z dx}{\int_L \phi_\theta^2 dx}$$

$$\zeta_{ae_{zz}} = \frac{\rho B^2}{4\tilde{m}_z}\frac{\omega_r}{\omega_z}H_1^* \frac{\int_{L_{\exp}} \phi_z^2 dx}{\int_L \phi_z^2 dx} \qquad \zeta_{ae_{z\theta}} = \frac{\rho B^3}{4\tilde{m}_z}\frac{\omega_r}{\omega_z}H_2^* \frac{\int_{L_{\exp}} \phi_z \phi_\theta dx}{\int_L \phi_z^2 dx}$$

$$\zeta_{ae_{\theta\theta}} = \frac{\rho B^4}{4\tilde{m}_\theta}\frac{\omega_r}{\omega_\theta}A_2^* \frac{\int_{L_{\exp}} \phi_\theta^2 dx}{\int_L \phi_\theta^2 dx} \qquad \zeta_{ae_{\theta z}} = \frac{\rho B^3}{4\tilde{m}_\theta}\frac{\omega_r}{\omega_\theta}A_1^* \frac{\int_{L_{\exp}} \phi_\theta \phi_z dx}{\int_L \phi_\theta^2 dx}$$

The solution procedure demands iterations, because the aerodynamic derivatives can only be read off if the outcome, ω_r and V_{cr}, are known. The theory of flutter was first presented by Theodorsen [28]. In cases where ω_θ / ω_z is larger than about 1.5, then Selberg's formula [22] may be used to provide a first estimate of the mean wind velocity that defines the flutter stability limit

$$V_{cr} = 0.6 B \omega_\theta \cdot \left\{ \left[1 - \left(\frac{\omega_z}{\omega_\theta} \right)^2 \right] \cdot \frac{\left(\tilde{m}_z \cdot \tilde{m}_\theta \right)^{1/2}}{\rho B^3} \right\}^{1/2} \tag{8.40}$$

Example 8.1

Let us consider a slender horizontal beam type of bridge with a cross section whose aerodynamic properties are close to those of an ideal flat plate, and set out to calculate the possible stability limits associated with the two mode shapes

$$\boldsymbol{\varphi}_1 = \begin{bmatrix} 0 & \phi_z & 0 \end{bmatrix}^T \quad \boldsymbol{\varphi}_2 = \begin{bmatrix} 0 & 0 & \phi_\theta \end{bmatrix}^T$$

with corresponding eigen-frequencies ω_z and ω_θ, and with modally equivalent and evenly distributed masses m_z and $\tilde{m}_\theta$. It is for simplicity assumed that $\phi_z \approx \phi_\theta$ and that $L_{\exp} = L$. Let us allot the following values to the necessary structural quantities

ρ (kg/m^3)	B (m)	$\tilde{m}_z$ (kg/m)	$\tilde{m}_\theta$ (kgm^2/m)	ω_z (rad/s)	$\zeta_z = \zeta_\theta$
1.25	20	10^4	$6 \cdot 10^5$	0.8	0.005

We wish to investigate the properties of the instability limits at various values of the frequency ratio ω_θ / ω_z, and thus it is assumed that $\omega_z = 0.8$ rad/s while ω_θ is arbitrary between

$\omega_\theta/\omega_z = 1$ and $\omega_\theta/\omega_z = 3$. To simplify the relevant expressions, let us introduce the following notation:

$$\beta_z = \frac{\rho B^2}{\tilde{m}_z} = 0.05 \qquad \beta_\theta = \frac{\rho B^4}{\tilde{m}_\theta} = 0.33 \quad \gamma = \frac{\omega_\theta}{\omega_z} \quad \text{and} \qquad \hat{\omega}_r = \frac{\omega_r}{\omega_\theta}$$

Due to the flat plate type of aerodynamic properties it is in this particular case only static divergence and flutter that may occur. The flat plate aerodynamic derivatives are given in Eq. 5.27 (and shown in Fig. 5.3), i.e.:

$$\begin{bmatrix} H_1^* & A_1^* \\ H_2^* & A_2^* \\ H_3^* & A_3^* \\ H_4^* & A_4^* \end{bmatrix} = \begin{bmatrix} -2\pi F\hat{V} & -\frac{\pi}{2}F\hat{V} \\ \frac{\pi}{2}\left(1+F+4G\hat{V}\right)\hat{V} & -\frac{\pi}{8}\left(1-F-4G\hat{V}\right)\hat{V} \\ 2\pi\left(F\hat{V}-G/4\right)\hat{V} & \frac{\pi}{2}\left(F\hat{V}-G/4\right)\hat{V} \\ \frac{\pi}{2}\left(1+4G\hat{V}\right) & \frac{\pi}{2}G\hat{V} \end{bmatrix}$$

where $\hat{V} = V/\left[B\omega_i(V)\right]$ is the reduced velocity, and where

$$F\left(\frac{\hat{\omega}_i}{2}\right) = \frac{J_1\cdot(J_1+Y_0)+Y_1\cdot(Y_1-J_0)}{(J_1+Y_0)^2+(Y_1-J_0)^2} \qquad \text{and} \qquad G\left(\frac{\hat{\omega}_i}{2}\right) = -\frac{J_1\cdot J_0+Y_1\cdot Y_0}{(J_1+Y_0)^2+(Y_1-J_0)^2}$$

are the real and imaginary parts of the so-called Theodorsen's circulatory function. $J_n\left(\hat{\omega}_i/2\right)$ and $Y_n\left(\hat{\omega}_i/2\right)$, $n = 0$ or 1, are first and second kind of Bessel functions with order n. $\hat{\omega}_i$ is the non-dimensional resonance frequency, i.e. $\hat{\omega}_i = B\omega_i(V)/V$. For an ideal flat plate type of cross section $C_M' = \pi/2$ (see quasi static solution given in Eq. 5.29). Thus, the stability limit with respect to static divergence is identified by

$$\frac{V_{cr}}{B\omega_\theta} = \left(\frac{2\tilde{m}_\theta}{\rho B^4}\cdot\frac{1}{C_M'}\right)^{1/2} \quad \Rightarrow \frac{V_{cr}}{B\omega_\theta} = \frac{2}{\sqrt{\beta_\theta\pi}} \approx 1.96$$

With respect to the flutter stability limit an approximate solution can be obtained from Eq. 8.40 (the Selberg formula), rendering

$$\frac{V_{cr}}{B\omega_\theta} = 0.6\cdot\left\{\left[1-\left(\frac{\omega_z}{\omega_\theta}\right)^2\right]\cdot\frac{\left(\tilde{m}_z\cdot\tilde{m}_\theta\right)^{1/2}}{\rho B^3}\right\}^{1/2} \approx 1.67\cdot\sqrt{1-\gamma^{-2}}$$

An exact solution can only be obtained from the simultaneous solution of Eqs. 8.38 and 8.39. Introducing the simplifications that $L_{\exp} = L$ and $\phi_z \approx \phi_\theta$ and the abbreviations for β_z, β_θ, γ and $\hat{\omega}_r$ defined above, then

$$\kappa_{ae_{zz}} = \frac{\rho B^2}{2\tilde{m}_z}\left(\frac{\omega_r}{\omega_z}\right)^2 H_4^* = \frac{\beta_z}{2} H_4^* \gamma^2 \hat{\omega}_r^2 \qquad \kappa_{ae_{z\theta}} = \frac{\rho B^3}{2\tilde{m}_z}\left(\frac{\omega_r}{\omega_z}\right)^2 H_3^* = \frac{\beta_z}{2} B H_3^* \gamma^2 \hat{\omega}_r^2$$

$$\kappa_{ae_{\theta\theta}} = \frac{\rho B^4}{2\tilde{m}_\theta}\left(\frac{\omega_r}{\omega_\theta}\right)^2 A_3^* = \frac{\beta_\theta}{2} A_3^* \hat{\omega}_r^2 \qquad \kappa_{ae_{\theta z}} = \frac{\rho B^3}{2\tilde{m}_\theta}\left(\frac{\omega_r}{\omega_\theta}\right)^2 A_4^* = \frac{\beta_\theta}{2}\frac{1}{B} A_4^* \hat{\omega}_2^2$$

$$\zeta_{ae_{zz}} = \frac{\rho B^2}{4\tilde{m}_z}\frac{\omega_r}{\omega_z} H_1^* = \frac{\beta_z}{4} H_1^* \gamma \hat{\omega}_r \qquad \zeta_{ae_{z\theta}} = \frac{\rho B^3}{4\tilde{m}_z}\frac{\omega_r}{\omega_z} H_2^* = \frac{\beta_z}{4} B H_2^* \gamma \hat{\omega}_r$$

$$\zeta_{ae_{\theta\theta}} = \frac{\rho B^4}{4\tilde{m}_\theta}\frac{\omega_r}{\omega_\theta} A_2^* = \frac{\beta_\theta}{4} A_2^* \hat{\omega}_r \qquad \zeta_{ae_{\theta z}} = \frac{\rho B^3}{4\tilde{m}_\theta}\frac{\omega_r}{\omega_\theta} A_1^* = \frac{\beta_\theta}{4}\frac{1}{B} A_1^* \hat{\omega}_r$$

Thus, Eqs. 8.29 and 8.30 are reduced to

$$\begin{aligned} \mathrm{Re}\left(\det\left(\hat{\mathbf{E}}_\eta\right)\right) = 1 - \left(1 + \gamma^2 + 4\gamma\zeta_z\zeta_\theta + \frac{\beta_z}{2}\gamma^2 H_4^* + \frac{\beta_\theta}{2} A_3^*\right)\hat{\omega}_r^2 + \gamma\left(\zeta_\theta\beta_z\gamma H_1^* + \zeta_z\beta_\theta A_2^*\right)\hat{\omega}_r^3 \\ + \gamma^2\left[1 + \frac{\beta_z}{2} H_4^* + \frac{\beta_\theta}{2} A_3^* + \frac{\beta_z\beta_\theta}{4}\left(A_1^* H_2^* - A_2^* H_1^* + A_3^* H_4^* - A_4^* H_3^*\right)\right]\hat{\omega}_r^4 = 0 \end{aligned}$$

$$\begin{aligned} \mathrm{Im}\left(\det\left(\hat{\mathbf{E}}_\eta\right)\right) = 2\hat{\omega}_r\left\{\zeta_z\gamma + \zeta_\theta - \frac{1}{4}\left(\beta_z\gamma^2 H_1^* + \beta_\theta A_2^*\right)\hat{\omega}_r\right. \\ - \left[\zeta_z\left(\frac{\beta_\theta}{2} A_3^* + \gamma\right) + \zeta_\theta\gamma^2\left(\frac{\beta_z}{2} H_4^* + 1\right)\right]\hat{\omega}_r^2 \\ \left. + \gamma^2\left[\frac{\beta_z\beta_\theta}{8}\left(H_1^* A_3^* - H_2^* A_4^* - H_3^* A_1^* + H_4^* A_2^*\right) + \frac{1}{4}\left(\beta_z H_1^* + \beta_\theta A_2^*\right)\right]\hat{\omega}_r^3\right\} = 0 \end{aligned}$$

It is seen that the solution of these equations requires the search for the lowest identical roots in a fourth and a third degree polynomial. Adopting ideal flat plate aerodynamic derivatives (see expressions above) the solution is shown in the upper diagram in Fig. 8.2 (together with the approximate solution given by Selberg's formula). The corresponding values of $\hat{\omega}_r$ are shown in the lower diagram in Fig. 8.2. At $\omega_\theta / \omega_z = 2$ the development of $\mathrm{Im}\left(\det\left(\hat{\mathbf{E}}_\eta\right)\right)$ and $\mathrm{Re}\left(\det\left(\hat{\mathbf{E}}_\eta\right)\right)$ with increasing values of $V/(B\omega_\theta)$ is shown in Fig. 8.1.

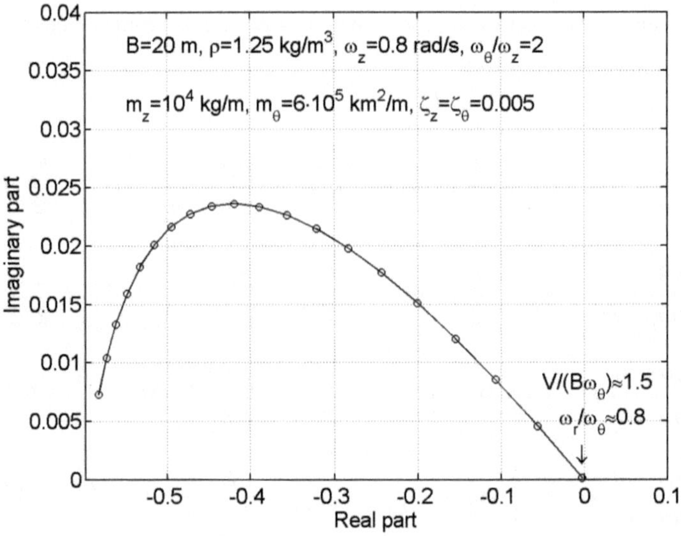

Fig. 8.1 *Development of imaginary and real parts at increasing values of* $V/(B\omega_\theta)$

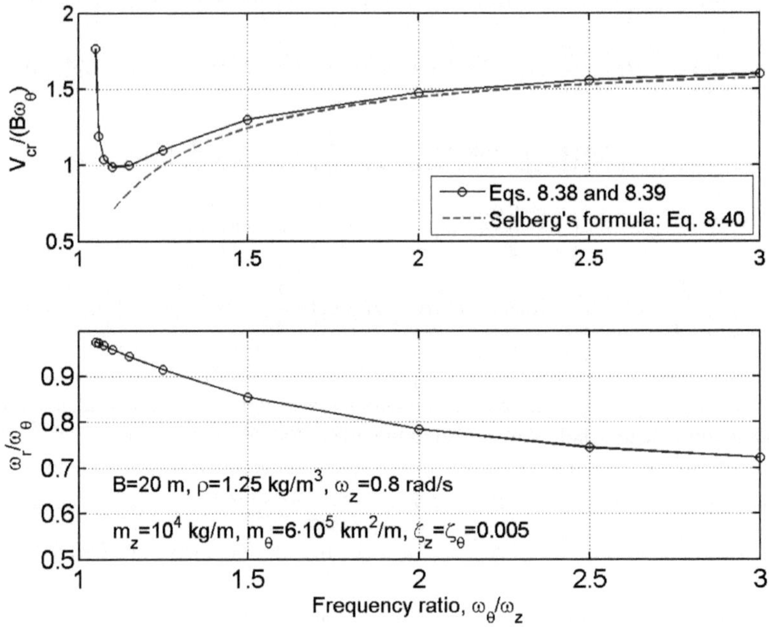

Fig. 8.2 *Upper diagram: flutter stability limit, lower diagram: corresponding frequency of resonant motion*

Chapter 9

THE BUFFETING THEORY IN A FINITE ELEMENT FORMAT

9.1 Introduction

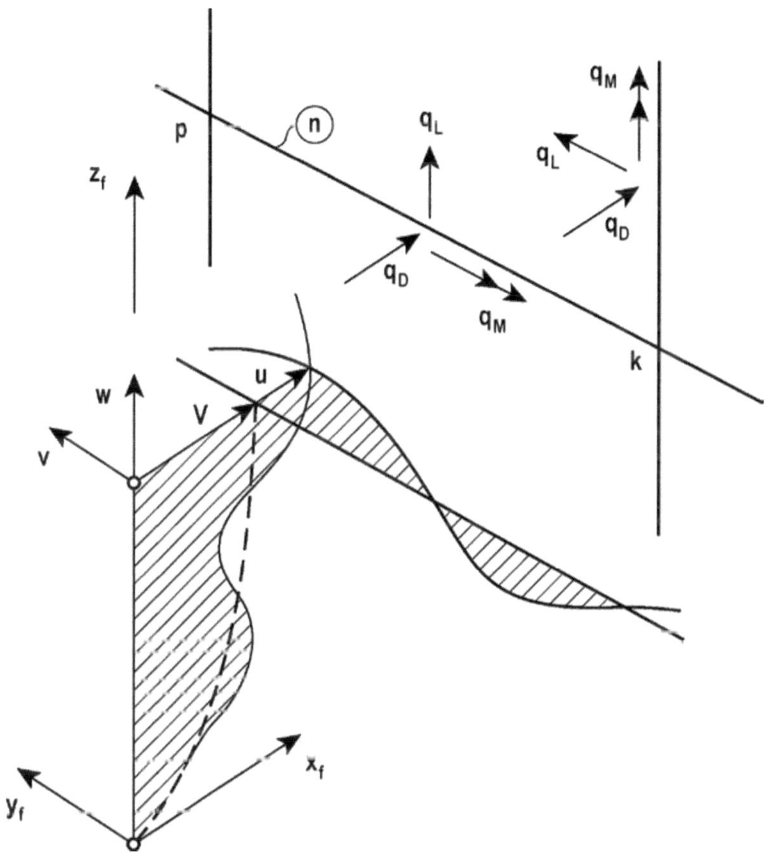

Fig. 9.1 *A line-like structure in a turbulent wind field*

In order to focus on the most important aspects of wind induced dynamic response the calculations presented in Chapter 6 (and the ensuing calculations of cross sectional forces presented in Chapter 7) have been performed in a modal format where each displacement component has been given a separate representation. These calculations may also be performed in a finite element format. Such a procedure is presented below. It has the advantage that it will comply with the computational methods usually applied elsewhere in structural mechanics. Thus, from a computer programming point of view, all the well known stiffness and mass properties from other types of structural dynamics problem will be applicable. The only difference is that the wind and motion induced loads need special attention. On the other hand, it should be noted that due to the fairly short correlation lengths and sharply dropping coherence properties of the turbulence components there will be demanding requirements for the choice of largest element length, i.e., the number of degrees of freedom in the finite element system may become cumbersome from a computational point of view. The same applies to the choice of time stepping increment in a time domain solution. In general, convergence should be checked. The overall problem of a structural system of line like members in a turbulent wind field, defined by the wind velocity components $V\left(z_f\right)+u\left(y_f,z_f,t\right)$, $v\left(y_f,z_f,t\right)$ and $w\left(y_f,z_f,t\right)$, is illustrated in Fig. 9.1. At an arbitrary position on an element n (between nodes p and k) the wind field and the interaction between flow and structural motion will generate three load components, one in the direction of drag, one in the across wind direction (vertical or horizontal depending on the orientation of the element in relation to the flow) and one torsion (pitching) moment. Adopting a system of six degrees of freedom in each node, there is a load vector $\left[R_1 \quad R_2 \quad R_3 \quad R_4 \quad R_5 \quad R_6\right]^T$ and a corresponding displacement vector $\left[r_1 \quad r_2 \quad r_3 \quad r_4 \quad r_5 \quad r_6\right]^T$ in each node, as shown in Fig. 9.2. It is taken for granted that the global axis X, Y and Z coincide with the flow axis $-y_f$, x_f and z_f, i.e. that the structural system is two-dimensional and perpendicular to the main wind flow direction. Unfortunately, a two-dimensional system is at the moment a necessary restriction as the experimental support of the wind load on an element at an arbitrary attitude in the flow is insufficient. Strictly speaking, the theory below is only sufficiently supported by experimental data if the elements of the system are either horizontal or vertical. Nonetheless, the possibility of a yaw angle has been included below. It should be noted that experimental data from structural aerodynamics usually complies with the force and displacement definitions given in Fig. 1.3, where pitching moment and cross sectional rotation are defined by windward edge up. It is in the following assumed that this definition applies to all the aerodynamic data (e.g. load coefficients, aerodynamic derivatives, etc.) that are adopted for numerical calculations. However, in the finite element theory it has been chosen to strictly comply with the usual convention that all external and internal forces and displacement degrees of freedom are vectors in global as well as local coordinates (see Figs. 9.2 and 9.3).

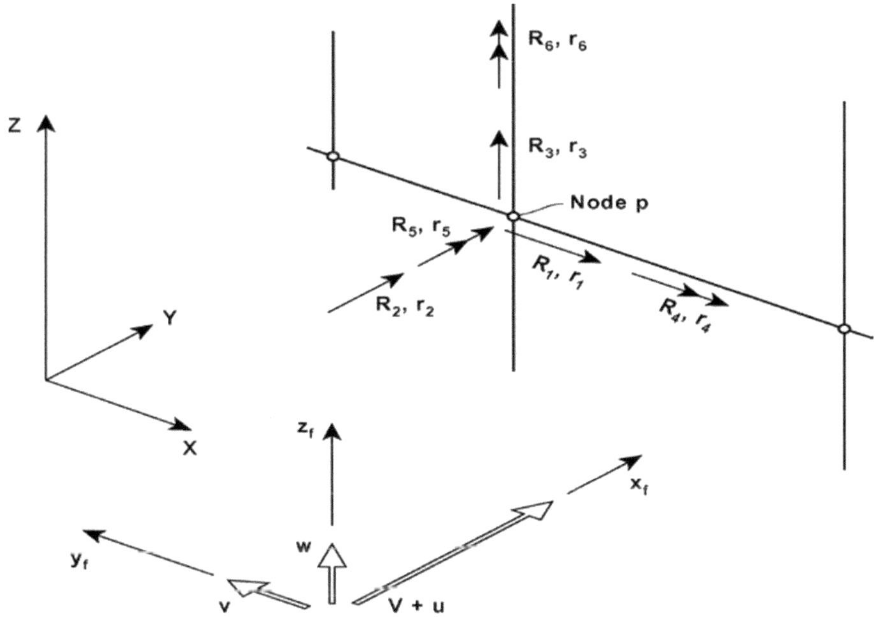

Fig. 9.2 *Nodal forces from external loads*

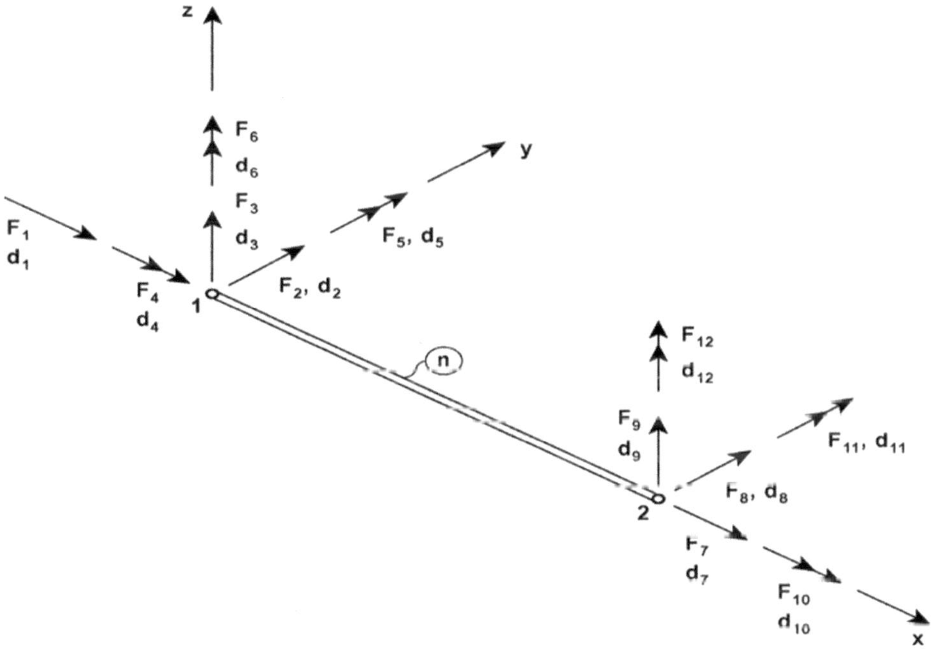

Fig. 9.3 *Element degrees of freedom and element end forces*

9.2 The element mechanical properties

A free body diagram of an arbitrary beam (line-like) type element n, with local axis x, y and z is illustrated in Fig. 9.3. It is taken for granted that all displacements as well as forces comprise a time-invariant mean value (the static part) and a stationary fluctuating (dynamic) part. At position x along its span the cross sectional displacements and rotation (torsion) are defined by

$$\mathbf{r}_{el_{tot}}(x,t)=\begin{bmatrix} r_x & r_y & r_z & r_\theta \end{bmatrix}_{el_{tot}}^T=\bar{\mathbf{r}}_{el}(x)+\mathbf{r}_{el}(x,t) \tag{9.1}$$

where index el indicates quantities at element level. At ends 1 and 2 it has the element nodal forces

$$\mathbf{F}_{tot}(t)=\begin{bmatrix}\mathbf{F}_1\\ \mathbf{F}_2\end{bmatrix}_{tot}=\bar{\mathbf{F}}+\mathbf{F}(t) \text{ where } \begin{cases}\mathbf{F}_{1_{tot}}=\begin{bmatrix}F_1 & F_2 & F_3 & F_4 & F_5 & F_6\end{bmatrix}_{tot}^T\\ \mathbf{F}_{2_{tot}}=\begin{bmatrix}F_7 & F_8 & F_9 & F_{10} & F_{11} & F_{12}\end{bmatrix}_{tot}^T\end{cases} \tag{9.2}$$

and corresponding local displacements

$$\mathbf{d}_{tot}(t)=\begin{bmatrix}\mathbf{d}_1\\ \mathbf{d}_2\end{bmatrix}_{tot}=\bar{\mathbf{d}}+\mathbf{d}(t) \text{ where } \begin{cases}\mathbf{d}_{1_{tot}}=\begin{bmatrix}d_1 & d_2 & d_3 & d_4 & d_5 & d_6\end{bmatrix}_{tot}^T\\ \mathbf{d}_{2_{tot}}=\begin{bmatrix}d_7 & d_8 & d_9 & d_{10} & d_{11} & d_{12}\end{bmatrix}_{tot}^T\end{cases} \tag{9.3}$$

It is assumed that the cross sectional displacement vector $\mathbf{r}_{el_{tot}}(x,t)$ with sufficient accuracy may be described by the product of a shape function matrix

$$\mathbf{N}(x)=\begin{bmatrix} N_1 & 0 & 0 & 0 & 0 & 0 & N_7 & 0 & 0 & 0 & 0 & 0\\ 0 & N_2 & 0 & 0 & 0 & N_6 & 0 & N_8 & 0 & 0 & 0 & N_{12}\\ 0 & 0 & N_3 & 0 & N_5 & 0 & 0 & 0 & N_9 & 0 & N_{11} & 0\\ 0 & 0 & 0 & N_4 & 0 & 0 & 0 & 0 & 0 & N_{10} & 0 & 0 \end{bmatrix} \tag{9.4}$$

and the nodal displacement vector $\mathbf{d}_{tot}(t)$, i.e. that

$$\mathbf{r}_{el_{tot}}(x,t)=\mathbf{N}(x)\cdot\mathbf{d}_{tot}(t) \tag{9.5}$$

where the twelve shape function N_i, $i=1-12$, are given in Fig. 9.4. These are identical to the shape functions commonly used elsewhere in structural mechanics. Since they are polynomial, it should be noted that they will represent an accurate solution to the time-invariant (static) part of the response but not to the dynamic part, as they will not fully satisfy the spanwise differential equation of motion.

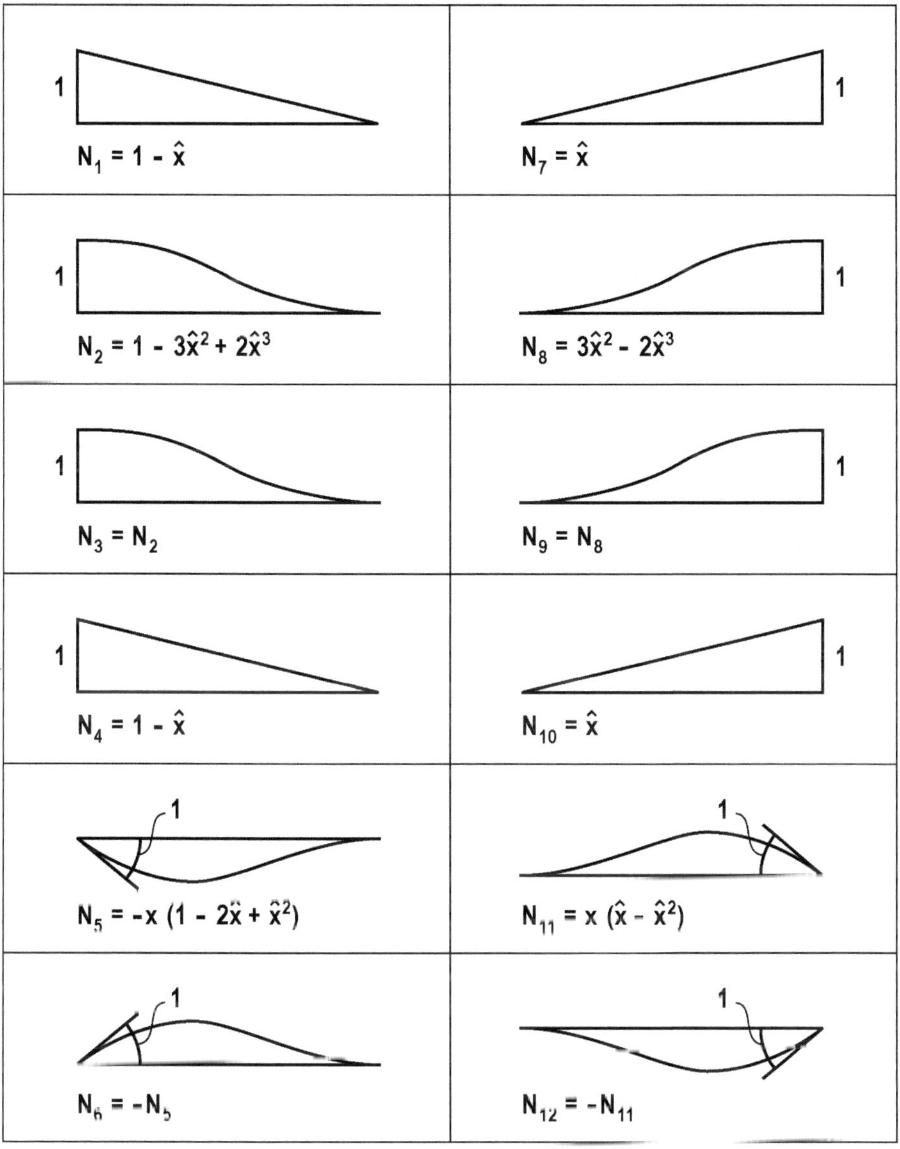

Fig. 9.4 *Shape functions* N_i, $i = 1-12$

Applying d'Alambert's principle at a position of external and internal equilibrium defined by $\mathbf{r}_{el_{tot}}(x,t)$, and let the system be subject to an incremental virtual displacement

$$\delta\mathbf{r}_{el} = \begin{bmatrix} \delta r_x & \delta r_y & \delta r_z & \delta r_\theta \end{bmatrix}^T \tag{9.6}$$

compatible with

$$\delta\mathbf{d} = \begin{bmatrix} \delta\mathbf{d}_1 \\ \delta\mathbf{d}_2 \end{bmatrix} \text{ where } \begin{cases} \delta\mathbf{d}_1 = [\delta d_1 \quad \delta d_2 \quad \delta d_3 \quad \delta d_4 \quad \delta d_5 \quad \delta d_6]^T \\ \delta\mathbf{d}_2 = [\delta d_7 \quad \delta d_8 \quad \delta d_9 \quad \delta d_{10} \quad \delta d_{11} \quad \delta d_{12}]^T \end{cases} \tag{9.7}$$

such that

$$\delta\mathbf{r}_{el} = \mathbf{N}(x) \cdot \delta\mathbf{d} \tag{9.8}$$

Then the external and internal works performed during this motion are given by:

$$W_{ext} = \delta\mathbf{d}^T \cdot \mathbf{F}_{tot} + \int_0^L \delta\mathbf{r}_{el}^T \cdot (-\mathbf{m}_0 \ddot{\mathbf{r}}_{el})dx \tag{9.9}$$

and

$$W_{\text{int}} = \int_0^L \int_A \delta\boldsymbol{\varepsilon}^T \cdot \mathbf{s} \cdot (dxdA) + \int_0^L \delta\mathbf{r}^T \cdot (\mathbf{c}_0 \dot{\mathbf{r}}_{el})dx \tag{9.10}$$

where (assuming shear centre axis)

$$\left.\begin{aligned} \mathbf{m}_0 &= diag\begin{bmatrix} m_x & m_y & m_z & m_\theta \end{bmatrix} \\ \mathbf{c}_0 &= diag\begin{bmatrix} c_x & c_y & c_z & c_\theta \end{bmatrix} \end{aligned}\right\} \tag{9.11}$$

are diagonal matrices containing the distributed mass and damping properties of the element, and where

$$\left.\begin{aligned} \mathbf{s} &= \begin{bmatrix} s_x & s_y & s_z & s_{yz} \end{bmatrix}_{el_{tot}}^T \\ \delta\boldsymbol{\varepsilon} &= \begin{bmatrix} \delta\varepsilon_x & \delta\varepsilon_y & \delta\varepsilon_z & \delta\gamma_\theta \end{bmatrix}^T \end{aligned}\right\} \tag{9.12}$$

s_x, s_y and s_z are cross sectional stress contributions from elastic beam stretching in the x direction and bending in the y and z directions, while s_{yz} is the cross sectional shear stress due to torsion. $\delta\varepsilon_x$, $\delta\varepsilon_y$, $\delta\varepsilon_z$ and $\delta\gamma_\theta$ are the corresponding virtual strain quantities. From basic theory of elasticity (Hook's law and Navier's hypothesis, see e.g. Chen & Atsuta [27]) $\mathbf{s}$ and $\delta\boldsymbol{\varepsilon}$ are given by

$$\left.\begin{aligned} \mathbf{s} &= \begin{bmatrix} Er'_x & -Eyr''_y & -Ezr''_z & Gr_p r'_\theta \end{bmatrix}^T_{el_{tot}} \\ \delta\boldsymbol{\varepsilon} &= \begin{bmatrix} \delta r'_x & -y\delta r''_y & -z\delta r''_z & r_p r'_\theta \end{bmatrix}^T \end{aligned}\right\} \tag{9.13}$$

where primes indicate derivation with respect to x and where E is the modulus of elasticity, G is the corresponding shear modulus and r_p is a cross sectional coordinate used to identify the St Venant torsion constant (r_p should be perceived as a symbolic representative as it strictly spoken is only applicable to a circular cross section). Defining

$$\hat{\mathbf{N}}(x) = \begin{bmatrix} N'_1 & 0 & 0 & 0 & 0 & 0 & N'_7 & 0 & 0 & 0 & 0 & 0 \\ 0 & N''_2 & 0 & 0 & 0 & N''_6 & 0 & N''_8 & 0 & 0 & 0 & N''_{12} \\ 0 & 0 & N''_3 & 0 & N''_5 & 0 & 0 & 0 & N''_9 & 0 & N''_{11} & 0 \\ 0 & 0 & 0 & N'_4 & 0 & 0 & 0 & 0 & 0 & N'_{10} & 0 & 0 \end{bmatrix} \tag{9.14}$$

and

$$\left.\begin{aligned} \mathbf{e} &= diag\begin{bmatrix} E & E & E & G \end{bmatrix} \\ \mathbf{f} &= \begin{bmatrix} 1 & -y & -z & r_p \end{bmatrix} \end{aligned}\right\} \tag{9.15}$$

then the internal work is given by

$$\begin{aligned} W_{\text{int}} &= \int_0^L \int_A \left(\mathbf{f}\cdot\hat{\mathbf{N}}\cdot\delta\mathbf{d}\right)^T \cdot \left(\mathbf{f}\cdot\mathbf{e}\cdot\hat{\mathbf{N}}\cdot\mathbf{d}_{tot}\right)\cdot dA\cdot dx + \int_0^L \left(\mathbf{N}\cdot\delta\mathbf{d}\right)^T\cdot\left(\mathbf{c}_0\cdot\mathbf{N}\cdot\dot{\mathbf{d}}\right)\cdot dx \\ &= \delta\mathbf{d}^T\cdot\int_0^L \hat{\mathbf{N}}^T\cdot\left(\int_A \mathbf{f}^T\cdot\mathbf{f}\cdot dA\cdot\mathbf{e}\right)\cdot\hat{\mathbf{N}}\cdot dx\cdot\mathbf{d}_{tot} + \delta\mathbf{d}^T\cdot\int_0^L \mathbf{N}^T\cdot\mathbf{c}_0\mathbf{N}\cdot dx\cdot\dot{\mathbf{d}} \end{aligned} \tag{9.16}$$

Defining $A = \int_A dA$, $I_z = \int_A y^2 dA$, $I_y = \int_A z^2 dA$, $I_t = \int_A r_p^2 dA$ and

$$\mathbf{k}_0 = \int_A \mathbf{f}^T\cdot\mathbf{f}\cdot dA\cdot\mathbf{e} = diag\begin{bmatrix} EA & EI_z & EI_y & GI_t \end{bmatrix} \tag{9.17}$$

then the internal work is given by

$$W_{\text{int}} = \delta\mathbf{d}^T\cdot\left(\mathbf{k}\cdot\mathbf{d}_{tot} + \mathbf{c}\cdot\dot{\mathbf{d}}\right) \tag{9.18}$$

where $\mathbf{k}$ and $\mathbf{c}$ are the element stiffness and damping matrices, defined by

$$\left.\begin{aligned}\mathbf{k} &= \int_0^L \hat{\mathbf{N}}^T \mathbf{k}_0 \hat{\mathbf{N}} dx \\ \mathbf{c} &= \int_0^L \mathbf{N}^T \mathbf{c}_0 \mathbf{N} dx\end{aligned}\right\} \tag{9.19}$$

By introducing $\delta\mathbf{r} = \mathbf{N} \cdot \delta\mathbf{d}$ and $\ddot{\mathbf{r}}_{el}(x,t) = \mathbf{N}(x) \cdot \ddot{\mathbf{d}}(t)$ then the external work (see Eq. 9.9) is given by:

$$W_{ext} = \delta\mathbf{d}^T \cdot \mathbf{F}_{tot} - \int_0^L (\mathbf{N} \cdot \delta\mathbf{d})^T \mathbf{m}_0 \left(\mathbf{N} \cdot \ddot{\mathbf{d}}\right) dx = \delta\mathbf{d}^T \cdot \left(\mathbf{F}_{tot} - \int_0^L \mathbf{N}^T \mathbf{m}_0 \mathbf{N} dx \cdot \ddot{\mathbf{d}}\right) \tag{9.20}$$

Thus, introducing the element mass matrix

$$\mathbf{m} = \int_0^L \mathbf{N}^T \mathbf{m}_0 \mathbf{N} dx \tag{9.21}$$

then the external work is given by

$$W_{ext} = \delta\mathbf{d}^T \cdot \left(\mathbf{F}_{tot} - \mathbf{m} \cdot \ddot{\mathbf{d}}\right) \tag{9.22}$$

By setting $W_{\text{int}} = W_{ext}$

$$\delta\mathbf{d}^T \cdot \left(\mathbf{k} \cdot \mathbf{d}_{tot} + \mathbf{c} \cdot \dot{\mathbf{d}}\right) = \delta\mathbf{d}^T \cdot \left(\mathbf{F}_{tot} - \mathbf{m} \cdot \ddot{\mathbf{d}}\right) \tag{9.23}$$

cancelling out $\delta\mathbf{d}$ and rearranging, then the following element equilibrium condition is obtained:

$$\mathbf{m}\ddot{\mathbf{d}} + \mathbf{c}\dot{\mathbf{d}} + \mathbf{k}\mathbf{d}_{tot} = \mathbf{F}_{tot} \tag{9.24}$$

Since $\mathbf{d}_{tot} = \bar{\mathbf{d}} + \mathbf{d}(t)$ and $\mathbf{F}_{tot} = \bar{\mathbf{F}} + \mathbf{F}(t)$ this equation may be split into a mean (static) part and a fluctuating (dynamic) part:

$$\left.\begin{aligned}\mathbf{k}\bar{\mathbf{d}} &= \bar{\mathbf{F}} \\ \mathbf{m}\ddot{\mathbf{d}} + \mathbf{c}\dot{\mathbf{d}} + \mathbf{k}\mathbf{d} &= \mathbf{F}\end{aligned}\right\} \tag{9.25}$$

This is the static and dynamic equilibrium conditions at element level. The element stiffness, damping and mass matrices are defined in Eqs. 9.19 and 9.21. By introducing the shape functions in Fig. 9.4 into Eqs. 9.4 and 9.14, then the following is obtained:

$$\mathbf{k}=\begin{bmatrix}\mathbf{k}_{11} & \mathbf{k}_{12}\\ \mathbf{k}_{21} & \mathbf{k}_{22}\end{bmatrix} \qquad \text{where} \qquad \mathbf{k}_{21}=\mathbf{k}_{12}^T \tag{9.26}$$

and

$$\mathbf{k}_{11}=\int_0^L \begin{bmatrix} EAN_1'^2 & 0 & 0 & 0 & 0 & 0\\ 0 & EI_zN_2''^2 & 0 & 0 & 0 & EI_zN_2''N_6''\\ 0 & 0 & EI_yN_3''^2 & 0 & EI_yN_3''N_5'' & 0\\ 0 & 0 & 0 & GI_tN_4'^2 & 0 & 0\\ 0 & 0 & EI_yN_5''N_3'' & 0 & EI_yN_5''^2 & 0\\ 0 & EI_zN_6''N_2'' & 0 & 0 & 0 & EI_zN_6''^2 \end{bmatrix} dx$$

$$= \begin{bmatrix} EA/L & 0 & 0 & 0 & 0 & 0\\ 0 & 12EI_z/L^3 & 0 & 0 & 0 & 6EI_z/L^2\\ 0 & 0 & 12EI_y/L^3 & 0 & -6EI_y/L^2 & 0\\ 0 & 0 & 0 & GI_t/L & 0 & 0\\ 0 & 0 & -6EI_y/L^2{}_3 & 0 & 4EI_y/L & 0\\ 0 & 6EI_z/L^2 & 0 & 0 & 0 & 4EI_z/L \end{bmatrix} \tag{9.27}$$

$$\mathbf{k}_{12}=\int_0^L \begin{bmatrix} EAN_1'N_7' & 0 & 0 & 0 & 0 & 0\\ 0 & EI_zN_2''N_8'' & 0 & 0 & 0 & EI_zN_2''N_{12}''\\ 0 & 0 & EI_yN_3''N_9'' & 0 & EI_yN_3''N_{11}'' & 0\\ 0 & 0 & 0 & GI_tN_4'N_{10}' & 0 & 0\\ 0 & 0 & EI_yN_5''N_9'' & 0 & EI_yN_5''N_{11}'' & 0\\ 0 & EI_zN_6''N_8'' & 0 & 0 & 0 & EI_zN_6''N_{12}'' \end{bmatrix} dx$$

$$= \begin{bmatrix} -EA/L & 0 & 0 & 0 & 0 & 0\\ 0 & -12EI_z/L^3 & 0 & 0 & 0 & 6EI_z/L^2\\ 0 & 0 & 12EI_y/L^3 & 0 & -6EI_y/L^2 & 0\\ 0 & 0 & 0 & -GI_t/L & 0 & 0\\ 0 & 0 & 6EI_y/L^2 & 0 & 2EI_y/L & 0\\ 0 & -6EI_z/L^2 & 0 & 0 & 0 & 2EI_z/L \end{bmatrix} \tag{9.28}$$

$$
\mathbf{k}_{22} = \int_0^L \begin{bmatrix} EAN_7'^2 & 0 & 0 & 0 & 0 & 0 \\ 0 & EI_z N_8''^2 & 0 & 0 & 0 & EI_z N_8'' N_{12}'' \\ 0 & 0 & EI_y N_9''^2 & 0 & EI_y N_9'' N_{11}'' & 0 \\ 0 & 0 & 0 & GI_t N_{10}'^2 & 0 & 0 \\ 0 & 0 & EI_y N_{11}'' N_9'' & 0 & EI_y N_{11}''^2 & 0 \\ 0 & EI_z N_{12}'' N_8'' & 0 & 0 & 0 & EI_z N_{12}''^2 \end{bmatrix} dx
$$

$$
= \begin{bmatrix} EA/L & 0 & 0 & 0 & 0 & 0 \\ 0 & 12EI_z/L^3 & 0 & 0 & 0 & -6EI_z/L^2 \\ 0 & 0 & 12EI_y/L^3 & 0 & 6EI_y/L^2 & 0 \\ 0 & 0 & 0 & GI_t/L & 0 & 0 \\ 0 & 0 & 6EI_y/L^2{}_3 & 0 & 4EI_y/L & 0 \\ 0 & -6EI_z/L^2 & 0 & 0 & 0 & 4EI_z/L \end{bmatrix} \tag{9.28}
$$

The element damping matrix is given by:

$$
\mathbf{c} = \begin{bmatrix} \mathbf{c}_{11} & \mathbf{c}_{12} \\ \mathbf{c}_{21} & \mathbf{c}_{22} \end{bmatrix} \quad \text{where} \quad \mathbf{c}_{21} = \mathbf{c}_{12}^T \tag{9.29}
$$

and

$$
\mathbf{c}_{11} = \int_0^L \begin{bmatrix} c_x N_1^2 & 0 & 0 & 0 & 0 & 0 \\ 0 & c_y N_2^2 & 0 & 0 & 0 & c_y N_2 N_6 \\ 0 & 0 & c_z N_3^2 & 0 & c_z N_3 N_5 & 0 \\ 0 & 0 & 0 & c_\theta N_4^2 & 0 & 0 \\ 0 & 0 & c_z N_5 N_3 & 0 & c_z N_5^2 & 0 \\ 0 & c_y N_6 N_2 & 0 & 0 & 0 & c_y N_6^2 \end{bmatrix} dx
$$

$$
= \frac{L}{420} \begin{bmatrix} 140c_x & 0 & 0 & 0 & 0 & 0 \\ 0 & 156c_y & 0 & 0 & 0 & 22c_y L \\ 0 & 0 & 156c_z & 0 & -22c_z L & 0 \\ 0 & 0 & 0 & 140c_\theta & 0 & 0 \\ 0 & 0 & -22c_z L & 0 & 4c_z L^2 & 0 \\ 0 & 22c_y L & 0 & 0 & 0 & 4c_y L^2 \end{bmatrix} \tag{9.30}
$$

$$
\mathbf{c}_{12} = \int_0^L \begin{bmatrix} c_x N_1 N_7 & 0 & 0 & 0 & 0 & 0 \\ 0 & c_y N_2 N_8 & 0 & 0 & 0 & c_y N_2 N_{12} \\ 0 & 0 & c_z N_3 N_9 & 0 & c_z N_3 N_{11} & 0 \\ 0 & 0 & 0 & c_\theta N_4 N_{10} & 0 & 0 \\ 0 & 0 & c_z N_5 N_9 & 0 & c_z N_5 N_{11} & 0 \\ 0 & c_y N_6 N_8 & 0 & 0 & 0 & c_y N_6 N_{12} \end{bmatrix} dx
$$

$$
= \frac{L}{420} \begin{bmatrix} 70c_x & 0 & 0 & 0 & 0 & 0 \\ 0 & 54c_y & 0 & 0 & 0 & -13c_y L \\ 0 & 0 & 54c_z & 0 & 13c_z L & 0 \\ 0 & 0 & 0 & 70c_\theta & 0 & 0 \\ 0 & 0 & -13c_z L & 0 & -3c_z L^2 & 0 \\ 0 & 13c_y L & 0 & 0 & 0 & -3c_y L^2 \end{bmatrix} \tag{9.31}
$$

$$
\mathbf{c}_{22} = \int_0^L \begin{bmatrix} c_x N_7^2 & 0 & 0 & 0 & 0 & 0 \\ 0 & c_y N_8^2 & 0 & 0 & 0 & c_y N_8 N_{12} \\ 0 & 0 & c_z N_9^2 & 0 & c_z N_9 N_{11} & 0 \\ 0 & 0 & 0 & c_\theta N_{10}^2 & 0 & 0 \\ 0 & 0 & c_z N_{11} N_9 & 0 & c_z N_{11}^2 & 0 \\ 0 & c_y N_{12} N_8 & 0 & 0 & 0 & c_y N_{12}^2 \end{bmatrix} dx
$$

$$
= \frac{L}{420} \begin{bmatrix} 140c_x & 0 & 0 & 0 & 0 & 0 \\ 0 & 156c_y & 0 & 0 & 0 & -22c_y L \\ 0 & 0 & 156c_z & 0 & 22c_z L & 0 \\ 0 & 0 & 0 & 140c_\theta & 0 & 0 \\ 0 & 0 & 22c_z L & 0 & 4c_z L^2 & 0 \\ 0 & -22c_y L & 0 & 0 & 0 & 4c_y L^2 \end{bmatrix} \tag{9.32}
$$

Similarly, the element mass matrix is given by:

$$
\mathbf{m} = \begin{bmatrix} \mathbf{m}_{11} & \mathbf{m}_{12} \\ \mathbf{m}_{21} & \mathbf{m}_{22} \end{bmatrix} \qquad \text{where} \quad \mathbf{m}_{21} = \mathbf{m}_{12}^T \tag{9.33}
$$

The necessary integrations will involve the same combinations of shape functions as those that have been applied to obtain the damping matrix, and thus

$$\mathbf{m}_{11} = \frac{L}{420}\begin{bmatrix} 140m_x & 0 & 0 & 0 & 0 & 0 \\ 0 & 156m_y & 0 & 0 & 0 & 22m_yL \\ 0 & 0 & 156m_z & 0 & -22m_zL & 0 \\ 0 & 0 & 0 & 140m_\theta & 0 & 0 \\ 0 & 0 & -22m_zL & 0 & 4m_zL^2 & 0 \\ 0 & 22m_yL & 0 & 0 & 0 & 4m_yL^2 \end{bmatrix} \tag{9.34}$$

$$\mathbf{m}_{12} = \frac{L}{420}\begin{bmatrix} 70m_x & 0 & 0 & 0 & 0 & 0 \\ 0 & 54m_y & 0 & 0 & 0 & -13m_yL \\ 0 & 0 & 54m_z & 0 & 13m_zL & 0 \\ 0 & 0 & 0 & 70m_\theta & 0 & 0 \\ 0 & 0 & -13m_zL & 0 & -3m_zL^2 & 0 \\ 0 & 13m_yL & 0 & 0 & 0 & -3m_yL^2 \end{bmatrix} \tag{9.35}$$

$$\mathbf{m}_{22} = \frac{L}{420}\begin{bmatrix} 140m_x & 0 & 0 & 0 & 0 & 0 \\ 0 & 156m_y & 0 & 0 & 0 & -22m_yL \\ 0 & 0 & 156m_z & 0 & 22m_zL & 0 \\ 0 & 0 & 0 & 140m_\theta & 0 & 0 \\ 0 & 0 & 22m_zL & 0 & 4m_zL^2 & 0 \\ 0 & -22m_yL & 0 & 0 & 0 & 4m_yL^2 \end{bmatrix} \tag{9.36}$$

The element property matrices given above may be found in many text books, see e.g. Hughes [25] and Cook et.al. [29]. They have been included here mainly for the sake of completeness. It should be noted that the development of damping properties at element level is not necessarily a rational choice. Alternatively, damping properties may be introduced at a structural global level (i.e. associated directly with the global degrees of freedom), e.g. in the form of Rayleigh damping or simply a diagonal type of modal damping matrix.

9.3 The wind load

All the necessary equations for the determination of the mean (static) as well as the fluctuation (dynamic) wind load at element level has previously been developed in Chapter 5. It is now simply a matter of implementing this theory into the framework of a finite element approach, i.e. to expand Eqs. 5.8 – 5.14 and 5.24 – 5.25 into the twelve degrees of freedom force and displacement system adopted for a finite element approach.

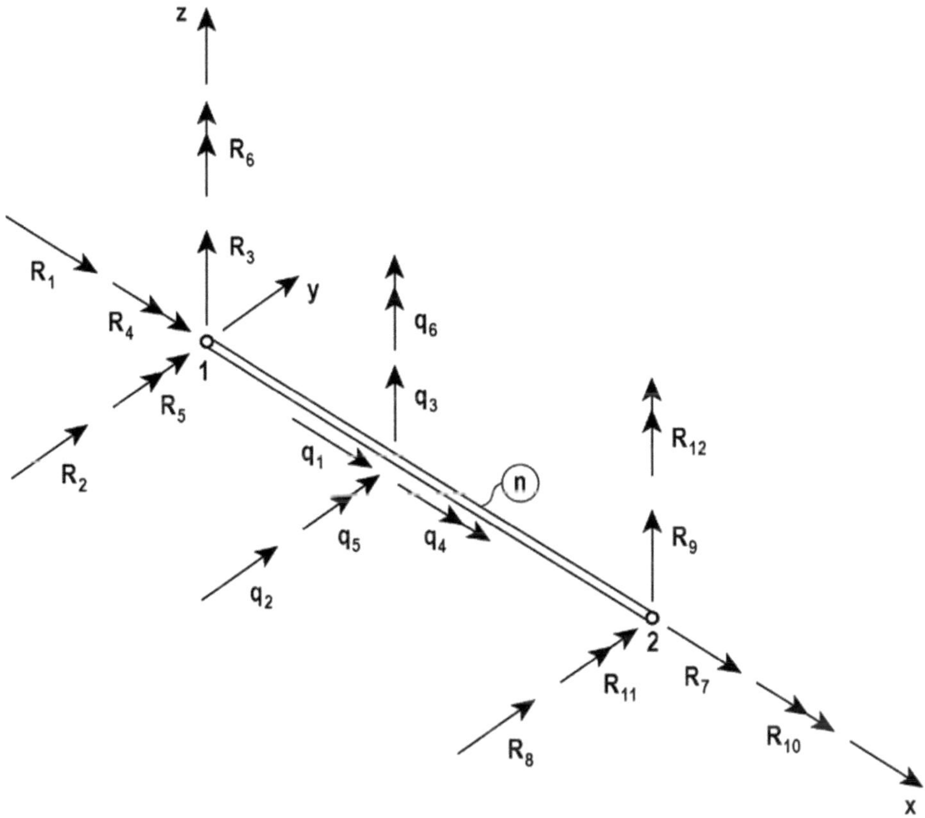

Fig. 9.5 *External load and load vector at element level*

At a global level the load vector is illustrated in Fig. 9.2. At element level (i.e. associated with an arbitrary element n) the load vector is illustrated on Fig. 9.5. It is defined by

$$\mathbf{R}_{n_{tot}}(t)=\begin{bmatrix}\mathbf{R}_1\\ \mathbf{R}_2\end{bmatrix}_{n_{tot}} \quad \text{where} \quad \begin{cases}\mathbf{R}_{1_{n_{tot}}}=\begin{bmatrix}R_1 & R_2 & R_3 & R_4 & R_5 & R_6\end{bmatrix}_{n_{tot}}^T\\ \mathbf{R}_{2_{n_{tot}}}=\begin{bmatrix}R_7 & R_8 & R_9 & R_{10} & R_{11} & R_{12}\end{bmatrix}_{n_{tot}}^T\end{cases} \tag{9.37}$$

These are external nodal forces (in element axis) caused by the distributed element load vector (see Fig. 9.5 and Eq. 5.8)

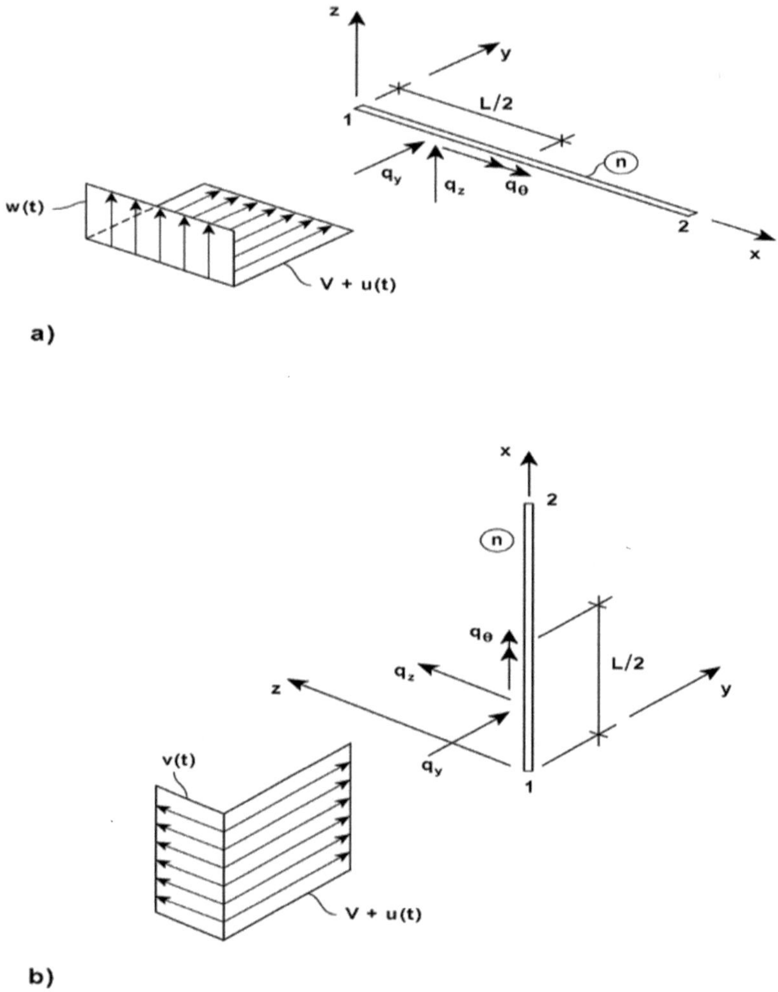

Fig. 9.6 *Wind load effects at horizontal and vertical positions of the element*

$$\mathbf{q}_{tot}(x,t) = \begin{bmatrix} q_1 \\ q_2 \\ q_3 \\ q_4 \\ q_5 \\ q_6 \end{bmatrix}_{tot} = \begin{bmatrix} 0 \\ q_y \\ q_z \\ q_\theta \\ 0 \\ 0 \end{bmatrix} = \begin{bmatrix} 0 \\ \bar{q}_y(x) \\ \bar{q}_z(x) \\ \bar{q}_\theta(x) \\ 0 \\ 0 \end{bmatrix} + \begin{bmatrix} 0 \\ q_y(x,t) \\ q_z(x,t) \\ q_\theta(x,t) \\ 0 \\ 0 \end{bmatrix} + \begin{bmatrix} 0 \\ q_{y_{ae}}(x,t) \\ q_{z_{ae}}(x,t) \\ q_{\theta_{ae}}(x,t) \\ 0 \\ 0 \end{bmatrix} \tag{9.38}$$

Thus, the load comprise a time invariant mean part (static load), a dynamic part caused by wind turbulence (buffeting load) and a dynamic part caused by interaction between the flow and structural motion (motion induced load), i.e.:

$$\mathbf{q}_{tot}(x,t) = \bar{\mathbf{q}}(x) + \mathbf{q}(x,t) + \mathbf{q}_{ae}(x,t) \tag{9.39}$$

As indicated in Fig. 9.6 it is for simplicity assumed that the element length (L) is short such that it may with sufficient accuracy be assumed that the flow components $V + u(x,t)$, $v(x,t)$ and $w(x,t)$ are constant along half the element span.

The mean (static) load vector:

The static load is given by (see Eq. 5.11)

$$\bar{\mathbf{q}}(x) = \frac{\rho V^2}{2} \cdot \begin{bmatrix} 0 & D\bar{C}_D & B\bar{C}_L & -B^2\bar{C}_M & 0 & 0 \end{bmatrix}^T \tag{9.40}$$

Thus, the mean (static) load vector at element level is given by

$$\bar{\mathbf{R}}_n = \begin{bmatrix} \bar{\mathbf{Q}}_1 \\ \bar{\mathbf{Q}}_2 \end{bmatrix}_n \tag{9.41}$$

where

$$\bar{Q}_{i_n}(Z_i) = \left\{ \frac{\rho}{2} \left[V(Z_i) \right]^2 \frac{L}{2} \cdot \begin{bmatrix} 0 & D\bar{C}_D & B\bar{C}_L & B^2\bar{C}_M & 0 & 0 \end{bmatrix}^T \right\}_n \tag{9.42}$$

and where i refers to element end 1 or 2, i.e. $i = 1$ or 2.

The turbulence induced (buffeting) load vector:

As illustrated in Fig. 9.6, the buffeting load will depend on orientation of the element in the flow, i.e. whether its position is horizontal or vertical, affecting the appropriate interpretation of the flow components $u(x,t), v(x,t)$ and $w(x,t)$. Thus, the buffeting load is given by (see Eq. 5.12)

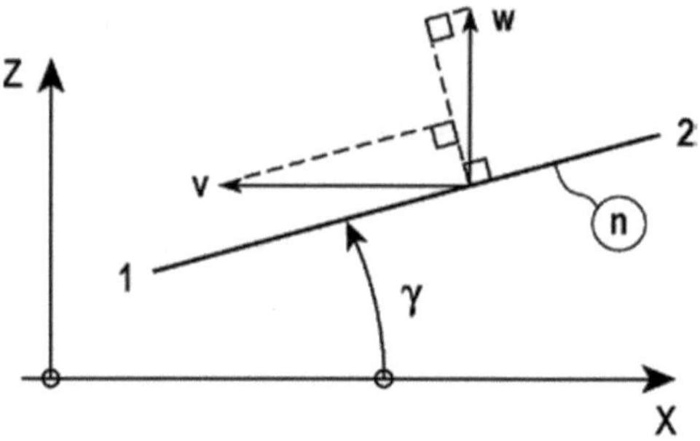

Fig. 9.7 *Definition of roll angle* γ

$$\mathbf{q}(x,t) = \frac{\rho V^2}{2} \begin{bmatrix} 0 & 0 & 0 \\ 2D\bar{C}_D I_u & (DC_D' - B\bar{C}_L) I_v & (DC_D' - B\bar{C}_L) I_w \\ 2B\bar{C}_L I_u & (BC_L' + D\bar{C}_D) I_v & (BC_L' + D\bar{C}_D) I_w \\ -2B^2\bar{C}_M I_u & -B^2 C_M' I_v & -B^2 C_M' I_w \\ 0 & 0 & 0 \\ 0 & 0 & 0 \end{bmatrix} \cdot \boldsymbol{\psi}_0 \hat{\mathbf{v}}_0 \quad (9.43)$$

where I_u, I_v and I_w are the turbulence intensities, $\boldsymbol{\psi}_0$ is a three by three matrix taking account of the element orientation (roll angle γ) in the flow (see Fig. 9.7)

$$\boldsymbol{\psi}_0 = \begin{bmatrix} 1 & 0 & 0 \\ 0 & \sin\gamma & 0 \\ 0 & 0 & \cos\gamma \end{bmatrix} \quad (9.44)$$

and where $\hat{\mathbf{v}}_0$ is the reduced turbulence velocity vector

$$\hat{\mathbf{v}}_0 = \begin{bmatrix} \dfrac{u(x,t)}{\sigma_u} & \dfrac{v(x,t)}{\sigma_v} & \dfrac{w(x,t)}{\sigma_w} \end{bmatrix}^T \quad (9.45)$$

Thus, the turbulence induced (buffeting) load vector at element level is given by:

$$\mathbf{R}_n(t) = \begin{bmatrix} \mathbf{Q}_1 \\ \mathbf{Q}_2 \end{bmatrix}_n \quad (9.46)$$

where

$$\mathbf{Q}_{i_n}(Z_i,t) = \left\{\mathbf{B}_{Q0}(Z_i)\cdot\boldsymbol{\psi}_0\cdot\hat{\mathbf{v}}_0\right\}_n \tag{9.47}$$

where $i = 1$ or 2 and where

$$\mathbf{B}_{Q0} = \frac{\rho\left[V(Z_i)\right]^2 L}{4}\begin{bmatrix} 0 & 0 & 0 \\ 2D\bar{C}_D I_u(Z_i) & \left(DC_D' - B\bar{C}_L\right)I_v(Z_i) & \left(DC_D' - B\bar{C}_L\right)I_w(Z_i) \\ 2B\bar{C}_L I_u(Z_i) & \left(BC_L' + D\bar{C}_D\right)I_v(Z_i) & \left(BC_L' + D\bar{C}_D\right)I_w(Z_i) \\ -2B^2\bar{C}_M I_u(Z_i) & -B^2 C_M' I_v(Z_i) & -B^2 C_M' I_w(Z_i) \\ 0 & 0 & 0 \\ 0 & 0 & 0 \end{bmatrix} \tag{9.48}$$

Thus,

$$\mathbf{R}_n(t) = \mathbf{B}_{Q_n}\cdot\boldsymbol{\psi}_n\cdot\hat{\mathbf{v}}_n(t) \tag{9.49}$$

where

$$\left.\begin{aligned} \mathbf{B}_{Q_n} &= \begin{bmatrix} \mathbf{B}_{Q0}(Z_1) & \mathbf{0} \\ \mathbf{0} & \mathbf{B}_{Q0}(Z_2) \end{bmatrix}_n \\ \boldsymbol{\psi}_n &= \begin{bmatrix} \boldsymbol{\psi}_0 & \mathbf{0} \\ \mathbf{0} & \boldsymbol{\psi}_0 \end{bmatrix}_n \\ \hat{\mathbf{v}}_n &= \begin{bmatrix} \hat{\mathbf{v}}_0(Z_1) \\ \hat{\mathbf{v}}_0(Z_2) \end{bmatrix}_n \end{aligned}\right\} \tag{9.50}$$

and, as mentioned above, where indices 1 and 2 refers to element ends.

<u>*The motion induced (aerodynamic) load vector:*</u>

As shown in Eq. 5.8, the motion induced load will comprise two contributions, one which is proportional to the element velocity and one which is proportional to its dynamic displacements. Thus,

$$\mathbf{q}_{ae}(x,t) = \mathbf{c}_{ae0}\begin{bmatrix} 0 \\ \dot{r}_{y_{el}} \\ \dot{r}_{z_{el}} \\ \dot{r}_{\theta_{el}} \\ 0 \\ 0 \end{bmatrix} + \mathbf{k}_{ae0}\begin{bmatrix} 0 \\ r_{y_{el}} \\ r_{z_{el}} \\ r_{\theta_{el}} \\ 0 \\ 0 \end{bmatrix} \tag{9.51}$$

If a quasi-static approach is adopted, then $\mathbf{c}_{ae0}$ and $\mathbf{k}_{ae0}$ are given by (see Eqs. 5.13 and 5.14):

$$\mathbf{c}_{ae0} = \frac{\rho V}{2}\begin{bmatrix} 0 & 0 & 0 & 0 & 0 & 0 \\ 0 & -2D\bar{C}_D & -\left(DC_D' - B\bar{C}_L\right) & 0 & 0 & 0 \\ 0 & -2B\bar{C}_L & -\left(BC_L' + D\bar{C}_D\right) & 0 & 0 & 0 \\ 0 & 2B^2\bar{C}_M & B^2C_M' & 0 & 0 & 0 \\ 0 & 0 & 0 & 0 & 0 & 0 \\ 0 & 0 & 0 & 0 & 0 & 0 \end{bmatrix} \tag{9.52}$$

$$\mathbf{k}_{ae0} = \frac{\rho V^2}{2}\begin{bmatrix} 0 & 0 & 0 & 0 & 0 & 0 \\ 0 & 0 & 0 & -DC_D' & 0 & 0 \\ 0 & 0 & 0 & -BC_L' & 0 & 0 \\ 0 & 0 & 0 & B^2C_M' & 0 & 0 \\ 0 & 0 & 0 & 0 & 0 & 0 \\ 0 & 0 & 0 & 0 & 0 & 0 \end{bmatrix} \tag{9.53}$$

If a modal approach with aerodynamic derivatives is adopted, then (see Eqs. 5.24 and 5.25)

$$\mathbf{c}_{ae0} = \frac{\rho B^2 \omega_i}{2}\begin{bmatrix} 0 & 0 & 0 & 0 & 0 & 0 \\ 0 & P_1^* & P_5^* & -BP_2^* & 0 & 0 \\ 0 & H_5^* & H_1^* & -BH_2^* & 0 & 0 \\ 0 & -BA_5^* & -BA_1^* & B^2A_2^* & 0 & 0 \\ 0 & 0 & 0 & 0 & 0 & 0 \\ 0 & 0 & 0 & 0 & 0 & 0 \end{bmatrix} \tag{9.54}$$

$$\mathbf{k}_{ae0} = \frac{\rho B^2 \omega_i^2}{2}\begin{bmatrix} 0 & 0 & 0 & 0 & 0 & 0 \\ 0 & P_4^* & P_6^* & -BP_3^* & 0 & 0 \\ 0 & H_6^* & H_4^* & -BH_3^* & 0 & 0 \\ 0 & -BA_6^* & -BA_4^* & B^2A_3^* & 0 & 0 \\ 0 & 0 & 0 & 0 & 0 & 0 \\ 0 & 0 & 0 & 0 & 0 & 0 \end{bmatrix} \tag{9.55}$$

where ω_i is the relevant frequency of motion (i.e. the mean wind velocity dependant resonance frequency of mode i). It should be noted that if a quasi-static approach is adopted then $\mathbf{c}_{ae_0}$ and $\mathbf{k}_{ae_0}$ from Eqs. 9.52 and 9.53 are applicable in time domain as well as in frequency domain, while an approach based on $\mathbf{c}_{ae_0}$ and $\mathbf{k}_{ae_0}$ containing aerodynamic derivatives given in Eqs. 9.54 and 9.55 is only applicable in a modal frequency domain approach. It is a physical requirement that loads obtained from aerodynamic derivatives should converge towards quasi-static loads as the frequency of motion is approaching zero, i.e. when the motion becomes very slow.

It should also be noted that a basic hypothesis behind the development of the buffeting theory in Chapter 5 was that fluctuations in the oncoming flow are instantaneously giving rise to corresponding fluctuations in the cross sectional loads. If a time domain solution in original coordinates (i.e. in the element degrees of freedom) is chosen then such a hypothesis will no longer be justifiable, in which case indicial memory functions will have to be introduced. Thus, it is necessary to introduce a dummy time history variable τ, a relative time $s = t - \tau$ (see Fig. 9.8) and a set of indicial functions

$$\Phi_{ij}(s) \quad \text{where} \quad \begin{cases} i = D, L \text{ or } M \\ j = \dot{y}, \dot{z} \text{ or } \theta \end{cases} \tag{9.56}$$

associated with interaction between drag, lift or moment forces and velocity of motion in y, z or θ directions. These functions describe how an incremental structural motion is giving rise to a corresponding change of motion induced load, i.e. the basic motion induced load hypothesis is given by

$$\frac{d\mathbf{q}_{ae}(x,t)}{d\tau} = \mathbf{c}_{ae_0}(s) \cdot \frac{d}{d\tau}\begin{bmatrix} 0 \\ \dot{r}_{y_{el}}(x,\tau) \\ \dot{r}_{z_{el}}(x,\tau) \\ \dot{r}_{\theta_{el}}(x,\tau) \\ 0 \\ 0 \end{bmatrix} + \mathbf{k}_{ae_0}(s) \cdot \frac{d}{d\tau}\begin{bmatrix} 0 \\ r_{y_{el}}(x,\tau) \\ r_{z_{el}}(x,\tau) \\ r_{\theta_{el}}(x,\tau) \\ 0 \\ 0 \end{bmatrix} \tag{9.57}$$

As time goes towards infinity it is a physical requirement that loads obtained from indicial functions will asymptotically be approaching quasi-static loads. Adopting functions $\Phi_{ij}(s) \xrightarrow[s \to \infty]{} 1$, i.e. similar to that which has been indicated in Fig. 9.8, then $\mathbf{c}_{ae_0}(s)$ and $\mathbf{k}_{ae_0}(s)$ are simply obtained by expanding the expressions in Eqs. 9.52 and 9.53 into

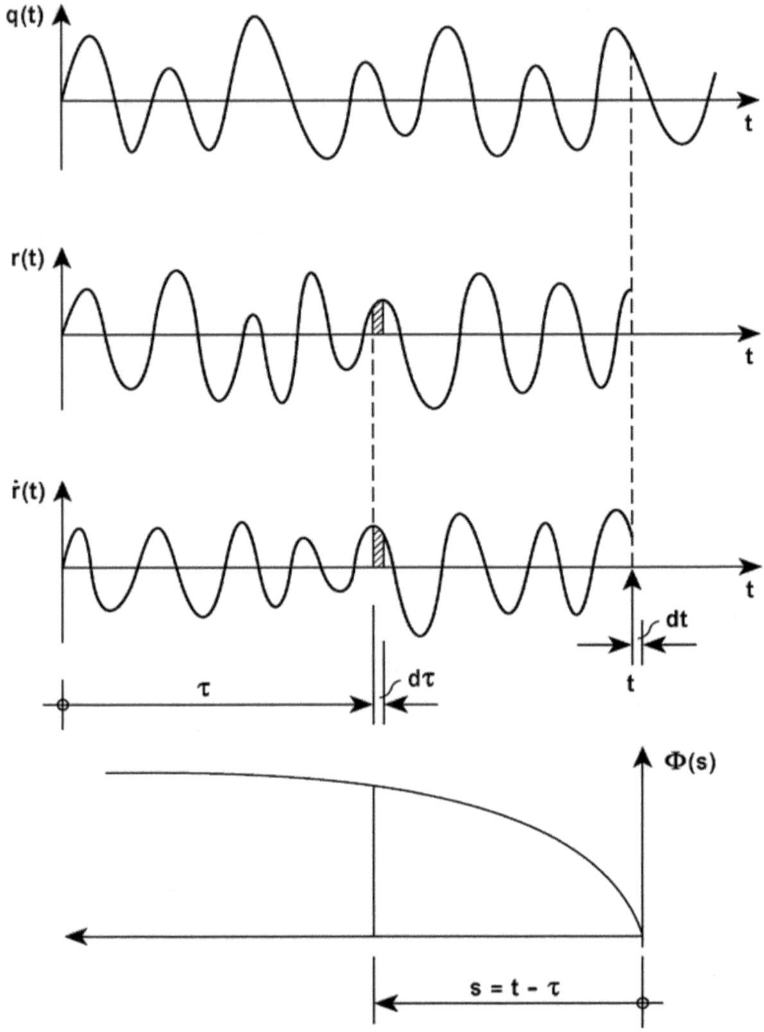

Fig. 9.8 *Relative time history integration of indicial functions*

$$\mathbf{c}_{ae_0}(s)=\frac{\rho V}{2}\begin{bmatrix} 0 & 0 & 0 & 0 & 0 & 0 \\ 0 & -2D\bar{C}_D\Phi_{D\dot{y}}(s) & -\left(DC_D'-B\bar{C}_L\right)\Phi_{D\dot{z}}(s) & 0 & 0 & 0 \\ 0 & -2B\bar{C}_L\Phi_{L\dot{y}}(s) & -\left(BC_L'+D\bar{C}_D\right)\Phi_{D\dot{z}}(s) & 0 & 0 & 0 \\ 0 & 2B^2\bar{C}_M\Phi_{M\dot{y}}(s) & B^2C_M'\Phi_{D\dot{z}}(s) & 0 & 0 & 0 \\ 0 & 0 & 0 & 0 & 0 & 0 \\ 0 & 0 & 0 & 0 & 0 & 0 \end{bmatrix} \tag{9.58}$$

and

$$\mathbf{k}_{ae_0}(s) = \frac{\rho V^2}{2} \begin{bmatrix} 0 & 0 & 0 & 0 & 0 & 0 \\ 0 & 0 & 0 & -DC_D' \Phi_{D\theta}(s) & 0 & 0 \\ 0 & 0 & 0 & -BC_L' \Phi_{L\theta}(s) & 0 & 0 \\ 0 & 0 & 0 & B^2 C_M' \Phi_{M\theta}(s) & 0 & 0 \\ 0 & 0 & 0 & 0 & 0 & 0 \\ 0 & 0 & 0 & 0 & 0 & 0 \end{bmatrix} \tag{9.59}$$

Thus, the motion induced load $\mathbf{q}_{ae}(x,t)$ may be obtained from history integration

$$\mathbf{q}_{ae}(x,t) = \int_0^t \left\{ \mathbf{c}_{ae_0}(s) \cdot \frac{d}{d\tau} \begin{bmatrix} 0 \\ \dot{r}_{y_{el}}(x,\tau) \\ \dot{r}_{z_{el}}(x,\tau) \\ \dot{r}_{\theta_{el}}(x,\tau) \\ 0 \\ 0 \end{bmatrix} + \mathbf{k}_{ae_0}(s) \cdot \frac{d}{d\tau} \begin{bmatrix} 0 \\ r_{y_{el}}(x,\tau) \\ r_{z_{el}}(x,\tau) \\ r_{\theta_{el}}(x,\tau) \\ 0 \\ 0 \end{bmatrix} \right\} d\tau \tag{9.60}$$

It is also a physical requirement that if the structural motion is harmonic then loads obtained from indicial functions must be equal to those obtained from aerodynamic derivatives. It is shown in Appendix D how this requirement may be used to determine indicial functions from known aerodynamic derivatives.

Thus, if Eq. 9.51 is applicable (i.e. a quasi-steady load hypothesis is adopted), then the motion induced load associated with element ends $i = 1$ or 2 is given by

$$\mathbf{Q}_{ae_{i_n}}(t) = \left\{ \frac{L}{2} \mathbf{c}_{ae_0}(Z_i) \cdot \dot{\mathbf{d}}_i(t) + \frac{L}{2} \mathbf{k}_{ae_0}(Z_i) \cdot \mathbf{d}_i(t) \right\}_n \tag{9.61}$$

where $\mathbf{d}_i$ is the is the six by one displacement vector at element ends $i = 1$ or 2.(the Z_i dependency comes from the variation of V with the elevation of element ends 1 and 2.) In this case the aerodynamic load vector at the level of element n is given by

$$\mathbf{R}_{ae_n}(t) = \begin{bmatrix} \mathbf{R}_{ae}(Z_1,t) \\ \mathbf{R}_{ae}(Z_2,t) \end{bmatrix}_n = \mathbf{c}_{ae_n}(Z_1,Z_2) \cdot \dot{\mathbf{d}}_n(t) + \mathbf{k}_{ae_n}(Z_1,Z_2) \cdot \mathbf{d}_n(t) \tag{9.62}$$

where $\mathbf{d}_n(t) = \begin{bmatrix} \mathbf{d}_1 & \mathbf{d}_2 \end{bmatrix}_n^T$ and

$$\mathbf{c}_{ae_n} = \begin{bmatrix} \frac{L}{2}\mathbf{c}_{ae_0}(Z_1) & \mathbf{0} \\ \mathbf{0} & \frac{L}{2}\mathbf{c}_{ae_0}(Z_2) \end{bmatrix}_n \text{ and } \mathbf{k}_{ae_n} = \begin{bmatrix} \frac{L}{2}\mathbf{k}_{ae_0}(Z_1) & \mathbf{0} \\ \mathbf{0} & \frac{L}{2}\mathbf{k}_{ae_0}(Z_2) \end{bmatrix}_n \tag{9.63}$$

where $\mathbf{c}_{ae_0}$ and $\mathbf{k}_{ae_0}$ are given in Eqs. 9.52 and 9.53.

On the other hand, if Eq. 9.60 is applicable (i.e. the concept of indicial functions is adopted), then

$$\mathbf{Q}_{ae_{i_n}}(t) = \left\{ \int_0^t \left[\frac{L}{2}\mathbf{c}_{ae_0}(Z_i,s)\cdot\dot{\mathbf{d}}_i(\tau) + \frac{L}{2}\mathbf{k}_{ae_0}(Z_i,s)\cdot\mathbf{d}_i(\tau) \right] d\tau \right\}_n \tag{9.64}$$

and the corresponding load vector at the level of element n is given by

$$\mathbf{R}_{ae_n}(t) = \begin{bmatrix} \mathbf{R}_{ae}(Z_1,t) \\ \mathbf{R}_{ae}(Z_2,t) \end{bmatrix}_n = \int_0^t \mathbf{c}_{ae_n}(Z_1,Z_2,s)\cdot\dot{\mathbf{d}}_n(\tau)d\tau + \int_0^t \mathbf{k}_{ae_n}(Z_1,Z_2,s)\cdot\mathbf{d}_n(\tau)d\tau \tag{9.65}$$

where

$$\mathbf{c}_{ae_n} = \begin{bmatrix} \frac{L}{2}\mathbf{c}_{ae_0}(Z_1,s) & \mathbf{0} \\ \mathbf{0} & \frac{L}{2}\mathbf{c}_{ae_0}(Z_2,s) \end{bmatrix}_n \text{ and } \mathbf{k}_{ae_n} = \begin{bmatrix} \frac{L}{2}\mathbf{k}_{ae_0}(Z_1,s) & \mathbf{0} \\ \mathbf{0} & \frac{L}{2}\mathbf{k}_{ae_0}(Z_2,s) \end{bmatrix}_n \tag{9.66}$$

where $\mathbf{c}_{ae_0}$ and $\mathbf{k}_{ae_0}$ are given in Eqs. 9.58 and 9.59.

Thus, all wind forces and motion induced forces at element level have been established. What remains before any response calculations can be performed is to establish equilibrium conditions expressed in global structural degrees of freedom.

9.4 The global analysis

As previously mentioned (see Chapter 9.2) it is taken for granted that displacements as well as forces at element level comprise a time-invariant mean value (the static part) and a stationary fluctuating (dynamic) part, i.e. $\mathbf{d}_{n_{tot}} = \bar{\mathbf{d}}_n + \mathbf{d}_n(t)$, $\mathbf{F}_{n_{tot}} = \bar{\mathbf{F}}_n + \mathbf{F}_n(t)$ and

$\mathbf{R}_{n_{tot}} = \bar{\mathbf{R}}_n + \mathbf{R}_n(t)$. This also applies at a global structural level, i.e. $\mathbf{r}_{tot} = \bar{\mathbf{r}} + \mathbf{r}(t)$ and $\mathbf{R}_{tot} = \bar{\mathbf{R}} + \mathbf{R}(t)$. Before proceeding it is necessary to define the connectivity matrix $\mathbf{A}_n$ describing the relationship between element degrees of freedom $\mathbf{d}_{n_{tot}}$ and global degrees of freedom $\mathbf{r}_{tot}$, i.e.:

$$\mathbf{d}_{n_{tot}} = \mathbf{A}_n \cdot \mathbf{r}_{tot} \tag{9.67}$$

Applying a set of virtual displacements to the system $\delta\mathbf{r}_{tot}$, then $\delta\mathbf{d}_{n_{tot}} = \mathbf{A}_n \cdot \delta\mathbf{r}_{tot}$, and since the virtual work exerted by the external forces (at global as well as at element level) must be equal to the sum of the virtual work of the internal forces, then

$$\mathbf{R}_{tot}^T \cdot \delta\mathbf{r} = \sum_{n=1}^{N} \mathbf{R}_{n_{tot}}^T \cdot \delta\mathbf{d}_{n_{tot}} = \sum_{n=1}^{N} \mathbf{F}_{n_{tot}}^T \cdot \delta\mathbf{d}_{n_{tot}} \tag{9.68}$$

where N is the total number of elements in the system. Introducing Eq. 9.67, then

$$\left(\sum_{n=1}^{N} \mathbf{R}_{n_{tot}}^T \mathbf{A}_n\right)^T \cdot \delta\mathbf{r}_{n_{tot}} = \left(\sum_{n=1}^{N} \mathbf{F}_{n_{tot}}^T \mathbf{A}_n\right)^T \cdot \delta\mathbf{r}_{n_{tot}} \tag{9.69}$$

and thus, it is seen that the equilibrium condition at a global structural level is given by

$$\sum_{n=1}^{N} \mathbf{A}_n^T \mathbf{R}_{n_{tot}} = \sum_{n=1}^{N} \mathbf{A}_n^T \mathbf{F}_{n_{tot}} \tag{9.70}$$

where N is the total number of elements in the system. As shown in Eq. 9.24

$$\mathbf{F}_{n_{tot}} = \mathbf{m}_n \ddot{\mathbf{d}}_n + \mathbf{c}_n \dot{\mathbf{d}}_n + \mathbf{k}\mathbf{d}_{n_{tot}} \tag{9.71}$$

and assuming that Eq 9.62 applies (i.e. that indicial functions are not in use), then

$$\mathbf{R}_{n_{tot}} = \bar{\mathbf{R}}_n + \mathbf{R}_n(t) + \mathbf{R}_{ae_n}(t) = \bar{\mathbf{R}}_n + \mathbf{R}_n(t) + \mathbf{c}_{ae_n} \dot{\mathbf{d}}_{tot} + \mathbf{k}_{ae_n} \mathbf{d}_{tot} \tag{9.72}$$

Introducing Eqs. 9.67, 9.71 and 9.72 into the equilibrium condition in Eq. 9.70 (and acknowledging that the time derivative of mean values are zero) then the following is obtained

$$\begin{aligned}&\sum_{n=1}^{N}\left(\mathbf{A}_n^T\bar{\mathbf{R}}_n+\mathbf{A}_n^T\mathbf{R}_n(t)+\mathbf{A}_n^T\mathbf{c}_{ae_n}\dot{\mathbf{r}}_{tot}+\mathbf{A}_n^T\mathbf{k}_{ae_n}\mathbf{r}_{tot}\right)\\&=\sum_{n=1}^{N}\left(\mathbf{A}_n^T\mathbf{m}_n\mathbf{A}_n\ddot{\mathbf{r}}+\mathbf{A}_n^T\mathbf{c}_n\mathbf{A}_n\dot{\mathbf{r}}+\mathbf{A}_n^T\mathbf{k}_n\mathbf{A}_n\mathbf{r}_{tot}\right)\end{aligned} \tag{9.73}$$

Defining the structural properties

$$\begin{bmatrix}\mathbf{M}\\\mathbf{C}\\\mathbf{K}\end{bmatrix}=\sum_{n=1}^{N}\begin{bmatrix}\mathbf{A}^T\mathbf{m}\mathbf{A}\\\mathbf{A}^T\mathbf{c}\mathbf{A}\\\mathbf{A}^T\mathbf{k}\mathbf{A}\end{bmatrix}_n \tag{9.74}$$

the motion induced properties

$$\begin{bmatrix}\mathbf{C}_{ae}\\\mathbf{K}_{ae}\end{bmatrix}=\sum_{n=1}^{N}\begin{bmatrix}\mathbf{A}^T\mathbf{c}_{ae}\mathbf{A}\\\mathbf{A}^T\mathbf{k}_{ae}\mathbf{A}\end{bmatrix}_n \tag{9.75}$$

and the load vectors

$$\begin{bmatrix}\bar{\mathbf{R}}\\\mathbf{R}(t)\end{bmatrix}=\sum_{n=1}^{N}\begin{bmatrix}\mathbf{A}^T\bar{\mathbf{R}}\\\mathbf{A}^T\mathbf{R}(t)\end{bmatrix}_n \tag{9.76}$$

then Eq. 9.73 may be written

$$\mathbf{M}\ddot{\mathbf{r}}_{tot}+\mathbf{C}\dot{\mathbf{r}}_{tot}+\mathbf{K}\mathbf{r}_{tot}=\bar{\mathbf{R}}+\mathbf{R}(t)+\mathbf{R}_{ae}(t) \tag{9.77}$$

where

$$\mathbf{R}_{ae}(t)=\mathbf{C}_{ae}\dot{\mathbf{r}}_{tot}(t)+\mathbf{K}_{ae}\mathbf{r}_{tot}(t) \tag{9.78}$$

Since $\dot{\mathbf{r}}_{tot}=\dot{\mathbf{r}}$ and $\ddot{\mathbf{r}}_{tot}=\ddot{\mathbf{r}}$ Eq. 9.77 may be split into a time invariant (mean static) equilibrium condition

$$(\mathbf{K}-\mathbf{K}_{ae})\bar{\mathbf{r}}=\bar{\mathbf{R}} \tag{9.79}$$

and a purely dynamic equilibrium condition

$$\mathbf{M}\,\ddot{\mathbf{r}}(t)+(\mathbf{C}-\mathbf{C}_{ae})\dot{\mathbf{r}}(t)+(\mathbf{K}-\mathbf{K}_{ae})\mathbf{r}(t)=\mathbf{R}(t) \tag{9.80}$$

This equation may be used to calculate dynamic response in time domain as well as in frequency domain. In a time domain approach it will be necessary to perform a time domain simulation of the fluctuating flow components $u(X_k,Z_k,t)$, $v(X_k,Z_k,t)$ and $w(X_k,Z_k,t)$ in every node k. In a frequency domain approach it will be necessary to introduce the stochastic properties of the flow components and to perform spanwise averaging to obtain the corresponding stochastic properties of the relevant response quantities.

If a time domain solution strategy with indicial functions is chosen, then everything above is still applicable, except that Eq. 9.65 must be introduced for the determination of motion induced loads. Thus, the aerodynamic load given in Eq. 9.78 must be replaced by:

$$\mathbf{R}_{ae}(t)=\sum_{n=1}^{N}\left\{\int_0^t \mathbf{A}_n^T\mathbf{c}_{ae_n}(s)\cdot\mathbf{A}_n\dot{\mathbf{r}}(\tau)d\tau+\int_0^t \mathbf{A}_n^T\mathbf{k}_{ae_n}(s)\cdot\mathbf{A}_n\mathbf{r}(\tau)d\tau\right\} \tag{9.81}$$

Recalling that $ds/d\tau=-1$ and assuming initial conditions $\mathbf{r}(\tau=0)=0$ and $\dot{\mathbf{r}}(\tau=0)=0$ then Eq. 9.81 can be further developed by integration by parts into

$$\begin{aligned}\mathbf{R}_{ae}(t)=&\sum_{n=1}^{N}\left\{\left[\mathbf{A}_n^T\mathbf{c}_{ae_n}(s)\cdot\mathbf{A}_n\dot{\mathbf{r}}(\tau)\right]_{\tau=0}^{\tau=t}-\int_0^t\mathbf{A}_n^T\cdot\frac{d\mathbf{c}_{ae_n}(s)}{ds}\cdot\frac{ds}{d\tau}\cdot\mathbf{A}_n\dot{\mathbf{r}}(\tau)d\tau\right.\\&\left.+\left[\mathbf{A}_n^T\mathbf{k}_{ae_n}(s)\cdot\mathbf{A}_n\mathbf{r}(\tau)\right]_{\tau=0}^{\tau=t}-\int_0^t\mathbf{A}_n^T\cdot\frac{d\mathbf{k}_{ae_n}(s)}{ds}\cdot\frac{ds}{d\tau}\cdot\mathbf{A}_n\mathbf{r}(\tau)d\tau\right\}\\=&\sum_{n=1}^{N}\left\{\mathbf{A}_n^T\mathbf{c}_{ae_n}(s=0)\cdot\mathbf{A}_n\dot{\mathbf{r}}(t)+\int_0^t\mathbf{A}_n^T\cdot\mathbf{c}'_{ae_n}(s)\cdot\mathbf{A}_n\dot{\mathbf{r}}(\tau)d\tau\right.\\&\left.+\mathbf{A}_n^T\mathbf{k}_{ae_n}(s=0)\cdot\mathbf{A}_n\mathbf{r}(t)+\int_0^t\mathbf{A}_n^T\cdot\mathbf{k}'_{ae_n}(s)\cdot\mathbf{A}_n\mathbf{r}(\tau)d\tau\right\}\end{aligned} \tag{9.82}$$

where the prime on $\mathbf{c}'_{ae_n}$ and $\mathbf{k}'_{ae_n}$ mean derivation with respect to s. Introducing

$$\begin{bmatrix}\mathbf{C}_{ae}(s=0)\\ \mathbf{K}_{ae}(s=0)\end{bmatrix}=\sum_{n=1}^{N}\begin{bmatrix}\mathbf{A}^T\cdot\mathbf{c}_{ae}(s=0)\cdot\mathbf{A}\\ \mathbf{A}^T\cdot\mathbf{k}_{ae}(s=0)\cdot\mathbf{A}\end{bmatrix}_n \tag{9.83}$$

and

$$\Delta\mathbf{R}_{ae}(t)=\sum_{n=1}^{N}\int_0^t\left[\mathbf{A}_n^T\cdot\mathbf{c}'_{ae_n}(s)\cdot\mathbf{A}_n\cdot\dot{\mathbf{r}}(\tau)+\mathbf{A}_n^T\cdot\mathbf{k}'_{ae_n}(s)\cdot\mathbf{A}_n\cdot\mathbf{r}(\tau)\right]d\tau \tag{9.84}$$

then the dynamic equilibrium condition is defined by

$$\mathbf{M}\,\ddot{\mathbf{r}}(t)+\left[\mathbf{C}-\mathbf{C}_{ae}(s=0)\right]\dot{\mathbf{r}}(t)+\left[\mathbf{K}-\mathbf{K}_{ae}(s=0)\right]\mathbf{r}(t)=\mathbf{R}(t)+\Delta\mathbf{R}_{ae}(t) \tag{9.85}$$

Since the motion induced load contributions are wind velocity dependant the system contains static as well as dynamic singularities, i.e. the static or the dynamic response will become infinitely large at some critical mean wind velocity.

9.5 *The time invariant static solution*

The time invariant static solution is given by Eq. 9.79, i.e.:

$$\bar{\mathbf{r}}=\left(\mathbf{K}-\mathbf{K}_{ae}\right)^{-1}\bar{\mathbf{R}} \tag{9.86}$$

The corresponding cross sectional element forces may be obtained from Eq. 9.25, i.e.:

$$\bar{\mathbf{F}}_n=\mathbf{k}_n\bar{\mathbf{d}}_n=\mathbf{k}_n\mathbf{A}_n\bar{\mathbf{r}} \tag{9.87}$$

9.6 *The quasi-static solution*

If the lowest eigen frequency of the structure is high, say beyond 4 Hz, then the structure is quasi-static, and the solution may be obtained as a sum of the time invariant solution $\bar{\mathbf{r}}$ (given in Eq. 9.83) and a slowly varying part

$$\mathbf{r}(t)=\left(\mathbf{K}-\mathbf{K}_{ae}\right)^{-1}\mathbf{R}(t) \tag{9.88}$$

where $\mathbf{R}(t)$ is the fluctuating load due to turbulence in the oncoming flow, i.e.:

$$\mathbf{R}(t)=\sum_{n=1}^{N}\mathbf{A}_n^T\mathbf{R}_n(t) \tag{9.89}$$

where $\mathbf{R}_n(t)$ is given in Eq. 9.49. There are two alternative solutions strategies. A time domain solution may be pursued, or alternatively a stochastic solution based on the covariance properties of the turbulence components may be chosen.

In a time domain solution it is necessary to perform a stochastic simulation of the stationary flow components $u(t)$, $v(t)$ and $w(t)$ contained in $\hat{\mathbf{v}}_n(t)$ (see Eqs. 9.45 and 9.50). Such a simulation procedure is presented in Appendix A.3. Thus, the solution is given by

$$\mathbf{r}_{tot}(t) = \overline{\mathbf{r}} + \mathbf{r}(t) = (\mathbf{K} - \mathbf{K}_{ae})^{-1}\left[\overline{\mathbf{R}} + \mathbf{R}(t)\right] \tag{9.90}$$

Since $u(t)$, $v(t)$ and $w(t)$ are stochastic, this solution will also be stochastic, i.e. it may be necessary to perform several calculations of $\mathbf{R}(t)$ such that statistics may be performed on a representative set of $\mathbf{r}(t)$ solutions (in general at least ten). Also, the simulations must be performed over a sufficiently large time window (e.g. T=10 min.). Thus, a time domain solution may numerically be quite demanding.

In a stochastic solution based on the covariance properties of the stationary turbulence components $u(t)$, $v(t)$ and $w(t)$ contained in $\hat{\mathbf{v}}_n(t)$ the solution is given by (see Eq. 6.2 and Fig. 6.2.a)

$$\mathbf{r}_{tot_{\max}} = \overline{\mathbf{r}} + \mathbf{r}_{\max} = \overline{\mathbf{r}} + k_p \boldsymbol{\sigma}_r \tag{9.91}$$

where $\overline{\mathbf{r}}$ is given in Eq. 9.86, k_p is a peak factor defined in Chapter 2.4 and $\boldsymbol{\sigma}_r$ is a vector containing all the standard deviations of the chosen set of displacement degrees of freedom in the system. $\boldsymbol{\sigma}_r$ may be extracted from the square root of the vector contained on the diagonal of the covariance matrix

$$\mathbf{Cov}_{rr} = E\left[\mathbf{r} \cdot \mathbf{r}^T\right] = \begin{bmatrix} \sigma_1^2 & \cdots & Cov_{1i} & \cdots & Cov_{1j} & \cdots & Cov_{1N_r} \\ \vdots & \ddots & \vdots & & \vdots & & \vdots \\ Cov_{i1} & \cdots & \sigma_i^2 & \cdots & Cov_{ij} & & \vdots \\ \vdots & & \vdots & \ddots & \vdots & & \vdots \\ Cov_{j1} & \cdots & Cov_{ji} & \cdots & \sigma_j^2 & & \vdots \\ \vdots & & & & & \ddots & \vdots \\ Cov_{N_r1} & \cdots & \cdots & \cdots & \cdots & \cdots & \sigma_{N_r}^2 \end{bmatrix} \tag{9.92}$$

where N_r is the total number of global degrees of freedom in the system. Since $\mathbf{r}(t) = (\mathbf{K} - \mathbf{K}_{ae})^{-1}\mathbf{R}(t)$ then

$$
\begin{aligned}
\mathbf{Cov}_{rr} &= E\left[\mathbf{r}\cdot\mathbf{r}^T\right] = E\left[\left\{\left(\mathbf{K}-\mathbf{K}_{ae}\right)^{-1}\mathbf{R}\right\}\cdot\left\{\left(\mathbf{K}-\mathbf{K}_{ae}\right)^{-1}\mathbf{R}\right\}^T\right] \\
&= \left(\mathbf{K}-\mathbf{K}_{ae}\right)^{-1}\cdot E\left[\mathbf{R}\cdot\mathbf{R}^T\right]\cdot\left(\left(\mathbf{K}-\mathbf{K}_{ae}\right)^{-1}\right)^T
\end{aligned}
\tag{9.93}
$$

Since $\mathbf{R}(t)=\sum_{n=1}^{N}\mathbf{A}^T\mathbf{R}_n(t)$ and $\mathbf{R}_n(t)=\mathbf{B}_{Q_n}\boldsymbol{\psi}_n\hat{\mathbf{v}}_n(t)$ (see Eqs. 9.49 and 9.76) then

$$
\begin{aligned}
E\left[\mathbf{R}\cdot\mathbf{R}^T\right] &= E\left[\left\{\sum_{n=1}^{N}\mathbf{A}_n^T\mathbf{R}_n\right\}\cdot\left\{\sum_{n=1}^{N}\mathbf{A}_n^T\mathbf{R}_n\right\}^T\right] = \sum_{n=1}^{N}\sum_{m=1}^{N}E\left[\left(\mathbf{A}_n^T\mathbf{R}_n\right)\cdot\left(\mathbf{A}_m^T\mathbf{R}_m\right)^T\right] \\
&= \sum_{n=1}^{N}\sum_{m=1}^{N}\mathbf{A}_n^T\cdot E\left[\mathbf{R}_n\cdot\mathbf{R}_m^T\right]\cdot\mathbf{A}_m \\
&= \sum_{n=1}^{N}\sum_{m=1}^{N}\mathbf{A}_n^T\cdot E\left[\left(\mathbf{B}_Q\boldsymbol{\psi}\hat{\mathbf{v}}\right)_n\cdot\left(\mathbf{B}_Q\boldsymbol{\psi}\hat{\mathbf{v}}\right)_m^T\right]\cdot\mathbf{A}_m \\
&= \sum_{n=1}^{N}\sum_{m=1}^{N}\mathbf{A}_n^T\left(\mathbf{B}_{Q_n}\boldsymbol{\psi}_n E\left[\hat{\mathbf{v}}_n\cdot\hat{\mathbf{v}}_m^T\right]\boldsymbol{\psi}_m^T\mathbf{B}_{Q_m}^T\right)\mathbf{A}_m \\
&= \sum_{n=1}^{N}\sum_{m=1}^{N}\mathbf{A}_n^T\cdot\left(\mathbf{B}_{Q_n}\cdot\boldsymbol{\psi}_n\cdot\mathbf{Cov}_{\hat{v}\hat{v}}\cdot\boldsymbol{\psi}_m^T\cdot\mathbf{B}_{Q_m}^T\right)\cdot\mathbf{A}_m
\end{aligned}
\tag{9.94}
$$

where $n=1,2,\ldots,N$ and $m=1,2,\ldots,N$ are dummy summation variables (N is the number of elements in the system) and where $\mathbf{Cov}_{\hat{v}\hat{v}}=E\left[\hat{\mathbf{v}}_n\cdot\mathbf{v}_m^T\right]$ is the covariance matrix of the reduced flow components contained in $\hat{\mathbf{v}}$, i.e.:

$$
\mathbf{Cov}_{\hat{v}\hat{v}} = E\left[\begin{bmatrix}\hat{u}_1(t)\\ \hat{v}_1(t)\\ \hat{w}_1(t)\\ \hat{u}_2(t)\\ \hat{v}_2(t)\\ \hat{w}_2(t)\end{bmatrix}_n\cdot\begin{bmatrix}\hat{u}_1(t) & \hat{v}_1(t) & \hat{w}_1(t) & \hat{u}_2(t) & \hat{v}_2(t) & \hat{w}_2(t)\end{bmatrix}_m\right]
\tag{9.95}
$$

where indices 1 and 2 refer to element ends 1 and 2. By performing the multiplications and the relevant statistical calculations, and assuming that all cross covariances between different flow components are negligible, i.e.

$$
Cov_{uv} = Cov_{uw} = Cov_{vw} \approx 0 \tag{9.96}
$$

then the following is obtained:

$$\mathbf{Cov}_{\hat{v}\hat{v}} =$$

$$\begin{bmatrix} \rho_{uu}(\Delta s_{1n1m}) & 0 & 0 & \rho_{uu}(\Delta s_{1n2m}) & 0 & 0 \\ 0 & \rho_{vv}(\Delta s_{1n1m}) & 0 & 0 & \rho_{vv}(\Delta s_{1n2m}) & 0 \\ 0 & 0 & \rho_{ww}(\Delta s_{1n1m}) & 0 & 0 & \rho_{ww}(\Delta s_{1n2m}) \\ \rho_{uu}(\Delta s_{2n1m}) & 0 & 0 & \rho_{uu}(\Delta s_{2n2m}) & 0 & 0 \\ 0 & \rho_{vv}(\Delta s_{2n1m}) & 0 & 0 & \rho_{vv}(\Delta s_{2n2m}) & 0 \\ 0 & 0 & \rho_{ww}(\Delta s_{2n1m}) & 0 & 0 & \rho_{ww}(\Delta s_{2n2m}) \end{bmatrix} \tag{9.97}$$

where Δs_{ij} is the separation in the global coordinate system between points $i = 1n$ or $2n$ and $j = 1m$ or $2m$, and where $\rho_{pp}(\Delta s_{ij})$, $p = u,v$ or w, are the covariance coefficients of the flow components, see Eq. 3.33, i.e.:

$$\rho_{pp}(\Delta s_{ij}) = \exp\left\{ -\left[\left(\frac{X_i - X_j}{{}^{x}L_p} \right)^2 + \left(\frac{Z_i - Z_j}{{}^{z}L_p} \right)^2 \right]^{\frac{1}{2}} \right\} \tag{9.98}$$

where ${}^{x}L_p$ and ${}^{z}L_p$ are the integral length scales of $p = u,v$ or w in X or Z directions (see Eq. 3.35). Thus (see Eq. 9.92 above),

$$\mathbf{Cov}_{rr} =$$

$$(\mathbf{K} - \mathbf{K}_{ae})^{-1} \cdot \left\{ \sum_{n=1}^{N} \sum_{m=1}^{N} \mathbf{A}_n^T \cdot \left(\mathbf{B}_n \cdot \boldsymbol{\psi}_n \cdot \mathbf{Cov}_{\hat{v}\hat{v}} \cdot \boldsymbol{\psi}_m^T \cdot \mathbf{B}_m^T \right) \cdot \mathbf{A}_m \right\} \cdot \left((\mathbf{K} - \mathbf{K}_{ae})^{-1} \right)^T \tag{9.99}$$

The extreme value of the response $\mathbf{r}_{tot_{\max}}$ may then readily be obtained from Eq. 9.91. Since this solution is quasi-static, the corresponding cross sectional element forces may be calculated from

$$\mathbf{F}_{tot_{n_{\max}}} = \mathbf{k}_n \mathbf{d}_{tot_{n_{\max}}} = \mathbf{k}_n \mathbf{A}_n \left(\bar{\mathbf{r}} + k_p \boldsymbol{\sigma}_r \right) - \mathbf{k}_n \boldsymbol{\Lambda}_n \mathbf{r}_{tot_{\max}} \tag{9.100}$$

9.7 Dynamic response calculations in frequency domain

If the lowest eigen frequency of the structure is in a region where dynamic effects can not be neglected, say below about 4 Hz, then a quasi-static solution shown in Chapter 9.6 above will no longer suffice and a full dynamic analysis will be required. However, the time invariant static solution shown in Chapter 9.5 is still valid, and thus, the total displacement response may be obtained as a sum of the time invariant solution $\bar{\mathbf{r}}$ (given in Eq. 9.86) and a purely dynamic part $\mathbf{r}(t)$, i.e. $\mathbf{r}_{tot}(t) = \bar{\mathbf{r}} + \mathbf{r}(t)$. However, in the following a stochastic solution in frequency domain will be shown, and from this only the statistical properties of the response will emerge, i.e. the result of the response calculation is the covariance matrix

$$\mathbf{Cov}_{rr} = E\left[\mathbf{r} \cdot \mathbf{r}^T\right] = \begin{bmatrix} \sigma_1^2 & \cdots & Cov_{1i} & \cdots & Cov_{1j} & \cdots & Cov_{1N_r} \\ \vdots & \ddots & \vdots & & \vdots & & \vdots \\ Cov_{i1} & \cdots & \sigma_i^2 & \cdots & Cov_{ij} & & \vdots \\ \vdots & & \vdots & \ddots & \vdots & & \vdots \\ Cov_{j1} & \cdots & Cov_{ji} & \cdots & \sigma_j^2 & & \vdots \\ \vdots & & & & & \ddots & \vdots \\ Cov_{N_r1} & \cdots & \cdots & \cdots & \cdots & \cdots & \sigma_{N_r}^2 \end{bmatrix} \tag{9.101}$$

where N_r is the number of degrees of freedom in the system. Thus, extreme values of displacement events are given by

$$\mathbf{r}_{tot_{\max}} = \bar{\mathbf{r}} + \mathbf{r}_{\max} = \bar{\mathbf{r}} + k_p \boldsymbol{\sigma}_r \tag{9.102}$$

where $\bar{\mathbf{r}}$ is given in Eq. 9.86, k_p is a peak factor defined in Chapter 2.4 and $\boldsymbol{\sigma}_r$ is a vector containing all the standard deviations of the chosen set of displacement degrees of freedom in the system. $\boldsymbol{\sigma}_r$ may be extracted from the square root of the vector contained on the diagonal of the covariance matrix, see Eq. 9.101.

In a frequency domain approach it is a necessary requirement that all load and response quantities are stationary such that a Fourier transform will render predictable coefficients throughout the entire time window of the process. I.e. motion induced load contributions which evolve from the previous history of the process (indicial functions) can not be included in such a solution strategy. Thus, response calculations in frequency domain must be based on (see Eq. 9.80)

$$\mathbf{M}\ddot{\mathbf{r}}(t) + (\mathbf{C} - \mathbf{C}_{ae})\dot{\mathbf{r}}(t) + (\mathbf{K} - \mathbf{K}_{ae})\mathbf{r}(t) = \mathbf{R}(t) \tag{9.103}$$

where $\mathbf{C}_{ae}$ and $\mathbf{K}_{ae}$ arc defined in Eqs. 9.75 (see also 9.63). Once frequency domain has been entered, then the content of $\mathbf{C}_{ae}$ and $\mathbf{K}_{ae}$ may be taken from experimentally determined aerodynamic derivatives (see Eqs. 9. 54 and 9.55) as well as from quasi static properties (see Eqs. 9. 52 and 9.53).

Taking the Fourier transform throughout Eq. 9.103, i.e. setting

$$\mathbf{r}(t) = \sum_{\omega} \mathbf{a}_r(\omega) \cdot e^{i\omega t} \qquad \text{and} \qquad \mathbf{R}(t) = \sum_{\omega} \mathbf{a}_R(\omega) \cdot e^{i\omega t} \tag{9.104}$$

where $\mathbf{a}_r(\omega)$ and $\mathbf{a}_R(\omega)$ are the Fourier coefficients of the displacement and load processes and where $\sum_{\omega}$ indicates summation over all ω-settings, then Eq. 9.103 is satisfied for each ω-setting if

$$\left[-\mathbf{M}\omega^2 + (\mathbf{C} - \mathbf{C}_{ae}) i\omega + (\mathbf{K} - \mathbf{K}_{ae}) \right] \cdot \mathbf{a}_r = \mathbf{a}_R \tag{9.105}$$

Thus,

$$\Rightarrow \quad \mathbf{a}_r = \mathbf{H}_r(\omega) \cdot \mathbf{a}_R \tag{9.106}$$

where

$$\mathbf{H}_r(\omega) = \left[-\mathbf{M}\omega^2 + (\mathbf{C} - \mathbf{C}_{ae}) i\omega + (\mathbf{K} - \mathbf{K}_{ae}) \right]^{-1} \tag{9.107}$$

The cross spectral density matrix of the response quantities corresponding to the chosen degrees of freedom is defined by

$$\mathbf{S}_{rr}(\omega) = \lim_{T \to \infty} \frac{1}{\pi T} \left(\mathbf{a}_r^* \ \mathbf{a}_r^T \right) - \begin{bmatrix} S_{r_1 r_1} & \cdots & S_{r_1 r_i} & \cdots & S_{r_1 r_j} & \cdots & S_{r_1 N_r} \\ \vdots & \ddots & \vdots & & \vdots & & \vdots \\ S_{r_i r_1} & \cdots & S_{r_i r_i} & \cdots & S_{r_i r_j} & & \vdots \\ \vdots & & \vdots & \ddots & \vdots & & \vdots \\ S_{r_j r_1} & \cdots & S_{r_j r_i} & \cdots & S_{r_j r_j} & & \vdots \\ \vdots & & & & & \ddots & \vdots \\ S_{N_r r_1} & \cdots & \cdots & \cdots & \cdots & \cdots & S_{r_{N_r} r_{N_r}} \end{bmatrix} \tag{9.108}$$

Introducing Eq. 9.106, then

$$\begin{aligned}\mathbf{S}_{rr}(\omega) &= \lim_{T\to\infty}\frac{1}{\pi T}\left(\mathbf{a}_r^* \cdot \mathbf{a}_r^T\right) = \lim_{T\to\infty}\frac{1}{\pi T}\left[\left(\mathbf{H}_r\mathbf{a}_R\right)^* \cdot \left(\mathbf{H}_r\mathbf{a}_R\right)^T\right] \\ &= \mathbf{H}_r^* \cdot \lim_{T\to\infty}\frac{1}{\pi T}\left(\mathbf{a}_R^* \cdot \mathbf{a}_R^T\right)\cdot \mathbf{H}_r^T = \mathbf{H}_r^* \cdot \mathbf{S}_{RR} \cdot \mathbf{H}_r^T\end{aligned} \tag{9.109}$$

where $\mathbf{S}_{RR}(\omega) = \lim_{T\to\infty}\frac{1}{\pi T}\left(\mathbf{a}_R^* \cdot \mathbf{a}_R^T\right)$ is the cross spectral density matrix of the buffeting wind load. The response covariance matrix may then be obtained simply by integration throughout the frequency domain, i.e.

$$\mathbf{Cov}_{rr} = \begin{bmatrix} \sigma_1^2 & \cdots & Cov_{1i} & \cdots & Cov_{1j} & \cdots & Cov_{1N_r} \\ \vdots & \ddots & \vdots & & \vdots & & \vdots \\ Cov_{i1} & \cdots & \sigma_i^2 & \cdots & Cov_{ij} & & \vdots \\ \vdots & & \vdots & \ddots & \vdots & & \vdots \\ Cov_{j1} & \cdots & Cov_{ji} & \cdots & \sigma_j^2 & & \vdots \\ \vdots & & & & & \ddots & \vdots \\ Cov_{N_r1} & \cdots & \cdots & \cdots & \cdots & \cdots & \sigma_{N_r}^2 \end{bmatrix} = \int_0^\infty \mathbf{S}_{rr}(\omega)d\omega \tag{9.110}$$

What then remains is to develop $\mathbf{S}_{RR}$ into an expression of known quantities. As shown in Eqs. 9.49 and 9.76

$$\mathbf{R}(t) = \sum_{n=1}^{N}\mathbf{A}_n^T \cdot \mathbf{R}_n(t) = \sum_{n=1}^{N}\mathbf{A}_n^T \cdot \left[\mathbf{B}_{Q_n}\boldsymbol{\psi}_n\hat{\mathbf{v}}_n(t)\right] \tag{9.111}$$

and thus a Fourier transform of the load vector will render

$$\mathbf{a}_R(\omega) = \sum_{n=1}^{N}\mathbf{A}_n^T \cdot \left[\mathbf{B}_{Q_n}\boldsymbol{\psi}_n\mathbf{a}_{\hat{v}_n}(\omega)\right] \tag{9.112}$$

where

$$\mathbf{a}_{\hat{v}_n}(\omega) = \begin{bmatrix} a_{\hat{u}_1} & a_{\hat{v}_1} & a_{\hat{w}_1} & a_{\hat{u}_2} & a_{\hat{v}_2} & a_{\hat{w}_2} \end{bmatrix}_n^T \tag{9.113}$$

contains the Fourier amplitudes of the reduced turbulence components at the nodes of element n (i.e. at element ends 1 and 2, see Eqs. 9.45 and 9.50). The cross spectral density matrix of the load is then given by

$$\begin{aligned}\mathbf{S}_{RR}(\omega) &= \lim_{T\to\infty}\frac{1}{\pi T}\left\{\left[\sum_{n=1}^{N}\mathbf{A}_n^T\cdot\left(\mathbf{B}_{Q_n}\boldsymbol{\psi}_n\mathbf{a}_{\hat{v}_n}\right)\right]^*\cdot\left[\sum_{n=1}^{N}\mathbf{A}_n^T\cdot\left(\mathbf{B}_{Q_n}\boldsymbol{\psi}_n\mathbf{a}_{\hat{v}_n}\right)\right]^T\right\}\\ &= \sum_{n=1}^{N}\sum_{m=1}^{N}\mathbf{A}_n^T\cdot\left[\mathbf{B}_{Q_n}\boldsymbol{\psi}_n\cdot\lim_{T\to\infty}\frac{1}{\pi T}\left(\mathbf{a}_{\hat{v}_n}^*\cdot\mathbf{a}_{\hat{v}_m}^T\right)\cdot\boldsymbol{\psi}_m^T\mathbf{B}_{Q_m}^T\right]\mathbf{A}_m\\ &= \sum_{n=1}^{N}\sum_{m=1}^{N}\mathbf{A}_n^T\cdot\left[\mathbf{B}_{Q_n}\boldsymbol{\psi}_n\cdot\mathbf{S}_{\hat{v}\hat{v}}(\omega)\cdot\boldsymbol{\psi}_m^T\mathbf{B}_{Q_m}^T\right]\mathbf{A}_m\end{aligned} \tag{9.114}$$

where

$$\mathbf{S}_{\hat{v}\hat{v}}(\omega) = \lim_{T\to\infty}\frac{1}{\pi T}\left(\mathbf{a}_{\hat{v}_n}^*\cdot\mathbf{a}_{\hat{v}_m}^T\right)$$

$$= \lim_{T\to\infty}\frac{1}{\pi T}\left(\begin{bmatrix}a_{\hat{u}_1}^*\\ a_{\hat{v}_1}^*\\ a_{\hat{w}_1}^*\\ a_{\hat{u}_2}^*\\ a_{\hat{v}_2}^*\\ a_{\hat{w}_2}^*\end{bmatrix}\cdot\begin{bmatrix}a_{\hat{u}_1} & a_{\hat{v}_1} & a_{\hat{w}_1} & a_{\hat{u}_2} & a_{\hat{v}_2} & a_{\hat{w}_2}\end{bmatrix}_m\right) = \begin{bmatrix}\mathbf{S}_{\hat{v}\hat{v}_{11}} & \mathbf{S}_{\hat{v}\hat{v}_{12}}\\ \mathbf{S}_{\hat{v}\hat{v}_{21}} & \mathbf{S}_{\hat{v}\hat{v}_{22}}\end{bmatrix} \tag{9.115}$$

is the cross spectral density matrix of the reduced turbulence components. Assuming that all cross spectral densities between two different flow components are negligible, and then its content is given by

$$\left.\begin{aligned}\mathbf{S}_{\hat{v}\hat{v}_{11}} &= \begin{bmatrix}\hat{S}_{uu}\left(\Delta s_{1_n 1_m},\omega\right) & 0 & 0\\ 0 & \hat{S}_{vv}\left(\Delta s_{1_n 1_m},\omega\right) & 0\\ 0 & 0 & \hat{S}_{ww}\left(\Delta s_{1_n 1_m},\omega\right)\end{bmatrix}\\ \mathbf{S}_{\hat{v}\hat{v}_{22}} &= \begin{bmatrix}\hat{S}_{uu}\left(\Delta s_{2_n 2_m},\omega\right) & 0 & 0\\ 0 & \hat{S}_{vv}\left(\Delta s_{2_n 2_m},\omega\right) & 0\\ 0 & 0 & \hat{S}_{ww}\left(\Delta s_{2_n 2_m},\omega\right)\end{bmatrix}\end{aligned}\right\} \tag{9.116}$$

$$\mathbf{S}_{\hat{v}\hat{v}_{12}} = \begin{bmatrix}\hat{S}_{uu}\left(\Delta s_{1_n 2_m},\omega\right) & 0 & 0\\ 0 & \hat{S}_{vv}\left(\Delta s_{1_n 2_m},\omega\right) & 0\\ 0 & 0 & \hat{S}_{ww}\left(\Delta s_{1_n 2_m},\omega\right)\end{bmatrix} = \mathbf{S}_{\hat{v}\hat{v}_{21}} \tag{9.117}$$

where Δs_{ij} is the distance between points i and j, and where $\hat{S}_{pp}\left(\Delta s_{ij},\omega\right)$, $p = u,v$ or w, is the reduced cross spectral density between flow component p at point i and itself at point j (i.e. the cross spectral density between u,v or w at ends 1 or 2 of element n and u,v or w at ends 1 or 2 of element m). As shown in Eq. 2.88, $\hat{S}_{pp}$ may be expressed by the product of the reduced single point spectra at points i and j and the reduced co spectrum between the same points, i.e.

$$\hat{S}_{pp}\left(\Delta s_{ij},\omega\right) = \sqrt{\hat{S}_{p_i}(\omega)\cdot\hat{S}_{p_j}(\omega)}\cdot\hat{C}o_{pp}\left(\Delta s_{ij},\omega\right) \qquad (9.118)$$

E.g., adopting a Kaimal type of auto spectrum and simple exponential co-spectrum decay, then (see Eqs. 3.25 and 3.41)

$$\left.\begin{aligned}
\hat{S}_{p_i}(\omega) &= \frac{S_{p_i}(\omega)}{\sigma_{p_i}^2} = \frac{A_p\cdot {}^{xf}L_p/V(Z_i)}{\left[1+1.5A_p\omega\,{}^{xf}L_p/V(Z_i)\right]^{5/3}}\\
\hat{S}_{p_j}(\omega) &= \frac{S_{p_j}(\omega)}{\sigma_{p_j}^2} = \frac{A_p\cdot {}^{xf}L_p/V(Z_j)}{\left[1+1.5A_p\omega\,{}^{xf}L_p/V(Z_j)\right]^{5/3}}\\
\hat{C}o_{pp}\left(\Delta s_{ij},\omega\right) &= \exp\left\{-\frac{\omega\sqrt{\left[c_{px}\left(X_i-X_j\right)\right]^2+\left[c_{pz}\left(Z_i-Z_j\right)\right]^2}}{\bar{V}_{ij}}\right\}\\
\bar{V}_{ij} &= \frac{1}{2}\left[V(Z_i)+V(Z_j)\right]
\end{aligned}\right\}
\quad p=\begin{Bmatrix}u\\v\\w\end{Bmatrix}\quad i=\begin{Bmatrix}1_n\\2_n\\1_m\\2_m\end{Bmatrix}\quad j=\begin{Bmatrix}1_n\\2_n\\1_m\\2_m\end{Bmatrix} \qquad (9.119)$$

and where indices n or m refers to element numbers and 1 or 2 refers to element end numbers. By defining the reduced auto spectral density matrices associated with elements n and m

$$\left.\begin{aligned}
\hat{\mathbf{S}}_n(\omega) &= diag\left[\hat{S}_{u_{1_n}}\quad \hat{S}_{v_{1_n}}\quad \hat{S}_{w_{1_n}}\quad \hat{S}_{u_{2_n}}\quad \hat{S}_{v_{2_n}}\quad \hat{S}_{w_{2_n}}\right]\\
\hat{\mathbf{S}}_m(\omega) &= diag\left[\hat{S}_{u_{1_m}}\quad \hat{S}_{v_{1_m}}\quad \hat{S}_{w_{1_m}}\quad \hat{S}_{u_{2_m}}\quad \hat{S}_{v_{2_m}}\quad \hat{S}_{w_{2_m}}\right]
\end{aligned}\right\} \qquad (9.120)$$

and the reduced covariance matrix between corresponding element ends

$$\hat{\mathbf{Co}}_{nm}\left(\Delta s_{nm},\omega\right)=\begin{bmatrix}\hat{\mathbf{Co}}_{1n1m} & \hat{\mathbf{Co}}_{1n2m}\\ \hat{\mathbf{Co}}_{1n2m} & \hat{\mathbf{Co}}_{2n2m}\end{bmatrix} \tag{9.121}$$

where

$$\left.\begin{aligned}
\hat{\mathbf{Co}}_{1n1m} &= \begin{bmatrix}\hat{Co}_{uu}\left(\Delta s_{1_n1_m},\omega\right) & 0 & 0\\ 0 & \hat{Co}_{vv}\left(\Delta s_{1_n1_m},\omega\right) & 0\\ 0 & 0 & \hat{Co}_{ww}\left(\Delta s_{1_n1_m},\omega\right)\end{bmatrix}\\
\hat{\mathbf{Co}}_{2n2m} &= \begin{bmatrix}\hat{Co}_{uu}\left(\Delta s_{2_n2_m},\omega\right) & 0 & 0\\ 0 & \hat{Co}_{uu}\left(\Delta s_{2_n2_m},\omega\right) & 0\\ 0 & 0 & \hat{Co}_{uu}\left(\Delta s_{2_n2_m},\omega\right)\end{bmatrix}
\end{aligned}\right\} \tag{9.122}$$

and

$$\left.\begin{aligned}
\hat{\mathbf{Co}}_{1n2m} &= \begin{bmatrix}\hat{Co}_{uu}\left(\Delta s_{1_n2_m},\omega\right) & 0 & 0\\ 0 & \hat{Co}_{vv}\left(\Delta s_{1_n2_m},\omega\right) & 0\\ 0 & 0 & \hat{Co}_{ww}\left(\Delta s_{1_n2_m},\omega\right)\end{bmatrix}\\
\hat{\mathbf{Co}}_{2n1m} &= \begin{bmatrix}\hat{Co}_{uu}\left(\Delta s_{2_n1_m},\omega\right) & 0 & 0\\ 0 & \hat{Co}_{vv}\left(\Delta s_{2_n1_m},\omega\right) & 0\\ 0 & 0 & \hat{Co}_{ww}\left(\Delta s_{2_n1_m},\omega\right)\end{bmatrix}
\end{aligned}\right\} \tag{9.123}$$

then $\mathbf{S}_{\hat{v}\hat{v}}\left(\omega\right)$ in Eq. 9.114 is given by

$$\mathbf{S}_{\hat{v}\hat{v}}\left(\omega\right)=\sqrt{\hat{\mathbf{S}}_n\left(\omega\right)}\cdot\hat{\mathbf{Co}}_{nm}\left(\omega\right)\cdot\sqrt{\hat{\mathbf{S}}_m\left(\omega\right)} \tag{9.124}$$

Combining Eqs. 9.109, 9.114 and 9.124 then the following cross spectral density response matrix is obtained

$$\begin{aligned}
\mathbf{S}_{rr}\left(\omega\right) &= \mathbf{H}_r^{*}\cdot\mathbf{S}_{RR}\cdot\mathbf{H}_r^{T}\\
&= \mathbf{H}_r^{*}\cdot\left\{\sum_{n=1}^{N}\sum_{m=1}^{N}\mathbf{A}_n^{T}\cdot\left[\mathbf{B}_n\cdot\boldsymbol{\psi}_n\cdot\left(\hat{\mathbf{S}}_n^{1/2}\cdot\hat{\mathbf{Co}}_{nm}\cdot\hat{\mathbf{S}}_m^{1/2}\right)\cdot\boldsymbol{\psi}_m^{T}\cdot\mathbf{B}_m^{T}\right]\mathbf{A}_m\right\}\cdot\mathbf{H}_r^{T}
\end{aligned} \tag{9.125}$$

from which the response covariance matrix may readily be obtained by frequency domain integration, i.e.

$$\mathbf{Cov}_{rr} = E\left[\mathbf{r}\cdot\mathbf{r}^T\right] = \begin{bmatrix} \sigma_1^2 & \cdots & Cov_{1i} & \cdots & Cov_{1j} & \cdots & Cov_{1N_r} \\ \vdots & \ddots & \vdots & & \vdots & & \vdots \\ Cov_{i1} & \cdots & \sigma_i^2 & \cdots & Cov_{ij} & & \vdots \\ \vdots & & \vdots & \ddots & \vdots & & \vdots \\ Cov_{j1} & \cdots & Cov_{ji} & \cdots & \sigma_j^2 & & \vdots \\ \vdots & & & & & \ddots & \vdots \\ Cov_{N_r1} & \cdots & \cdots & \cdots & \cdots & \cdots & \sigma_{N_r}^2 \end{bmatrix} = \int_0^\infty \mathbf{S}_{rr}(\omega)d\omega \tag{9.126}$$

For the ensuing calculations of the statistical properties of cross sectional response forces at element end points it is necessary also to determine the covariance between the displacement response and its derivatives. The general solution to the problem of determining the covariance between a stationary process $x(t)$ and its derivatives $\dot{x}(t)$ and $\ddot{x}(t)$ has been shown in Chapter 2.9 (see Eq. 2.94). Recalling that for a stationary process $E[x\cdot\dot{x}] = E[\dot{x}\cdot\ddot{x}] = 0$, then

$$\begin{bmatrix} \mathbf{Cov}_{rr} & \mathbf{Cov}_{r\dot{r}} & \mathbf{Cov}_{r\ddot{r}} \\ \mathbf{Cov}_{\dot{r}r} & \mathbf{Cov}_{\dot{r}\dot{r}} & \mathbf{Cov}_{\dot{r}\ddot{r}} \\ \mathbf{Cov}_{\ddot{r}r} & \mathbf{Cov}_{\ddot{r}\dot{r}} & \mathbf{Cov}_{\ddot{r}\ddot{r}} \end{bmatrix} = \int_0^\infty \begin{bmatrix} 1 & 0 & -\omega^2 \\ 0 & \omega^2 & 0 \\ -\omega^2 & 0 & \omega^4 \end{bmatrix} \cdot \mathbf{S}_{rr} d\omega \tag{9.127}$$

Since the displacement response vector associated with element number n is given by $\mathbf{d}_n = \mathbf{A}_n \cdot \mathbf{r}$ it is seen that

$$\begin{bmatrix} \mathbf{Cov}_{d_n\dot{d}_n} \\ \mathbf{Cov}_{\dot{d}_n d_n} \\ \mathbf{Cov}_{\dot{d}_n\ddot{d}_n} \\ \mathbf{Cov}_{\ddot{d}_n\dot{d}_n} \end{bmatrix} = E\begin{bmatrix} \mathbf{d}\cdot\dot{\mathbf{d}}^T \\ \dot{\mathbf{d}}\cdot\mathbf{d}^T \\ \dot{\mathbf{d}}\cdot\ddot{\mathbf{d}}^T \\ \ddot{\mathbf{d}}\cdot\dot{\mathbf{d}}^T \end{bmatrix}_n = E\begin{bmatrix} (\mathbf{A}_n\mathbf{r})\cdot(\mathbf{A}_n\dot{\mathbf{r}})^T \\ (\mathbf{A}_n\dot{\mathbf{r}})\cdot(\mathbf{A}_n\mathbf{r})^T \\ (\mathbf{A}_n\dot{\mathbf{r}})\cdot(\mathbf{A}_n\ddot{\mathbf{r}})^T \\ (\mathbf{A}_n\ddot{\mathbf{r}})\cdot(\mathbf{A}_n\dot{\mathbf{r}})^T \end{bmatrix} = \mathbf{A}_n \cdot E\begin{bmatrix} \mathbf{r}\cdot\dot{\mathbf{r}}^T \\ \dot{\mathbf{r}}\cdot\mathbf{r}^T \\ \dot{\mathbf{r}}\cdot\ddot{\mathbf{r}}^T \\ \ddot{\mathbf{r}}\cdot\dot{\mathbf{r}}^T \end{bmatrix} \cdot \mathbf{A}_n^T = \begin{bmatrix} \mathbf{0} \\ \mathbf{0} \\ \mathbf{0} \\ \mathbf{0} \end{bmatrix} \tag{9.128}$$

while

$$\begin{bmatrix} \mathbf{Cov}_{d_n d_n} \\ \mathbf{Cov}_{d_n \ddot{d}_n} \\ \mathbf{Cov}_{\ddot{d}_n d_n} \\ \mathbf{Cov}_{\dot{d}_n \dot{d}_n} \\ \mathbf{Cov}_{\ddot{d}_n \ddot{d}_n} \end{bmatrix} = E \begin{bmatrix} \mathbf{d} \cdot \mathbf{d}^T \\ \mathbf{d} \cdot \ddot{\mathbf{d}}^T \\ \ddot{\mathbf{d}} \cdot \mathbf{d}^T \\ \dot{\mathbf{d}} \cdot \dot{\mathbf{d}}^T \\ \ddot{\mathbf{d}} \cdot \ddot{\mathbf{d}}^T \end{bmatrix}_n = E \begin{bmatrix} (\mathbf{A}_n \mathbf{r}) \cdot (\mathbf{A}_n \mathbf{r})^T \\ (\mathbf{A}_n \mathbf{r}) \cdot (\mathbf{A}_n \ddot{\mathbf{r}})^T \\ (\mathbf{A}_n \ddot{\mathbf{r}}) \cdot (\mathbf{A}_n \mathbf{r})^T \\ (\mathbf{A}_n \dot{\mathbf{r}}) \cdot (\mathbf{A}_n \dot{\mathbf{r}})^T \\ (\mathbf{A}_n \ddot{\mathbf{r}}) \cdot (\mathbf{A}_n \ddot{\mathbf{r}})^T \end{bmatrix} = \mathbf{A}_n \cdot E \begin{bmatrix} \mathbf{r} \cdot \mathbf{r}^T \\ \mathbf{r} \cdot \ddot{\mathbf{r}}^T \\ \ddot{\mathbf{r}} \cdot \mathbf{r}^T \\ \dot{\mathbf{r}} \cdot \dot{\mathbf{r}}^T \\ \ddot{\mathbf{r}} \cdot \ddot{\mathbf{r}}^T \end{bmatrix} \cdot \mathbf{A}_n^T = \mathbf{A}_n \cdot \begin{bmatrix} \mathbf{Cov}_{rr} \\ \mathbf{Cov}_{r\ddot{r}} \\ \mathbf{Cov}_{\ddot{r}r} \\ \mathbf{Cov}_{\dot{r}\dot{r}} \\ \mathbf{Cov}_{\ddot{r}\ddot{r}} \end{bmatrix} \cdot \mathbf{A}_n^T \tag{9.129}$$

By introducing the relevant expressions in Eq. 9.127 the following is obtained

$$\begin{bmatrix} \mathbf{Cov}_{d_n d_n} \\ \mathbf{Cov}_{d_n \ddot{d}_n} \\ \mathbf{Cov}_{\ddot{d}_n d_n} \\ \mathbf{Cov}_{\dot{d}_n \dot{d}_n} \\ \mathbf{Cov}_{\ddot{d}_n \ddot{d}_n} \end{bmatrix} = \mathbf{A}_n \cdot \left(\int_0^\infty \begin{bmatrix} 1 \\ -\omega^2 \\ -\omega^2 \\ \omega^2 \\ \omega^4 \end{bmatrix} \cdot \mathbf{S}_{rr}(\omega) d\omega \right) \cdot \mathbf{A}_n^T \tag{9.130}$$

The corresponding response force vector

$$\mathbf{F}_n(t) = \begin{bmatrix} \mathbf{F}_1 \\ \mathbf{F}_2 \end{bmatrix}_n \quad \text{where} \quad \begin{cases} \mathbf{F}_{1_n} = \begin{bmatrix} F_1 & F_2 & F_3 & F_4 & F_5 & F_6 \end{bmatrix}_n^T \\ \mathbf{F}_{2_n} = \begin{bmatrix} F_7 & F_8 & F_9 & F_{10} & F_{11} & F_{12} \end{bmatrix}_n^T \end{cases} \tag{9.131}$$

associated with element number n is defined by the local element dynamic equilibrium condition (see Eq. 9.25)

$$\mathbf{F}_n = \mathbf{m}_n \ddot{\mathbf{d}}_n + \mathbf{c}_n \dot{\mathbf{d}}_n + \mathbf{k}_n \mathbf{d}_n \tag{9.132}$$

and thus

$$\begin{aligned} \mathbf{Cov}_{F_n F_n} &= E\left[\mathbf{F}_n \cdot \mathbf{F}_n^T\right] = E\left[\left(\mathbf{m}_n \ddot{\mathbf{d}}_n + \mathbf{c}_n \dot{\mathbf{d}}_n + \mathbf{k}_n \mathbf{d}_n\right) \cdot \left(\mathbf{m}_n \ddot{\mathbf{d}}_n + \mathbf{c}_n \dot{\mathbf{d}}_n + \mathbf{k}_n \mathbf{d}_n\right)^T\right] \\ &= \mathbf{m}_n \cdot E\left[\ddot{\mathbf{d}}_n \cdot \ddot{\mathbf{d}}_n^T\right] \cdot \mathbf{m}_n^T + \mathbf{c}_n \cdot E\left[\dot{\mathbf{d}}_n \cdot \dot{\mathbf{d}}_n^T\right] \cdot \mathbf{c}_n^T + \mathbf{k}_n \cdot E\left[\mathbf{d}_n \cdot \mathbf{d}_n^T\right] \cdot \mathbf{k}_n^T \\ &\quad + \mathbf{k}_n \cdot E\left[\mathbf{d}_n \cdot \ddot{\mathbf{d}}_n^T\right] \cdot \mathbf{m}_n^T + \mathbf{m}_n \cdot E\left[\ddot{\mathbf{d}}_n \cdot \mathbf{d}_n^T\right] \cdot \mathbf{k}_n^T \end{aligned} \tag{9.133}$$

Introducing Eq. 9.128 the following is obtained

$$\mathbf{Cov}_{F_nF_n} = E\left[\mathbf{F}_n \cdot \mathbf{F}_n^T\right] = \begin{bmatrix} \sigma_{F_1}^2 & \cdots & Cov_{F_1F_i} & \cdots & Cov_{F_1F_{12}} \\ \vdots & \ddots & \vdots & \cdots & \vdots \\ Cov_{F_iF_1} & \cdots & \sigma_{F_i}^2 & \cdots & Cov_{F_iF_{12}} \\ \vdots & \cdots & \vdots & \ddots & \vdots \\ Cov_{F_{12}F_1} & \cdots & Cov_{F_{12}F_i} & \cdots & \sigma_{12}^2 \end{bmatrix} \tag{9.134}$$

$$= \mathbf{m}_n \cdot \mathbf{Cov}_{\ddot{d}_n\ddot{d}_n} \mathbf{m}_n^T + \mathbf{c}_n \cdot \mathbf{Cov}_{\dot{d}_n\dot{d}_n} \mathbf{c}_n^T + \mathbf{k}_n \cdot \mathbf{Cov}_{d_nd_n} \mathbf{k}_n^T$$
$$+\mathbf{m}_n \cdot \mathbf{Cov}_{\ddot{d}_nd_n} \mathbf{k}_n^T + \mathbf{k}_n \cdot \mathbf{Cov}_{d_n\ddot{d}_n} \mathbf{m}_n^T$$

where $\mathbf{Cov}_{\ddot{d}_n\ddot{d}_n}$, $\mathbf{Cov}_{\dot{d}_n\dot{d}_n}$, $\mathbf{Cov}_{d_nd_n}$, $\mathbf{Cov}_{\ddot{d}_nd_n}$ and $\mathbf{Cov}_{d_n\ddot{d}_n}$ are defined in Eq. 9.130. It should be noted that if damping has been defined at a global level (e.g. in the form of Rayleigh damping $\mathbf{C} = \alpha\mathbf{M} + \beta\mathbf{K}$), then the damping properties at element level should comply with the same choices of damping properties(i.e. $\mathbf{c}_n = \alpha\mathbf{m}_n + \beta\mathbf{k}_n$). It is also worth noting that if the chosen element length L_n is sufficiently small then the mass and damping terms above will be small, i.e. $\mathbf{Cov}_{F_nF_n} \approx \mathbf{k}_n \cdot \mathbf{Cov}_{d_nd_n} \mathbf{k}_n^T$.

9.8 Frequency domain response calculations in modal coordinates

The dynamic equilibrium condition in original discretised coordinates (see Eq. 9.80)

$$\mathbf{M}\,\ddot{\mathbf{r}}(t) + (\mathbf{C} - \mathbf{C}_{ae})\dot{\mathbf{r}}(t) + (\mathbf{K} - \mathbf{K}_{ae})\mathbf{r}(t) = \mathbf{R}(t) \tag{9.135}$$

may readily be transformed into a modal format by choosing

$$\mathbf{r}(t) = \mathbf{\Phi} \cdot \boldsymbol{\eta}(t) \tag{9.136}$$

where the vector $\boldsymbol{\eta}(t)$ contains N_{mod} modal coordinates

$$\boldsymbol{\eta}(t) = \begin{bmatrix} \eta_1 & \eta_2 & \cdots & \eta_i & \cdots & \eta_{N_{\text{mod}}} \end{bmatrix}^T \tag{9.137}$$

and $\mathbf{\Phi}$ contains the mode shapes

$$\mathbf{\Phi} = \begin{bmatrix} \boldsymbol{\varphi}_1 & \boldsymbol{\varphi}_2 & \cdots & \boldsymbol{\varphi}_i & \cdots & \boldsymbol{\varphi}_{N_{\text{mod}}} \end{bmatrix} \tag{9.138}$$

and where $\boldsymbol{\varphi}_i$ $(i = 1,2,\ldots,N_{\text{mod}})$ contains the mode shape values, each associated with the corresponding global degree of freedom number $k = 1,2,\ldots,N_r$, i.e.

$$\boldsymbol{\varphi}_i = \begin{bmatrix} \phi_1 & \phi_2 & \cdots & \phi_k & \cdots & \phi_{N_r} \end{bmatrix}^T \tag{9.139}$$

It should be noted that these eigen-modes are identical to those in Chapters 4, 6 and 7, but their mathematical description differs in the way their components are organised. While the eigen-modes in Chapters 4, 6 and 7 are based on a description of components associated with motion in horizontal, vertical and torsion directions, the eigen-modes in Eq. 9.137 contain displacement components associated with the global degrees of freedom in the finite element model of the system.

It should also be noted that because $\mathbf{K}_{ae}$ depends of the mean wind velocity, so will the total stiffness of the system, and thus, the resonance frequencies and the associated mode shapes will change with increasing mean wind velocities. These changes are not negligible with respect to the overall behaviour of the system in the close vicinity of an instability limit, where it will usually be necessary to update system quantities due to these effects (resonance frequencies in particular, but it can not be ruled out that in some cases it may also be necessary to update mode-shapes). Sufficiently far from an instability limit these effects are small such that the modal solution strategy of most wind engineering problems may be based on the eigen-frequencies ω_i and corresponding mode shapes $\boldsymbol{\varphi}_i$ as determined from the eigen-value problem in still air, i.e. from

$$\left[\mathbf{K} - \omega_i^2 \mathbf{M}\right] \boldsymbol{\varphi}_i = \mathbf{0} \tag{9.140}$$

Introducing Eq. 9.136 into Eq. 9.135, and pre-multiplying the entire equation by $\mathbf{\Phi}^T$

$$\mathbf{\Phi}^T \mathbf{M}\, \mathbf{\Phi} \ddot{\boldsymbol{\eta}}(t) + \left(\mathbf{\Phi}^T \mathbf{C} \mathbf{\Phi} - \mathbf{\Phi}^T \mathbf{C}_{ae} \mathbf{\Phi}\right) \dot{\boldsymbol{\eta}}(t) + \left(\mathbf{\Phi}^T \mathbf{K} \mathbf{\Phi} - \mathbf{\Phi}^T \mathbf{K}_{ae} \mathbf{\Phi}\right) \boldsymbol{\eta}(t) = \mathbf{\Phi}^T \mathbf{R}(t) \tag{9.141}$$

and defining the modal quantities

$$\left.\begin{aligned} \tilde{\mathbf{M}} &= \mathbf{\Phi}^T \mathbf{M} \mathbf{\Phi} \\ \tilde{\mathbf{C}} &= \mathbf{\Phi}^T \mathbf{C} \mathbf{\Phi} \\ \tilde{\mathbf{K}} &= \mathbf{\Phi}^T \mathbf{M} \mathbf{\Phi} \end{aligned}\right\} \quad \text{and} \quad \left.\begin{aligned} \tilde{\mathbf{C}}_{ae} &= \mathbf{\Phi}^T \mathbf{C}_{ae} \mathbf{\Phi} \\ \tilde{\mathbf{K}}_{ae} &= \mathbf{\Phi}^T \mathbf{K}_{ae} \mathbf{\Phi} \\ \tilde{\mathbf{R}} &= \mathbf{\Phi}^T \mathbf{R} \end{aligned}\right\} \tag{9.142}$$

then the following modal dynamic equilibrium condition is obtained

$$\tilde{\mathbf{M}}\ddot{\boldsymbol{\eta}} + \left(\tilde{\mathbf{C}} - \tilde{\mathbf{C}}_{ae}\right)\dot{\boldsymbol{\eta}} + \left(\tilde{\mathbf{K}} - \tilde{\mathbf{K}}_{ae}\right)\boldsymbol{\eta} = \tilde{\mathbf{R}} \tag{9.143}$$

Due to the orthogonal properties of the mode shapes all the off diagonal terms in $\tilde{\mathbf{M}}$, $\tilde{\mathbf{C}}$ and $\tilde{\mathbf{K}}$ are zeros. Thus

$$\left.\begin{array}{l} \tilde{\mathbf{M}} = diag\left[\tilde{M}_i\right] \\ \tilde{\mathbf{C}} = diag\left[\tilde{C}_i\right] \\ \tilde{\mathbf{K}} = diag\left[\tilde{K}_i\right] \end{array}\right\} \quad \text{where} \quad \left.\begin{array}{l} \tilde{M}_i = \boldsymbol{\varphi}_i^T \mathbf{M} \boldsymbol{\varphi}_i \\ C_i = \boldsymbol{\varphi}_i^T \mathbf{C} \boldsymbol{\varphi}_i \\ \tilde{K}_i = \boldsymbol{\varphi}_i^T \mathbf{K} \boldsymbol{\varphi}_i \end{array}\right\} \tag{9.144}$$

However, it is readily seen that $\tilde{\mathbf{K}}$ may more conveniently be determined from Eq. 9.140. Pre-multiplying $\left[\mathbf{K} - \omega_i^2 \mathbf{M}\right]\boldsymbol{\varphi}_i = \mathbf{0}$ by $\boldsymbol{\varphi}_i^T$, and thus it is seen that

$$\tilde{K}_i = \omega_i^2 \tilde{M}_i \tag{9.145}$$

Furthermore, it is common practice to introduce N_{mod} modal damping ratios ζ_i , each associated with the corresponding modal critical damping $2\tilde{M}_i\omega_i$, and thus

$$\tilde{C}_i = 2\tilde{M}_i \omega_i \zeta_i \tag{9.146}$$

It is seen from Eqs. 9.52 – 9.55 that $\mathbf{c}_{ae_0}$ and $\mathbf{k}_{ae_0}$ are not symmetric, and thus, nor are the corresponding element properties $\mathbf{c}_{ae_n}$ and $\mathbf{k}_{ae_n}$ defined in Eq. 9.63. Therefore, the aerodynamic damping and stiffness matrices $\mathbf{C}_{ae}$ and $\mathbf{K}_{ae}$ are non-symmetric, and the orthogonal mode shapes will not nullify off-diagonal terms in $\tilde{\mathbf{C}}_{ae}$ and $\tilde{\mathbf{K}}_{ae}$ above. Thus

$$\tilde{\mathbf{C}}_{ae} = \begin{bmatrix} \ddots & & \iddots \\ & \tilde{C}_{ae_{ij}} & \\ \iddots & & \ddots \end{bmatrix} \quad \text{where} \quad \tilde{C}_{ae_{ij}} = \boldsymbol{\varphi}_i^T \mathbf{C}_{ae} \boldsymbol{\varphi}_j \tag{9.147}$$

and

$$\tilde{\mathbf{K}}_{ae} = \begin{bmatrix} \ddots & & \iddots \\ & \tilde{K}_{ae_{ij}} & \\ \iddots & & \ddots \end{bmatrix} \quad \text{where} \quad \tilde{K}_{ae_{ij}} = \boldsymbol{\varphi}_i^T \mathbf{K}_{ae} \boldsymbol{\varphi}_j \tag{9.148}$$

Thus, all of Eq. 9.143 has been defined. A time domain solution of this equation is dealt with in chapter 9.9 below.

The procedure for a frequency domain solution in modal coordinates is identical to that which has been shown in original coordinates in chapter 9.6. Taking the Fourier transform throughout Eq. 9.143, i.e. setting

$$\boldsymbol{\eta}(t) = \sum_{\omega} \mathbf{a}_{\eta}(\omega) \cdot e^{i\omega t} \qquad \text{and} \qquad \tilde{\mathbf{R}}(t) = \sum_{\omega} \mathbf{a}_{\tilde{R}}(\omega) \cdot e^{i\omega t} \tag{9.149}$$

where N_{mod} by 1 vectors $\mathbf{a}_{\eta}(\omega)$ and $\mathbf{a}_{\tilde{R}}(\omega)$ contain the Fourier coefficients of the modal coordinates and the modal and load, and where $\sum_{\omega}$ indicates summation over all ω-settings. Then Eq. 9.141 is satisfied for each ω-setting if

$$\left[-\tilde{\mathbf{M}}\omega^2 + \left(\tilde{\mathbf{C}} - \tilde{\mathbf{C}}_{ae} \right) i\omega + \left(\tilde{\mathbf{K}} - \tilde{\mathbf{K}}_{ae} \right) \right] \cdot \mathbf{a}_{\eta} = \mathbf{a}_{\tilde{R}} \tag{9.150}$$

Thus,

$$\Rightarrow \qquad \mathbf{a}_{\eta} = \tilde{\mathbf{H}}_{\eta}(\omega) \cdot \mathbf{a}_{\tilde{R}} \tag{9.151}$$

where

$$\tilde{\mathbf{H}}_{\eta}(\omega) = \left[-\tilde{\mathbf{M}}\omega^2 + \left(\tilde{\mathbf{C}} - \tilde{\mathbf{C}}_{ae} \right) i\omega + \left(\tilde{\mathbf{K}} - \tilde{\mathbf{K}}_{ae} \right) \right]^{-1} \tag{9.152}$$

The cross spectral density matrix of the modal coordinates is defined by

$$\mathbf{S}_{\eta\eta}(\omega) = \lim_{T \to \infty} \frac{1}{\pi T} \left(\mathbf{a}_{\eta}^{*} \cdot \mathbf{a}_{\eta}^{T} \right) = \begin{bmatrix} S_{\eta_1\eta_1} & \cdots & S_{\eta_1\eta_j} & \cdots & S_{\eta_1\eta_{N_{\text{mod}}}} \\ \vdots & \ddots & \vdots & \ddots & \vdots \\ S_{\eta_i\eta_1} & \cdots & S_{\eta_i\eta_j} & \cdots & S_{\eta_i\eta_{N_{\text{mod}}}} \\ \vdots & \ddots & \vdots & \ddots & \vdots \\ S_{\eta_{N_{\text{mod}}}\eta_1} & \cdots & S_{\eta_{N_{\text{mod}}}\eta_j} & \cdots & S_{\eta_{N_{\text{mod}}}\eta_{N_{\text{mod}}}} \end{bmatrix} \tag{9.153}$$

Introducing Eq. 9.151, then

$$\begin{aligned} \mathbf{S}_{\eta\eta}(\omega) &= \lim_{T \to \infty} \frac{1}{\pi T} \left(\mathbf{a}_{\eta}^{*} \; \mathbf{a}_{\eta}^{T} \right) = \lim_{T \to \infty} \frac{1}{\pi T} \left[\left(\tilde{\mathbf{H}}_{\eta} \mathbf{a}_{\tilde{R}} \right)^{*} \cdot \left(\tilde{\mathbf{H}}_{\eta} \mathbf{a}_{\tilde{R}} \right)^{T} \right] \\ &= \tilde{\mathbf{H}}_{\eta}^{*} \cdot \lim_{T \to \infty} \frac{1}{\pi T} \left(\mathbf{a}_{\tilde{R}}^{*} \cdot \mathbf{a}_{\tilde{R}}^{T} \right) \cdot \tilde{\mathbf{H}}_{\eta}^{T} = \tilde{\mathbf{H}}_{\eta}^{*} \cdot \mathbf{S}_{\tilde{R}\tilde{R}} \cdot \tilde{\mathbf{H}}_{\eta}^{T} \end{aligned} \tag{9.154}$$

where $\mathbf{S}_{\tilde{R}\tilde{R}}(\omega) = \lim_{T\to\infty} \frac{1}{\pi T}\left(\mathbf{a}_{\tilde{R}}^{*} \cdot \mathbf{a}_{\tilde{R}}^{T}\right)$ is the cross spectral density matrix of the modal buffeting wind load. $\mathbf{R}(t)$ is given in Eq. 9.111, and thus

$$\tilde{\mathbf{R}}(t) = \mathbf{\Phi}^T \mathbf{R}(t) = \mathbf{\Phi}^T \cdot \sum_{n=1}^{N} \mathbf{A}_n^T \cdot \mathbf{R}_n(t) = \mathbf{\Phi}^T \cdot \sum_{n=1}^{N} \mathbf{A}_n^T \cdot \left[\mathbf{B}_{Q_n} \boldsymbol{\psi}_n \hat{\mathbf{v}}_n(t)\right] \tag{9.155}$$

Taking the Fourier transform throughout Eq. 9.155 renders

$$\mathbf{a}_{\tilde{R}}(\omega) = \mathbf{\Phi}^T \cdot \sum_{n=1}^{N} \mathbf{A}_n^T \cdot \left[\mathbf{B}_{Q_n} \boldsymbol{\psi}_n \mathbf{a}_{\hat{v}_n}(\omega)\right] \tag{9.156}$$

and thus, the cross spectral density matrix of the modal buffeting wind load is given by

$$\begin{aligned}
\mathbf{S}_{\tilde{R}\tilde{R}}(\omega) &= \lim_{T\to\infty} \frac{1}{\pi T} \left\{ \left[\mathbf{\Phi}^T \sum_{n=1}^{N} \mathbf{A}_n^T \left(\mathbf{B}_{Q_n} \boldsymbol{\psi}_n \mathbf{a}_{\hat{v}_n}^{*}\right)\right] \cdot \left[\mathbf{\Phi}^T \sum_{n=1}^{N} \mathbf{A}_n^T \left(\mathbf{B}_{Q_n} \boldsymbol{\psi}_n \mathbf{a}_{\hat{v}_n}\right)\right]^T \right\} \\
&= \mathbf{\Phi}^T \left\{ \sum_{n=1}^{N} \sum_{m=1}^{N} \mathbf{A}_n^T \left[\mathbf{B}_{Q_n} \boldsymbol{\psi}_n \cdot \lim_{T\to\infty} \frac{1}{\pi T} \left(\mathbf{a}_{\hat{v}_n}^{*} \cdot \mathbf{a}_{\hat{v}_m}^{T}\right) \cdot \boldsymbol{\psi}_m^T \mathbf{B}_{Q_m}^T \right] \mathbf{A}_m \right\} \mathbf{\Phi} \\
&= \mathbf{\Phi}^T \cdot \left\{ \sum_{n=1}^{N} \sum_{m=1}^{N} \mathbf{A}_n^T \cdot \left[\mathbf{B}_{Q_n} \boldsymbol{\psi}_n \cdot \hat{\mathbf{S}}_{\hat{v}\hat{v}}(\omega) \cdot \boldsymbol{\psi}_m^T \mathbf{B}_{Q_m}^T \right] \cdot \mathbf{A}_m \right\} \cdot \mathbf{\Phi}
\end{aligned} \tag{9.157}$$

where $\hat{\mathbf{S}}_{\hat{v}\hat{v}}(\omega) = \lim_{T\to\infty} \frac{1}{\pi T}\left(\mathbf{a}_{\hat{v}_n}^{*} \cdot \mathbf{a}_{\hat{v}_m}^{T}\right)$ is the cross spectral density matrix of the reduced turbulence velocity vector, previously developed in chapter 9.6, see Eq. 9.124. Thus

$$\begin{aligned}
\mathbf{S}_{\eta\eta}(\omega) &= \tilde{\mathbf{H}}_\eta^{*}(\omega) \cdot \mathbf{S}_{\tilde{R}\tilde{R}}(\omega) \cdot \tilde{\mathbf{H}}_\eta^{T}(\omega) = \tilde{\mathbf{H}}_\eta^{*}(\omega) \cdot \left[\mathbf{\Phi}^T \cdot \mathbf{S}_{RR}(\omega) \cdot \mathbf{\Phi}\right] \cdot \tilde{\mathbf{H}}_\eta^{T}(\omega) \\
&= \tilde{\mathbf{H}}_\eta^{*} \left(\mathbf{\Phi}^T \left\{ \sum_{n=1}^{N} \sum_{m=1}^{N} \mathbf{A}_n^T \left[\mathbf{B}_n \boldsymbol{\psi}_n \cdot \left(\hat{\mathbf{S}}_n^{1/2} \hat{\mathbf{C}}\mathbf{o}_{nm} \hat{\mathbf{S}}_m^{1/2}\right) \cdot \boldsymbol{\psi}_m^T \mathbf{B}_m^T \right] \mathbf{A}_m \right\} \mathbf{\Phi} \right) \tilde{\mathbf{H}}_\eta^{T}
\end{aligned} \tag{9.158}$$

where $\hat{\mathbf{C}}\mathbf{o}_{nm}(\Delta s_{nm}, \omega)$, $\hat{\mathbf{S}}_n(\omega)$ and $\hat{\mathbf{S}}_m(\omega)$ are defined in Eqs. 9.120 and 9.121. Finally, since $\mathbf{r}(t) = \mathbf{\Phi} \cdot \boldsymbol{\eta}(t)$, then $\mathbf{a}_r(\omega) = \mathbf{\Phi} \cdot \mathbf{a}_\eta(\omega)$, and thus, the cross spectral density matrix of the displacement response quantities is given by

$$\mathbf{S}_{rr}(\omega) = \lim_{T\to\infty}\frac{1}{\pi T}\left(\mathbf{a}_r^* \cdot \mathbf{a}_r^T\right) = \lim_{T\to\infty}\frac{1}{\pi T}\left[\left(\mathbf{\Phi a}_\eta\right)^* \cdot \left(\mathbf{\Phi a}_\eta\right)^T\right]$$

$$= \mathbf{\Phi} \cdot \lim_{T\to\infty}\frac{1}{\pi T}\left(\mathbf{a}_\eta^* \cdot \mathbf{a}_\eta^T\right)\cdot \mathbf{\Phi}^T = \mathbf{\Phi}\cdot\mathbf{S}_{\eta\eta}\cdot\mathbf{\Phi}^T \tag{9.159}$$

$$= \mathbf{\Phi}\cdot\left[\tilde{\mathbf{H}}_\eta^* \cdot \left(\mathbf{\Phi}^T \mathbf{S}_{RR}\mathbf{\Phi}\right)\cdot \tilde{\mathbf{H}}_\eta^T\right]\cdot\mathbf{\Phi}^T$$

where

$$\mathbf{S}_{RR}(\omega) = \sum_{n=1}^{N}\sum_{m=1}^{N}\mathbf{A}_n^T\left[\mathbf{B}_{Q_n}\boldsymbol{\psi}_n \cdot \left(\hat{\mathbf{S}}_n^{1/2}\hat{\mathbf{C}}\mathrm{o}_{nm}\hat{\mathbf{S}}_m^{1/2}\right)\cdot \boldsymbol{\psi}_m^T\mathbf{B}_{Q_m}^T\right]\mathbf{A}_m \tag{9.160}$$

The response covariance matrix may then be obtained simply by integration throughout the frequency domain, i.e.

$$\mathbf{Cov}_{rr} = \begin{bmatrix} \sigma_1^2 & \cdots & Cov_{1i} & \cdots & Cov_{1j} & \cdots & Cov_{1N_r} \\ \vdots & \ddots & \vdots & & \vdots & & \vdots \\ Cov_{i1} & \cdots & \sigma_i^2 & \cdots & Cov_{ij} & & \vdots \\ \vdots & & \vdots & \ddots & \vdots & & \vdots \\ Cov_{j1} & \cdots & Cov_{ji} & \cdots & \sigma_j^2 & & \vdots \\ \vdots & & & & & \ddots & \vdots \\ Cov_{N_r 1} & \cdots & \cdots & \cdots & \cdots & \cdots & \sigma_{N_r}^2 \end{bmatrix} = \int_0^\infty \mathbf{S}_{rr}(\omega)d\omega \tag{9.161}$$

This solution strategy will render identical results to that which has been developed in chapter 6.3. The only difference is that the mode shapes have been arranged in a different order, i.e. in this chapter in an order which correspond to the chosen degrees of freedom in a finite element formulation.

Once $\mathbf{S}_{rr}(\omega)$ has been established, then the ensuing determination of the statistical properties of cross sectional forces is identical to that which has been shown in chapter 9.6 above (see Es. 9.127 – 9.134).

9.9 Dynamic response calculations in time domain

In a time domain solution the total displacement response may be obtained as a sum of the time invariant solution $\overline{\mathbf{r}}$ (given in Eq. 9.86) and a purely dynamic part $\mathbf{r}(t)$, i.e.

$$\mathbf{r}_{tot}(t) = \overline{\mathbf{r}} + \mathbf{r}(t) \tag{9.162}$$

A solution strategy may be pursued in the original finite element degrees of freedom (developed in Chapter 9.4) or in modal coordinates (see chapter 9.8). If a solution in original degrees of freedom is pursued, then the calculation of $\mathbf{r}(t)$ will require the solution of the dynamic equation given in Eq. 9.80 or 9.85, depending on the choice of motion induced load description. It is seen that either of these equations may be written on the general form

$$\mathbf{M}\,\ddot{\mathbf{r}}(t) + \mathbf{C}_{net}\dot{\mathbf{r}}(t) + \mathbf{K}_{net}\mathbf{r}(t) = \mathbf{R}_{dyn}(t) \tag{9.163}$$

where $\left.\begin{aligned} \mathbf{C}_{net} &= \mathbf{C} - \mathbf{C}_{ae} \\ \mathbf{K}_{net} &= \mathbf{K} - \mathbf{K}_{ae} \\ \mathbf{R}_{dyn}(t) &= \mathbf{R}(t) \end{aligned}\right\}$ if quasi-static properties are adopted,

$\left.\begin{aligned} \mathbf{C}_{net} &= \mathbf{C} - \mathbf{C}_{ae}(s=0) \\ \mathbf{K}_{net} &= \mathbf{K} - \mathbf{K}_{ae}(s=0) \\ \mathbf{R}_{dyn}(t) &= \mathbf{R}(t) + \Delta\mathbf{R}_{ae}(t) \end{aligned}\right\}$ if indicial functions are adopted.

If a solution in modal coordinates is pursued, then the calculation of $\mathbf{r}(t) = \mathbf{\Phi} \cdot \boldsymbol{\eta}(t)$ requires solution of the dynamic equation given in Eq. 9.143, i.e.

$$\tilde{\mathbf{M}}\ddot{\boldsymbol{\eta}}(t) + \left(\tilde{\mathbf{C}} - \tilde{\mathbf{C}}_{ae}\right)\dot{\boldsymbol{\eta}}(t) + \left(\tilde{\mathbf{K}} - \tilde{\mathbf{K}}_{ae}\right)\boldsymbol{\eta}(t) = \tilde{\mathbf{R}}(t) \tag{9.164}$$

where all quantities are defined in Eqs. 9.142, 9.144 – 9.148 and 9.155.

In any time domain solution it will be necessary to perform a stochastic simulation of the stationary flow components $u(t)$, $v(t)$ and $w(t)$ contained in the buffeting load vector $\mathbf{R}_{dyn}(t)$ or $\tilde{\mathbf{R}}(t)$. Such a simulation procedure is presented in Appendix A.3.

There are a number of iteration procedures available for a time domain solution strategy. Only a selected few are included below. In any case, as illustrated in Fig. 9.9 a time domain solution will involve some discretisation of the load processes $\mathbf{R}_{dyn}(t)$ or $\tilde{\mathbf{R}}(t)$ at time step t_k ($k = 1,2,,....,N_k$), and a stepwise calculation of the corresponding response ($\mathbf{r}$ or $\boldsymbol{\eta}$). Based on the knowledge of the response at time step t_k and the discretised values of the load, the task at hand is to calculate the response at time step t_{k+1}. Such a forward prediction routine is called explicit if it is based on the known response history alone. It is called implicit if it contains assumptions about the response situation or equilibrium condition in the unknown future of the system. I.e., in an explicit routine $\mathbf{r}(t_k + \Delta t)$ is a function of $\mathbf{r}$, $\dot{\mathbf{r}}$ and $\ddot{\mathbf{r}}$ at $t = t_k$, while an implicit routine contains some assumptions about the development of the motion in the time step between t_k and $t_k + \Delta t$. Obviously, the shorter time step, the easier it is to obtain a good solution. Among the implicit methods there are two main classes of solution strategies, i.e. the direct iteration procedures and the numeric integration methods. Some of these are presented below.

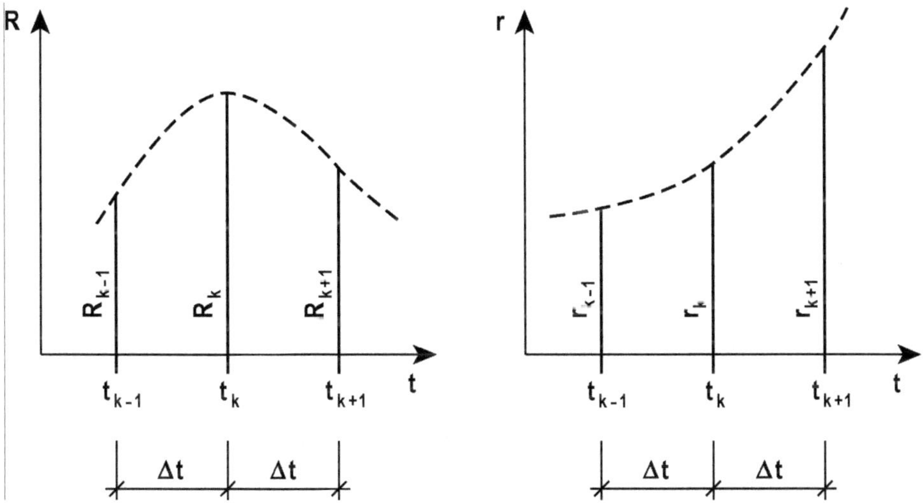

Fig. 9.9 *Time domain*

<u>*The second central difference method*</u>

Consider the situation at time step t_{k-1}, t_k and t_{k+1}. A Taylor series expansion of $\mathbf{r}_{k+1}$ and $\mathbf{r}_{k-1}$ is given by:

$$\left.\begin{aligned} \mathbf{r}_{k+1} &= \mathbf{r}_k + \Delta t \cdot \dot{\mathbf{r}}_k + \frac{\Delta t^2}{2} \cdot \ddot{\mathbf{r}}_k + \cdots \\ \mathbf{r}_{k-1} &= \mathbf{r}_k - \Delta t \cdot \dot{\mathbf{r}}_k + \frac{\Delta t^2}{2} \cdot \ddot{\mathbf{r}}_k - \cdots \end{aligned}\right\} \tag{9.165}$$

Thus, considering only the three first terms of the Taylor series expansion, $\mathbf{r}_{k+1} - \mathbf{r}_{k-1}$ renders

$$\dot{\mathbf{r}}_k \approx \frac{1}{2\Delta t} \cdot \left(\mathbf{r}_{k+1} - \mathbf{r}_{k-1}\right) \tag{9.166}$$

while $\mathbf{r}_{k+1} + \mathbf{r}_{k-1}$ renders

$$\ddot{\mathbf{r}}_k \approx \frac{1}{\Delta t^2} \cdot \left(\mathbf{r}_{k+1} - 2\mathbf{r}_k + \mathbf{r}_{k-1} \right) \tag{9.167}$$

Dynamic equilibrium at t_k is given by

$$\mathbf{M}\,\ddot{\mathbf{r}}_k + \mathbf{C}_{net}\dot{\mathbf{r}}_k + \mathbf{K}_{net}\mathbf{r}_k = \mathbf{R}_{dyn_k} \tag{9.168}$$

Introducing $\dot{\mathbf{r}}_k$ and $\ddot{\mathbf{r}}_k$ from Eqs. 9.166 and 9.167

$$\mathbf{M}\,\frac{1}{\Delta t^2} \cdot \left(\mathbf{r}_{k+1} - 2\mathbf{r}_k + \mathbf{r}_{k-1} \right) + \mathbf{C}_{net}\,\frac{1}{2\Delta t} \cdot \left(\mathbf{r}_{k+1} - \mathbf{r}_{k-1} \right) + \mathbf{K}_{net}\mathbf{r}_k = \mathbf{R}_{dyn_k} \tag{9.169}$$

and solving for $\mathbf{r}_{k+1}$

$$\mathbf{r}_{k+1} = \left(\mathbf{M} + \frac{\Delta t}{2}\mathbf{C}_{net} \right)^{-1} \left[\Delta t^2 \mathbf{R}_{dyn_k} + \left(2\mathbf{M} - \Delta t^2 \mathbf{K}_{net} \right)\mathbf{r}_k - \left(\mathbf{M} - \frac{\Delta t}{2}\mathbf{C}_{net} \right)\mathbf{r}_{k-1} \right] \tag{9.170}$$

Thus, it is seen that $\mathbf{r}_{k+1}$ may be estimated based on knowledge about the load and response quantise at t_k and t_{k-1}.

For the establishment of initial conditions at $t = 0$ before the iteration procedure can start it is necessary to define (choose) $\mathbf{r}_0$ and $\dot{\mathbf{r}}_0$. Dynamic equilibrium at $t = 0$ will then render the corresponding acceleration

$$\ddot{\mathbf{r}}_0 = \mathbf{M}^{-1} \cdot \left(\mathbf{R}_{dyn_0} - \mathbf{C}_{net}\dot{\mathbf{r}}_0 - \mathbf{K}_{net}\mathbf{r}_0 \right) \tag{9.171}$$

while eliminating $\mathbf{r}_1$ from (see Eqs. 9.166 and 9.167)

$$\left. \begin{aligned} \dot{\mathbf{r}}_0 &\approx \frac{1}{2\Delta t} \cdot \left(\mathbf{r}_1 - \mathbf{r}_{-1} \right) \\ \ddot{\mathbf{r}}_0 &\approx \frac{1}{\Delta t^2} \cdot \left(\mathbf{r}_1 - 2\mathbf{r}_0 + \mathbf{r}_{-1} \right) \end{aligned} \right\} \tag{9.172}$$

renders

$$\mathbf{r}_{-1} = \mathbf{r}_0 - \Delta t \cdot \dot{\mathbf{r}}_0 + \frac{\Delta t^2}{2}\ddot{\mathbf{r}}_0 \tag{9.173}$$

Thus,

$$\mathbf{r}_1 = \left(\mathbf{M} + \frac{\Delta t}{2}\mathbf{C} \right)^{-1} \left[\Delta t^2 \mathbf{R}_0 + \left(2\mathbf{M} - \Delta t^2 \mathbf{K} \right)\mathbf{r}_0 - \left(\mathbf{M} - \frac{\Delta t}{2}\mathbf{C} \right)\mathbf{r}_{-1} \right] \quad (9.174)$$

The stability of the second central difference method may be evaluated by considering an undamped and unloaded single-degree-of freedom system (se Eqs. 9.167 and 9.168) with mass M_i, stiffness K_i and eigen-frequency $\omega_i^2 = K_i / M_i$

$$\left. \begin{aligned} & M_i \frac{1}{\Delta t^2}\left(r_{k+1} - 2r_k + r_{k-1} \right) + K_i r_k = 0 \\ & \Rightarrow r_{k+1} + \left(\Delta t^2 \omega_i^2 - 2 \right) r_k + r_{k-1} = 0 \end{aligned} \right\} \quad (9.175)$$

which, for a harmonic motion $r = ae^{\lambda t}$ (where a is the amplitude), will render

$$\left. \begin{aligned} & ae^{\lambda(t_k + \Delta t)} + \left(\Delta t^2 \omega_i^2 - 2 \right) ae^{\lambda t_k} + ae^{\lambda(t_k - \Delta t)} = 0 \\ \Rightarrow \quad & \left(e^{\lambda \Delta t} \right)^2 + \left(\Delta t^2 \omega_i^2 - 2 \right) e^{\lambda \Delta t} + 1 = 0 \\ \Rightarrow \quad & \left. \begin{aligned} \left(e^{\lambda \Delta t} \right)_1 \\ \left(e^{\lambda \Delta t} \right)_2 \end{aligned} \right\} = \frac{1}{2}\left[2 - \Delta t^2 \omega_i^2 \pm \Delta t \omega_i \sqrt{\Delta t^2 \omega_i^2 - 4} \right] \end{aligned} \right\} \quad (9.176)$$

and thus the response at t_{k+1} is given by

$$r_{k+1} = ae^{\lambda(t_k + \Delta t)} = ae^{\lambda t_k} e^{\lambda \Delta t} = ae^{\lambda t_k} \left[c_1 \left(e^{\lambda \Delta t} \right)_1 + c_1 \left(e^{\lambda \Delta t} \right)_2 \right] \quad (9.177)$$

where c_1 and c_2 are constants dependant on initial conditions. In a second order equation $\alpha x^2 + \beta x + \gamma = 0$ the product of the roots $x_1 \cdot x_2 = \gamma / \alpha$. Thus, it is seen from Eq. 9.176 that $\left(e^{\lambda \Delta t} \right)_1 \cdot \left(e^{\lambda \Delta t} \right)_2 = 1$. It may be taken for granted that both roots are distinct. A positive radicand in the solution in Eq. 9.176 will render two real roots, and since the product of the two roots is unity one of them must be larger that one, and thus, the solution is consistently growing, i.e. it is unstable. A negative radicand on the other hand will render complex roots, and the product of the two roots can only be unity if they are complex conjugates and both has an absolute value equal to one. Thus, the solution is numerically stable if

$$\Delta t^2 \omega_i^2 - 4 \le 0 \qquad \Rightarrow \qquad \Delta t \le \frac{2}{\omega_i} \tag{9.178}$$

where (obviously), ω_i is the largest eigen-frequency expected to play any significant role in the response behaviour of the system.

In any case Δt should not be chosen larger than about $1/\left(2\omega_{R_{\max}}\right)$ where $\omega_{R_{\max}}$ is the largest frequency contained in the load.

<u>Numeric integration methods</u>

The numeric integration methods are based on the assumption that higher order quantities may be obtained from the situation at the onset of the time step and integration of the response development between time steps, i.e. that

$$\left.\begin{aligned} \dot{\mathbf{r}}_{k+1} &= \dot{\mathbf{r}}_k + \int_0^{\Delta t} \ddot{\mathbf{r}}(\tau) d\tau \\ \mathbf{r}_{k+1} &= \mathbf{r}_k + \int_0^{\Delta t} \dot{\mathbf{r}}(\tau) d\tau \end{aligned}\right\} \quad \text{where} \quad 0 \le \tau \le \Delta t \tag{9.179}$$

As illustrated in Fig. 9.10 the approximation is that the acceleration variation within the time step is either assumed constant and equal to its initial value, it is assumed equal to the average acceleration over the time step or it is assumed linear across the time step. If constant initial acceleration is assumed (see Fig. 9.10.a), then

$$\ddot{\mathbf{r}}(\tau) = \ddot{\mathbf{r}}_k \qquad \Rightarrow \qquad \dot{\mathbf{r}}(\tau) = \dot{\mathbf{r}}_k + \int_0^{\tau} \ddot{\mathbf{r}}_k d\tau = \dot{\mathbf{r}}_k + \ddot{\mathbf{r}}_k \cdot \tau \tag{9.180}$$

and thus

$$\left.\begin{aligned} \dot{\mathbf{r}}_{k+1} &= \dot{\mathbf{r}}_k + \int_0^{\Delta t} \ddot{\mathbf{r}}_k d\tau = \dot{\mathbf{r}}_k + \Delta t\, \ddot{\mathbf{r}}_k \\ \mathbf{r}_{k+1} &= \mathbf{r}_k + \int_0^{\Delta t} \left(\dot{\mathbf{r}}_k + \tau \cdot \ddot{\mathbf{r}}_k\right) d\tau = \mathbf{r}_k + \Delta t \cdot \dot{\mathbf{r}}_k + \frac{\Delta t^2}{2} \ddot{\mathbf{r}}_k \end{aligned}\right\} \tag{9.181}$$

If the concept of a constant average acceleration is adopted (see Fig. 9.10.b), then

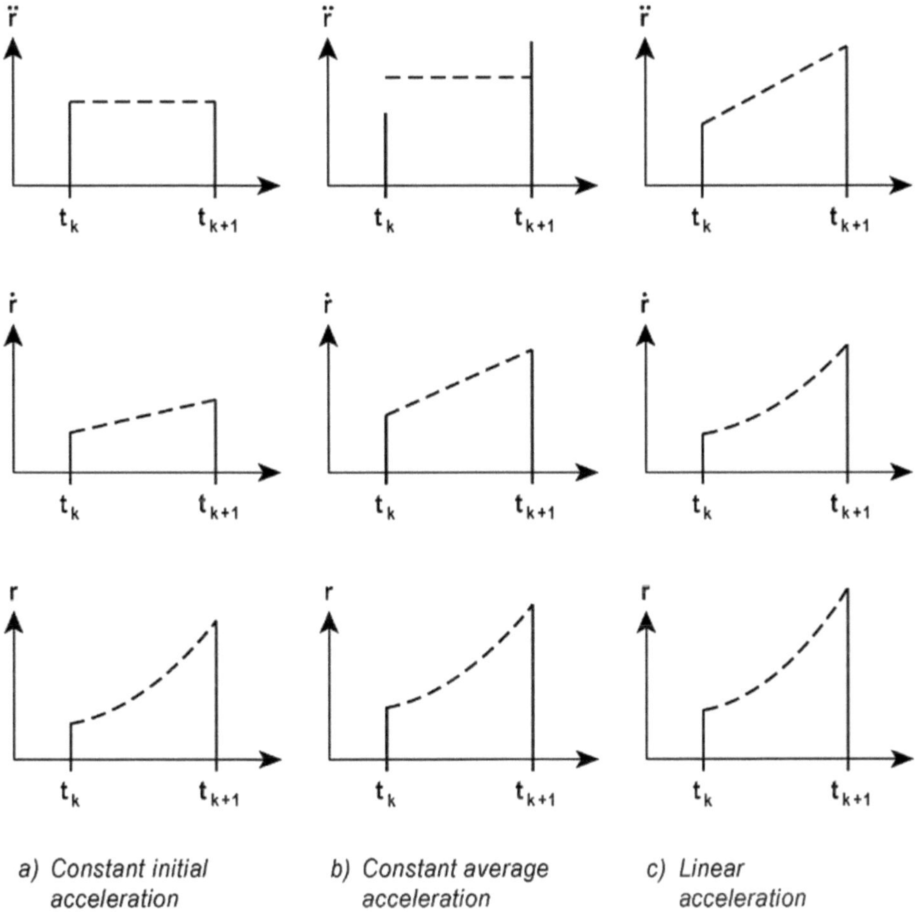

Fig. 9.10 *Numeric integration assumptions*

$$\ddot{\mathbf{r}}(\tau) = \frac{1}{2}\left(\ddot{\mathbf{r}}_k + \ddot{\mathbf{r}}_{k+1}\right) \quad \Rightarrow \quad \dot{\mathbf{r}}(\tau) = \dot{\mathbf{r}}_k + \int_0^{\tau} \frac{1}{2}\left(\ddot{\mathbf{r}}_k + \ddot{\mathbf{r}}_{k+1}\right) d\tau = \dot{\mathbf{r}}_k + \frac{\tau}{2}\left(\ddot{\mathbf{r}}_k + \ddot{\mathbf{r}}_{k+1}\right) \tag{9.182}$$

and thus

$$\left.\begin{aligned} \dot{\mathbf{r}}_{k+1} &= \dot{\mathbf{r}}_k + \int_0^{\Delta t} \frac{1}{2}\left(\ddot{\mathbf{r}}_k + \ddot{\mathbf{r}}_{k+1}\right) d\tau = \dot{\mathbf{r}}_k + \frac{\Delta t}{2}\left(\ddot{\mathbf{r}}_k + \ddot{\mathbf{r}}_{k+1}\right) \\ \mathbf{r}_{k+1} &= \mathbf{r}_k + \int_0^{\Delta t} \left[\dot{\mathbf{r}}_k + \frac{\tau}{2}\left(\ddot{\mathbf{r}}_k + \ddot{\mathbf{r}}_{k+1}\right)\right] d\tau = \mathbf{r}_k + \Delta t \cdot \dot{\mathbf{r}}_k + \frac{\Delta t^2}{4}\left(\ddot{\mathbf{r}}_k + \ddot{\mathbf{r}}_{k+1}\right) \end{aligned}\right\} \tag{9.183}$$

If the concept of a linear acceleration is adopted (see Fig. 9.10.c), then

$$\ddot{\mathbf{r}}(\tau) = \ddot{\mathbf{r}}_k + (\ddot{\mathbf{r}}_{k+1} - \ddot{\mathbf{r}}_k)\frac{\tau}{\Delta t} \tag{9.184}$$

in which case

$$\left.\begin{aligned} \dot{\mathbf{r}}(\tau) &= \dot{\mathbf{r}}_k + \int_0^\tau \left[\ddot{\mathbf{r}}_k + (\ddot{\mathbf{r}}_{k+1} - \ddot{\mathbf{r}}_k)\frac{\tau}{\Delta t}\right] d\tau = \dot{\mathbf{r}}_k + \ddot{\mathbf{r}}_k \tau + (\ddot{\mathbf{r}}_{k+1} - \ddot{\mathbf{r}}_k)\frac{\tau^2}{2\Delta t} \\ \mathbf{r}(\tau) &= \mathbf{r}_k + \int_0^\tau \left[\dot{\mathbf{r}}_k + \ddot{\mathbf{r}}_k \tau + (\ddot{\mathbf{r}}_{k+1} - \ddot{\mathbf{r}}_k)\frac{\tau^2}{2\Delta t}\right] d\tau = \mathbf{r}_k + \dot{\mathbf{r}}_k \tau + \ddot{\mathbf{r}}_k \frac{\tau^2}{2} + (\ddot{\mathbf{r}}_{k+1} - \ddot{\mathbf{r}}_k)\frac{\tau^3}{6\Delta t} \end{aligned}\right\} \tag{9.185}$$

and thus

$$\left.\begin{aligned} \dot{\mathbf{r}}_{k+1} &= \dot{\mathbf{r}}_k + \Delta t \ddot{\mathbf{r}}_k + \frac{\Delta t}{2}(\ddot{\mathbf{r}}_{k+1} - \ddot{\mathbf{r}}_k) = \dot{\mathbf{r}}_k + \frac{\Delta t}{2}(\ddot{\mathbf{r}}_k + \ddot{\mathbf{r}}_{k+1}) \\ \mathbf{r}_{k+1} &= \mathbf{r}_k + \Delta t \cdot \dot{\mathbf{r}}_k + \frac{\Delta t^2}{2}\ddot{\mathbf{r}}_k + \frac{\Delta t^2}{6}(\ddot{\mathbf{r}}_{k+1} - \ddot{\mathbf{r}}_k) = \mathbf{r}_k + \Delta t \cdot \dot{\mathbf{r}}_k + \frac{\Delta t^2}{3}\ddot{\mathbf{r}}_k + \frac{\Delta t^2}{6}\ddot{\mathbf{r}}_{k+1} \end{aligned}\right\} \tag{9.186}$$

The concept of integrating an assumed variation of the acceleration between t_k and t_{k+1} presented above may all be generalised into the formulation first suggested by Newmark [32]:

$$\left.\begin{aligned} \dot{\mathbf{r}}_{k+1} &= \dot{\mathbf{r}}_k + (1-\gamma) \cdot \Delta t \cdot \ddot{\mathbf{r}}_k + \gamma \cdot \Delta t \cdot \ddot{\mathbf{r}}_{k+1} \\ \mathbf{r}_{k+1} &= \mathbf{r}_k + \Delta t \cdot \dot{\mathbf{r}}_k + \left(\frac{1}{2} - \beta\right) \cdot \Delta t^2 \cdot \ddot{\mathbf{r}}_k + \beta \cdot \Delta t^2 \cdot \ddot{\mathbf{r}}_{k+1} \end{aligned}\right\} \tag{9.187}$$

where γ and β are weighting parameters, each to be chosen according to prescribed requirements regarding numerical stability and accuracy. From the second expression in Eq. 9.187 the acceleration at t_{k+1}

$$\ddot{\mathbf{r}}_{k+1} = \frac{1}{\beta \Delta t^2}(\mathbf{r}_{k+1} - \mathbf{r}_k) - \left[\frac{1}{\beta \Delta t}\dot{\mathbf{r}}_k + \left(\frac{1}{2\beta} - 1\right) \cdot \ddot{\mathbf{r}}_k\right] \tag{9.188}$$

is obtained, which, combined with the first expression in Eq. 9.187, renders

$$\dot{\mathbf{r}}_{k+1} = \frac{\gamma}{\beta \Delta t}(\mathbf{r}_{k+1} - \mathbf{r}_k) - \left(\frac{\gamma}{\beta} - 1\right) \cdot \dot{\mathbf{r}}_k - \left(\frac{\gamma}{2\beta} - 1\right)\Delta t \cdot \ddot{\mathbf{r}}_k \tag{9.189}$$

It is seen that the conditions at t_k may be defined by

$$\left.\begin{aligned} \mathbf{a}_k &= \frac{1}{\beta \Delta t^2}\mathbf{r}_k + \frac{1}{\beta \Delta t}\dot{\mathbf{r}}_k + \left(\frac{1}{2\beta} - 1\right)\cdot \ddot{\mathbf{r}}_k \\ \mathbf{b}_k &= \frac{\gamma}{\beta \Delta t}\mathbf{r}_k + \left(\frac{\gamma}{\beta} - 1\right)\cdot \dot{\mathbf{r}}_k + \left(\frac{\gamma}{2\beta} - 1\right)\Delta t \cdot \ddot{\mathbf{r}}_k \end{aligned}\right\} \tag{9.190}$$

in which case $\ddot{\mathbf{r}}_{k+1}$ and $\dot{\mathbf{r}}_{k+1}$ simplifies into

$$\left.\begin{aligned} \ddot{\mathbf{r}}_{k+1} &= \frac{1}{\beta \Delta t^2}\mathbf{r}_{k+1} - \mathbf{a}_k \\ \dot{\mathbf{r}}_{k+1} &= \frac{1}{\beta \Delta t}\mathbf{r}_{k+1} - \mathbf{b}_k \end{aligned}\right\} \tag{9.191}$$

Introducing this into the dynamic equilibrium equation at t_{k+1}

$$\mathbf{M}\,\ddot{\mathbf{r}}_{k+1} + \mathbf{C}_{net}\dot{\mathbf{r}}_{k+1} + \mathbf{K}_{net}\mathbf{r}_{k+1} = \mathbf{R}_{dyn_{k+1}} \tag{9.192}$$

will then render

$$\left(\frac{1}{\beta \Delta t^2}\mathbf{M} + \frac{\gamma}{\beta \Delta t}\mathbf{C}_{net} + \mathbf{K}_{net}\right)\cdot \mathbf{r}_{k+1} = \mathbf{R}_{dyn_{k+1}} + \mathbf{M}\cdot \mathbf{a}_k + \mathbf{C}_{net}\cdot \mathbf{b}_k \tag{9.193}$$

Defining

$$\left.\begin{aligned} \mathbf{K}_{eff_{k+1}} &= \frac{1}{\beta \Delta t^2}\mathbf{M} + \frac{\gamma}{\beta \Delta t}\mathbf{C}_{net} + \mathbf{K}_{net} \\ \mathbf{R}_{eff_{k+1}} &= \mathbf{R}_{dyn_{k+1}} + \mathbf{M}\cdot \mathbf{a}_k + \mathbf{C}_{net}\cdot \mathbf{b}_k \end{aligned}\right\} \tag{9.194}$$

and thus

$$\mathbf{r}_{k+1} = \mathbf{K}_{eff_{k+1}}^{-1}\cdot \mathbf{R}_{eff_{k+1}} \tag{9.195}$$

It is seen that the response at time step t_{k+1} is calculated from the load at t_{k+1} as well as the displacement, velocity and acceleration response at t_k. If the system is entirely linear, then $\mathbf{K}_{eff}$ is constant throughout the calculations.

Hilber, Hughes & Taylor [33] have suggested an extension of Newmark's method by the introduction of the numerical damping coefficient $\alpha \le 0$ into the dynamic equilibrium condition

$$\mathbf{M}\,\ddot{\mathbf{r}}_{k+1} + (1+\alpha)\mathbf{C}_{net}\dot{\mathbf{r}}_{k+1} - \alpha\mathbf{C}_{net}\dot{\mathbf{r}}_{k} + (1+\alpha)\mathbf{K}_{net}\mathbf{r}_{k+1} - \alpha\mathbf{K}_{net}\mathbf{r}_{k} = \mathbf{R}_{dyn_\alpha} \tag{9.196}$$

and accordingly, evaluate the dynamic load at a $(1+\alpha)t_{k+1} - \alpha t_k = t_{k+1} + \alpha\Delta t$. I.e., if load linearity within the time step is adopted, then

$$\mathbf{R}_\alpha = (1+\alpha)\mathbf{R}_{k+1} - \alpha\mathbf{R}_k \tag{9.197}$$

Combining Eqs. 9.191 and 9.196 and solving for $\mathbf{r}_{k+1}$ will then again render $\mathbf{r}_{k+1} = \mathbf{K}_{eff_{k+1}}^{-1} \cdot \mathbf{R}_{eff_{k+1}}$, but now $\mathbf{K}_{eff_{k+1}}$ and $\mathbf{R}_{eff_{k+1}}$ are extended into

$$\left.\begin{aligned} \mathbf{K}_{eff_{k+1}} &= \frac{1}{\beta\Delta t^2}\mathbf{M} + (1+\alpha)\frac{\gamma}{\beta\Delta t}\mathbf{C}_{net} + (1+\alpha)\mathbf{K}_{net} \\ \mathbf{R}_{eff_{k+1}} &= (1+\alpha)\mathbf{R}_{k+1} - \alpha\mathbf{R}_k + \mathbf{M}\cdot\mathbf{a}_k + \mathbf{C}_{net}\cdot\mathbf{b}_k + \alpha\mathbf{c}_k \end{aligned}\right\} \tag{9.198}$$

where

$$\mathbf{c}_k = \left(\frac{\gamma}{\beta\Delta t}\mathbf{C}_{net} + \mathbf{K}_{net}\right)\cdot\mathbf{r}_k + \mathbf{C}_{net}\frac{\gamma}{\beta}\dot{\mathbf{r}}_k + \mathbf{C}_{net}\left(\frac{\gamma}{2\beta} - 1\right)\Delta t \cdot \ddot{\mathbf{r}}_k \tag{9.199}$$

For the numeric integration methods the establishment of initial conditions at $t = 0$ before the iteration procedure can start requires the choice of $\mathbf{r}_0$ and $\dot{\mathbf{r}}_0$. Dynamic equilibrium at $t = 0$ renders the corresponding acceleration

$$\ddot{\mathbf{r}}_0 = \mathbf{M}^{-1}\cdot\left(\mathbf{R}_{dyn_0} - \mathbf{C}_{net}\dot{\mathbf{r}}_0 - \mathbf{K}_{net}\mathbf{r}_0\right) \tag{9.200}$$

and thus, iteration may commence. Stability may be evaluated from the properties of a single degree of freedom system (or a modal approach) similar to that which has been shown for the central difference method shown above. In general, the Newmark method is unconditionally stable if

$$\gamma \ge \gamma_0 = 1/2 \quad \text{and} \qquad \beta \ge \beta_0 = \frac{1}{4}\left(\gamma + \frac{1}{2}\right)^2 \tag{9.201}$$

For β-values below β_0 it is only conditionally stable. The stability limit is then given by

$$\Delta t \le \Delta t_{cr} = \frac{1}{\omega_i \sqrt{\beta_0 - \beta}} \tag{9.202}$$

For $\gamma = \gamma_0$ there is no numeric (artificial) damping in the system. With $\gamma > \gamma_0$ or $\gamma < \gamma_0$ positive or negative numeric damping is introduced into the system. Positive numeric damping may be used as an effective tool to dampen out undesirable effects of higher modes in the system (which may also be obtained by adopting Hilber, Hughes & Taylor method with $-1/3 < \alpha < 0$). With $\gamma = 1/2$ and $\beta = 0$ Newmark's method becomes identical to the second central difference method, in which case the stability limit is given by $\Delta t_{cr} = 2/\omega_i$, where ω_i is the largest eigen-frequency contained in the system. If $\gamma = 1/2$ and $\beta = 1/4$ then Newmark's method becomes identical to a numerical integration method based on the assumption of a constant average acceleration, which is unconditionally stable. If $\gamma = 1/2$ and $\beta = 1/6$ then Newmark's method becomes identical to a numerical integration method based on the assumption of a linear variation of the acceleration, in which case the stability limit is given by $\Delta t_{cr} = \sqrt{12}/\omega_i$.

As previously mentioned Δt should never be chosen larger than about $1/\left(2\omega_{R_{\max}}\right)$ where $\omega_{R_{\max}}$ is the largest frequency contained in the load.

Tangent-stiffness approach:

For heavily non-linear displacement or material problems the stiffness may change considerably throughout the response process, in which case necessary accuracy may only be obtained by updating the stiffness from one time step to the next. In such cases a tangent-stiffness approach may be adopted. Assuming sufficiently short time steps and linearity within each step, then the change of internal forces from t_k to t_{k+1} is given by

$$\Delta \mathbf{R}_k^{\text{int}} = \mathbf{K}_k^{\tan} \cdot \Delta \mathbf{r} \tag{9.203}$$

where $\mathbf{K}_k^{\tan}$ is the updated tangent stiffness at t_k and $\Delta \mathbf{r} = \mathbf{r}_{k+1} - \mathbf{r}_k$. Thus, the internal force vector at t_{k+1} is

$$\mathbf{R}_{k+1}^{\text{int}} = \mathbf{R}_k^{\text{int}} + \mathbf{K}_k^{\tan} \cdot \Delta \mathbf{r} \tag{9.204}$$

The dynamic equilibrium condition at t_{k+1} is then given by (see Eq. 9.192)

$$\mathbf{M}\,\ddot{\mathbf{r}}_{k+1} + \mathbf{C}_{net}\dot{\mathbf{r}}_{k+1} + \mathbf{R}_k^{\mathrm{int}} + \mathbf{K}_{net_k}^{\tan}\Delta\mathbf{r} = \mathbf{R}_{dyn_{k+1}} \tag{9.205}$$

Introducing the Newmark iteration scheme given in Eqs. 9.188 and 9.189 (and that $\mathbf{r}_{k+1} - \mathbf{r}_k = \Delta\mathbf{r}$), then the following is obtained

$$\begin{aligned}&\left(\frac{1}{\beta\Delta t^2}\mathbf{M} + \frac{1}{\beta\Delta t}\mathbf{C}_{net} + \mathbf{K}_{net_k}^{\tan}\right)\cdot\Delta\mathbf{r} = \mathbf{R}_{dyn_{k+1}} - \mathbf{R}_k^{\mathrm{int}}\\ &+\mathbf{M}\cdot\left[\frac{1}{\beta\Delta t}\dot{\mathbf{r}}_k + \left(\frac{1}{2\beta}-1\right)\cdot\ddot{\mathbf{r}}_k\right] + \mathbf{C}_{net}\cdot\left[\left(\frac{\gamma}{\beta}-1\right)\cdot\dot{\mathbf{r}}_k + \left(\frac{1}{2\beta}-1\right)\cdot\Delta t\cdot\ddot{\mathbf{r}}_k\right]\end{aligned} \tag{9.206}$$

Thus

$$\Delta\mathbf{r} = \mathbf{K}_{eff_k}^{-1}\cdot\left(\mathbf{R}_{dyn_{k+1}} - \mathbf{R}_k^{\mathrm{int}} + \mathbf{M}\cdot\mathbf{a}_{eff_k} + \mathbf{C}_{net}\cdot\mathbf{b}_{eff_k}\right) \tag{9.207}$$

where

$$\left.\begin{aligned}\mathbf{K}_{eff_k} &= \frac{1}{\beta\Delta t^2}\mathbf{M} + \frac{1}{\beta\Delta t}\mathbf{C}_{net} + \mathbf{K}_{net_k}^{\tan}\\ \mathbf{a}_{eff_k} &= \frac{1}{\beta\Delta t}\dot{\mathbf{r}}_k + \left(\frac{1}{2\beta}-1\right)\cdot\ddot{\mathbf{r}}_k\\ \mathbf{b}_{eff_k} &= \left(\frac{\gamma}{\beta}-1\right)\cdot\dot{\mathbf{r}}_k + \left(\frac{1}{2\beta}-1\right)\cdot\Delta t\cdot\ddot{\mathbf{r}}_k\end{aligned}\right\} \tag{9.208}$$

Such a procedure will generally require error control. This may be obtained by minimising the estimated external load error $\Delta\mathbf{R}_{k+1}^{err}$, defined as the difference between the actual load at t_{k+1} and the corresponding load which can be calculated from the estimated displacements

$$\Delta\mathbf{R}_{k+1}^{err} = \mathbf{R}_{dyn_{k+1}} - \left(\mathbf{M}\,\ddot{\mathbf{r}}_{k+1} + \mathbf{C}_{net}\dot{\mathbf{r}}_{k+1} + \mathbf{R}_k^{\mathrm{int}}\right)_{est} \tag{9.209}$$

Thus, iterations until $\Delta\mathbf{R}_{k+1}^{err}$ is less than a specified limit will be required within each time step. Initial conditions and stability criteria are identical to those presented above for the numeric integration methods.

Appendix A

TIME DOMAIN SIMULATIONS

A.1 Introduction

It is in the following taken for granted that the stochastic space and time domain simulation of a process x implies the extraction of single point or simultaneous multiple point time series from known frequency domain cross spectral information about the process. The process may contain coherent or non-coherent properties in space and time. Thus, a multiple point representation is associated with the spatial occurrence of the process. For a non-coherent process there is no statistical connection between the simulated time series that occur at various positions in space, and thus, the simulation may be treated as a representation of independent single point time series. This type of simulation is shown in chapter A.2. For a coherent process there is a prescribed statistical connection between each of the spatial representatives within a set of M simulated time series. E.g., if the simulated time series represent the space and time distribution of a wind field, there will be a certain statistical connection between the instantaneous values $x_m(t), m = 1,2,....,M$ that matches the spatial properties of the wind field. Such a simulation is shown in chapter A.3. The simulation procedure presented below is taken from Shinozuka [23] and Deodatis [24].

Simulating time series from spectra is particularly useful for two reasons. First, there are some response calculations that render results which are more or less narrow banded (or contain beating effects), and thus, they do not necessarily comply with the assumptions behind the peak factor given in Eq. 2.45. These cases may require separate time domain simulations to establish an appropriate peak factor for the calculation of maximum response. This application will usually only require single point simulations. Secondly, if the relevant cross spectra of the wind field properties in frequency domain are known, there is always the possibility of a time domain simulation of the entire wind field, or those of the flow components that are deemed necessary. Together with the buffeting load theory in chapter 5.1 this is a tempting option, as time domain step-wise load effect integration may be performed, and thus, the response calculation may be carried out in time domain instead of the frequency domain approach that is shown in chapter 6. The mathematical procedure for such an approach may be found in many text books, see e.g. Hughes [25]. The main advantage is that such an approach may contain many of the non-linear effects that had to be simplified or discarded in the linear theory that was required for a frequency domain solution. The disadvantage is that motion induced load effects can only be fully included if a new set of indicial functions are introduced (see e.g. Scanlan [26]). These may not be readily available.

A.2 Simulation of single point time series

The mathematical development from a single time series to its auto-spectral density is presented in chapter 2.5. In principle, the process is illustrated on Fig. 2.11. A time domain simulation is obtained by the reverse process.

Let $S_x(\omega)$ be the single-sided single point auto spectral density of an arbitrary stochastic variable x, for simplicity with zero mean value. A time domain representative, $x(t)$, can then be obtained by subdividing S_x into N blocks along the frequency axis, each centred at ω_k ($k = 1,2,\ldots,N$) and covering a frequency segment $\Delta\omega_k$, as shown in Fig. A.1.

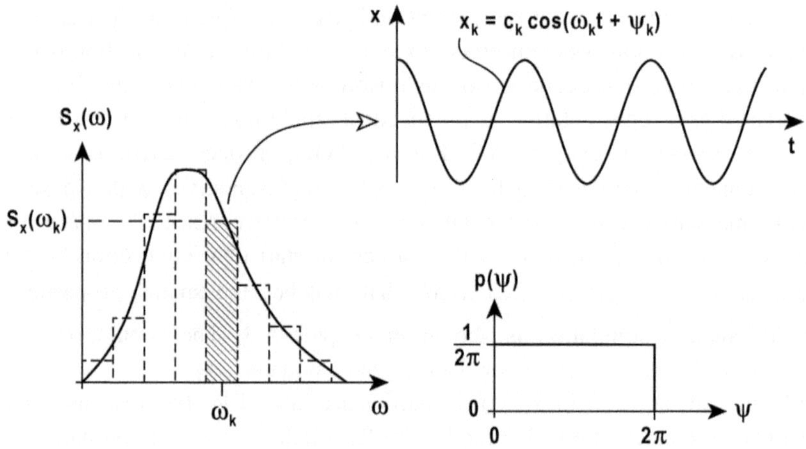

Fig. A.1 *Spectral decomposition*

On a discrete form $S_x(\omega_k)$ is the variance of each harmonic component per frequency segment, as defined in Eq. 2.53 (see also Fig. 2.11), i.e.

$$S_x(\omega_k) = c_k^2 / (2\Delta\omega_k) \tag{A.1}$$

A time series representative of x is then obtained by

$$x(t) = \sum_{k=1}^{N} c_k \cos(\omega_k t + \psi_k) \tag{A.2}$$

where

$$c_k = \left[2 \cdot S_x(\omega_k) \cdot \Delta\omega_k\right]^{1/2} \tag{A.3}$$

and where ψ_k are arbitrary phase angles between zero and 2π, one for each harmonic component. Alternatively, Eq. A.2 may be replaced by the exponential format (often encountered in the literature)

$$x(t) = \mathrm{Re}\left\{\sum_{k=1}^{N} c_k \cdot \exp\left[i\left(\omega_k t + \psi_k\right)\right]\right\} \tag{A.4}$$

The variance of $x(t)$ is $\sum_{k=1}^{N} \frac{c_k^2}{2}$, which in the limit of $\Delta\omega \to 0$ and $N \to \infty$,

$$\sigma_x^2 = \lim_{\substack{\Delta\omega \to 0 \\ N \to \infty}} \sum_{k=1}^{N} \frac{c_k^2}{2} = \int_0^\infty S_x(\omega) d\omega \tag{A.5}$$

I.e., if the discretization is sufficiently fine, then the variance of the simulated representative, $x(t)$, is equal to or close enough to the variance of the parent variable. The procedure is further illustrated in Example A.1 and Fig. A.2. Any number of such representatives may be simulated simply by changing the choice of phase angles. Obviously, the accuracy of such a simulation depends on the discretization fineness, but there is also the unfavourable possibility of aliasing. Let ω_c be the upper cut-off frequency, beyond which there is none or only negligible spectral information about the process. Assuming constant frequency segments

$$\Delta\omega = \omega_c / N \tag{A.6}$$

then each simulated time series will be periodic with period

$$T = 2\pi / \Delta\omega \tag{A.7}$$

Thus, time series without aliasing will be obtained if they are generated with a time step

$$\Delta t \le 2\pi / \left(2\omega_c\right) \tag{A.8}$$

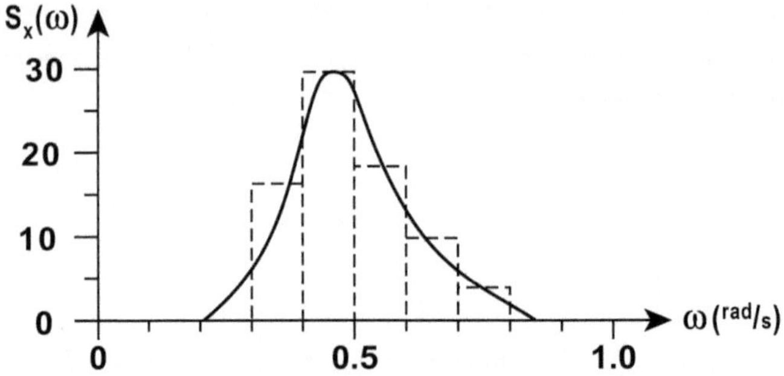

k	$S_x(\omega_k)$	ω_k	$\Delta\omega$	$c_k = \sqrt{2S_x\Delta\omega}$
1	16	0.35	0.1	1.8
2	30	0.45	0.1	2.4
3	18	0.55	0.1	1.9
4	10	0.65	0.1	1.4
5	4	0.75	0.1	0.9

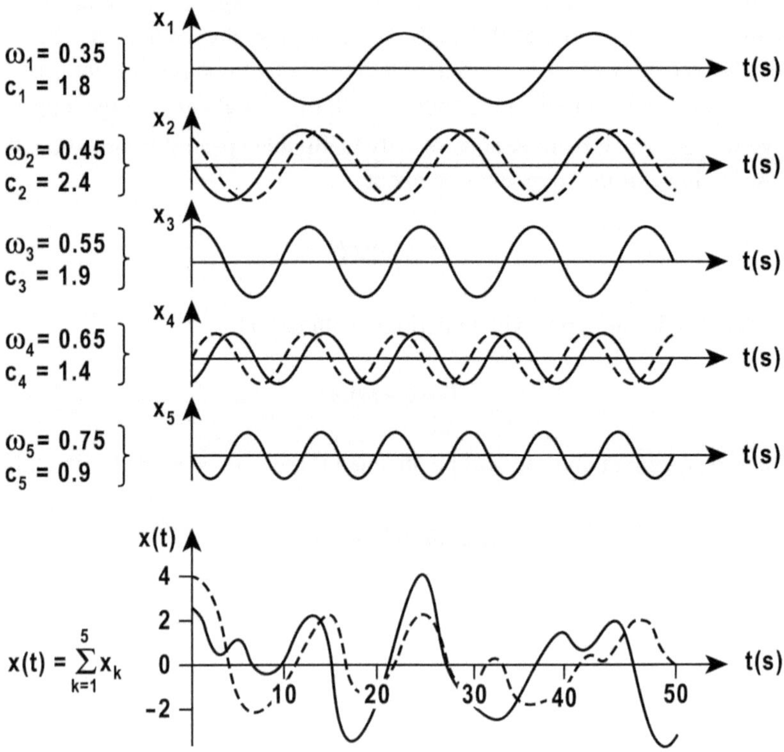

Fig. A.2 *Simulation of single point time series*

Example A.1
The top diagram in Fig. A.2 shows the single point single sided spectrum of a process x of which we wish to portray two representatives in time domain. As shown, the frequency span of the spectrum is first divided into five equal frequency segments, and the corresponding values ω_k and $S_x(\omega_k)$, $k = 1,2,...,5$, are read off. Thus the process is represented by five harmonic components whose amplitudes $c_k = \sqrt{2 \cdot S_x(\omega_k) \cdot \Delta\omega}$ are given in the far right hand side column in the table of Fig. A.2. Thus

$$x(t) = \sum_{k=1}^{5} \sqrt{2S_x(\omega_k)\Delta\omega} \cdot \cos(\omega_k t + \psi_k)$$

What then remains is to choose five arbitrary value of ψ_k. In Fig. A.2 the five cosine components are first shown by fully drawn lines, representing a certain choice of ψ_k values. The sum of these components shown in the lower diagram in Fig. A.2 is an arbitrary representation of the process $x(t)$. If the second and the fourth of these components are moved an arbitrary time shift, then together with the remaining unchanged components they sum up to become another arbitrary representation of the process shown by the broken line in Fig. A.2. As can be seen, the two simulated representatives look quite different in time domain, although they come from the same spectral density. What is important is that they both have zero mean and the same variance, i.e. they have identical statistical properties up to and including the variance.

A.3 Simulation of spatially non–coherent time series

While the procedure presented above may be used to simulate single point time series representatives of x, it is not applicable if we wish to simulate multiple point time series whose properties are expected to be distributed according to certain coherence properties. E.g., let us assume that we wish to simulate the turbulence components

$$x(y_f, z_f, t) \qquad x = \begin{cases} u \\ v \\ w \end{cases} \tag{A.9}$$

of a stationary and homogeneous wind field at a chosen number of points M in a plane perpendicular to the main flow direction. It is then important to capture the fact that these time series are representatives of simultaneous events, and therefore, they must contain the appropriate spatial coherence properties that are characteristic to the process.

For simplicity it is in the following assumed that cross spectra between the u , v and w components are negligible, i.e. that

$$S_{xy}(\omega,\Delta s) \approx 0 \quad \left.\begin{matrix} x \\ y \end{matrix}\right\} = u,v,w \tag{A.10}$$

where Δs is the spatial separation in the $y_f - z_f$ plane. We will then only need information about the cross spectra of the turbulence components themselves, $S_{xx}(\omega,\Delta s)$. Let $Cov_{x_m x_n}(\tau)$ be the covariance and $S_{x_m x_n}(\omega)$ the corresponding cross spectral density between two arbitrary points m and n. As shown in chapter 2.6 these quantities constitute a Fourier transform pair. An M by M cross spectral density matrix

$$\mathbf{S}_{xx}(\omega) = \begin{bmatrix} S_{x_1x_1} & \cdots & S_{x_1x_n} & \cdots & S_{x_1x_M} \\ \vdots & \ddots & \vdots & ⋰ & \vdots \\ S_{x_mx_1} & \cdots & S_{x_mx_n} & \cdots & S_{x_mx_M} \\ \vdots & ⋰ & \vdots & \ddots & \vdots \\ S_{x_Mx_1} & \cdots & S_{x_Mx_n} & \cdots & S_{x_Mx_M} \end{bmatrix} \tag{A.11}$$

will then contain all the space and frequency domain information that is necessary for a time domain simulation of M time series with the correct statistical properties for a special representation of the process. It follows from the assumptions of stationarity and homogeneity that

$$Cov_{x_m x_n} = Cov_{x_n x_m} \tag{A.12}$$

and thus,

$$S_{x_m x_n} = S^*_{x_n x_m} \tag{A.13}$$

This implies that $\mathbf{S}_{xx}(\omega)$ is Hermitian and non–negative definite. A Cholesky decomposition of $\mathbf{S}_{xx}$ will then render a lower triangular matrix

$$\mathbf{G}_{xx}(\omega) = \begin{bmatrix} G_{x_1x_1} & 0 & 0 & 0 & \cdots & 0 & \cdots & 0 \\ G_{x_2x_1} & G_{x_2x_2} & 0 & 0 & \cdots & 0 & \cdots & 0 \\ \vdots & \vdots & & \vdots & & \vdots & & \vdots \\ G_{x_mx_1} & G_{x_mx_2} & \cdots & G_{x_mx_n} & \cdots & G_{x_mx_m} & 0 & 0 \\ \vdots & \vdots & & \vdots & & \vdots & & \vdots \\ G_{x_Mx_1} & G_{x_Mx_2} & \cdots & G_{x_Mx_n} & \cdots & G_{x_Mx_m} & \cdots & G_{x_Mx_M} \end{bmatrix} \tag{A.14}$$

whose properties are such that

$$\mathbf{S}_{xx}(\omega) = \mathbf{G}_{xx} \cdot \mathbf{G}_{xx}^{*T} \tag{A.15}$$

Assuming a frequency segmentation of N equidistant points, the simulated simultaneous time series at $m = 1,2,....,M$ are then given by

$$x_m(t) = \sum_{n=1}^{m}\sum_{j=1}^{N}\left|G_{mn}(\omega_j)\right| \cdot \sqrt{2\Delta\omega} \cdot \cos(\omega_j \cdot t + \psi_{nj}) \tag{A.16}$$

where j is the frequency segment number and ψ_{nj} is an arbitrary phase angle between zero and 2π. In most cases of a homogeneous wind field (see Eq. 2.87)

$$S_{xx}(\omega,\Delta s) = S_x(\omega) \cdot \hat{S}_{xx}(\omega,\Delta s) \tag{A.17}$$

where S_x is the single-point spectral density of the process, $\Delta s = \left|s_{x_m} - s_{x_n}\right|$ is the spatial separation between points x_m and x_n, and where

$$\hat{S}_{xx}(\omega,\Delta s) = \sqrt{Coh_{xx}(\omega,\Delta s)} \cdot \exp\left[i\varphi_{xx}(\omega)\right] \tag{A.18}$$

Thus, defining a Cholesky decomposition $\hat{\mathbf{S}}_{xx}(\omega) = \hat{\mathbf{G}}_{xx} \cdot \hat{\mathbf{G}}_{xx}^{*T}$, then the time series at $m = 1,2,....,M$ are given by

$$x_m(t) = \sum_{n=1}^{m}\sum_{j=1}^{N}\left|\hat{G}_{mn}(\omega_j)\right| \cdot \sqrt{2S_x(\omega_j) \cdot \Delta\omega} \cdot \cos(\omega_j \cdot t + \psi_{nj}) \tag{A.19}$$

where $\hat{G}_{mn}$ is the content of $\hat{\mathbf{G}}_{xx}$ (i.e. the reduced versions of G_{mn} in Eq. A.14)

$$\hat{\mathbf{G}}_{xx}(\omega) = \begin{bmatrix} \hat{G}_{11} & 0 & 0 & 0 & \cdots & 0 & \cdots & 0 \\ \hat{G}_{21} & \hat{G}_{22} & 0 & 0 & \cdots & 0 & \cdots & 0 \\ \vdots & \vdots & & \vdots & & \vdots & & \vdots \\ \hat{G}_{m1} & \hat{G}_{m2} & \cdots & \hat{G}_{mn} & \cdots & \hat{G}_{mm} & 0 & 0 \\ \vdots & \vdots & & \vdots & & \vdots & & \vdots \\ \hat{G}_{M1} & \hat{G}_{M2} & \cdots & \hat{G}_{Mn} & \cdots & \hat{G}_{Mm} & & \hat{G}_{MM} \end{bmatrix} \tag{A.20}$$

and where a Cholesky decomposition will render

$$\hat{G}_{11}(\omega_j) = \left[\hat{S}_{xx}(\omega_j, 0)\right]^{1/2} \tag{A.21}$$

$$\hat{G}_{mm}(\omega_j) = \left[\hat{S}_{xx}(\omega_j, 0) - \sum_{k=1}^{m-1} \hat{G}_{mk}^2(\omega_j)\right]^{1/2} \tag{A.22}$$

$$\hat{G}_{mn}(\omega_j) = \frac{\hat{S}_{xx}(\omega_j, \Delta s) - \sum_{k=1}^{n-1} \hat{G}_{mk}(\omega_j) \cdot \hat{G}_{nk}(\omega_j)}{\hat{G}_{nn}(\omega_j)} \tag{A.23}$$

Example A.2

A process x is statistically distributed in time and space. Its cross-spectrum $S_{xx}(\omega, \Delta s)$ is defined by the product between the single point spectrum $S_x(\omega)$ shown in Fig. A.3 and its root-coherence function $\sqrt{Coh_{xx}(\omega, \Delta s)}$ shown in Fig. A.4. I.e.,

$$S_{xx}(\omega, \Delta s) = S_x(\omega) \cdot \sqrt{Coh_{xx}(\omega, \Delta s)}$$

The phase spectrum $\exp\left[i\varphi_{xx}(\omega)\right]$ is assumed equal to unity for all relevant values of ω and Δs. Let us set out to simulate the process at three points in space, each a distance 10 m apart. Thus,

$$\mathbf{\Delta s} = \begin{bmatrix} \Delta s_1 & \Delta s_2 & \Delta s_3 \end{bmatrix}^T = \begin{bmatrix} 0 & 10 & 20 \end{bmatrix}^T$$

Let us for simplicity settle with the three point frequency segmentation shown in Fig. A.3. I.e.

$$\boldsymbol{\omega} = \begin{bmatrix} \omega_1 & \omega_2 & \omega_3 \end{bmatrix}^T = \begin{bmatrix} 0.3 & 0.7 & 1.1 \end{bmatrix}^T \qquad \text{and} \qquad \Delta\omega = 0.4$$

(It should be noted that this frequency segmentation is only justified by the wish of obtaining mathematical expressions with reasonable length, such that a complete solution may be presented. For any practical purposes such a coarse segmentation will most often render unduly inaccurate results.) The single point spectrum at these frequency settings are then (see Fig. A.3)

$$\mathbf{S}_x = \begin{bmatrix} S_x(\omega_1) & S_x(\omega_2) & S_x(\omega_3) \end{bmatrix}^T = \begin{bmatrix} 4.0 & 7.6 & 3.0 \end{bmatrix}^T$$

while the corresponding values of the root coherence function are given by (see Fig. A.4)

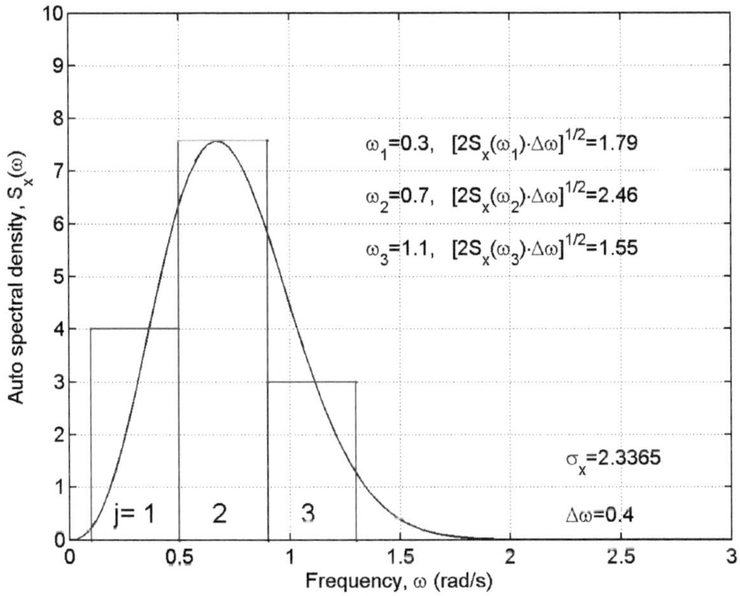

Fig. A.3 *Single point spectrum*

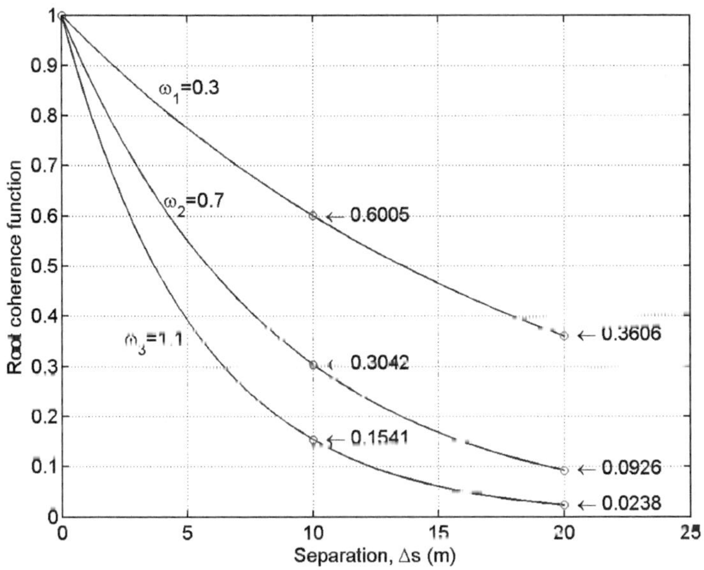

Fig. A.4 *Root coherence function at* $\omega = 0.3,\ 0.7$ and 1.1

$\sqrt{Coh_{xx}(\omega,\Delta s)}$:		Δs		
		0	10	20
ω	0.3	1.0	0.6005	0.3606
	0.7	1.0	0.3042	0.0926
	1.1	1.0	0,1541	0.0238

Thus,

$$\hat{\mathbf{S}}_{xx}\left(\omega_j=0.3,\Delta s_{mn}\right)=\begin{bmatrix}1 & & sym.\\ 0.6005 & 1 & \\ 0.3606 & 0.6005 & 1\end{bmatrix}$$

$$\hat{\mathbf{S}}_{xx}\left(\omega_j=0.7,\Delta s_{mn}\right)=\begin{bmatrix}1 & & sym.\\ 0.3042 & 1 & \\ 0.0926 & 0.3042 & 1\end{bmatrix}$$

$$\hat{\mathbf{S}}_{xx}\left(\omega_j=1.1,\Delta s_{mn}\right)=\begin{bmatrix}1 & & sym.\\ 0.1541 & 1 & \\ 0.0238 & 0.1541 & 1\end{bmatrix}$$

$$\hat{\mathbf{G}}_{xx}\left(\omega_j\right)=\begin{bmatrix}\hat{G}_{11} & 0 & 0\\ \hat{G}_{21} & \hat{G}_{22} & 0\\ \hat{G}_{31} & \hat{G}_{32} & \hat{G}_{33}\end{bmatrix} \text{ is defined such that } \hat{\mathbf{S}}_{xx}\left(\omega_j,\Delta s_n\right)=\hat{\mathbf{G}}_{xx}\cdot\hat{\mathbf{G}}_{xx}^T$$

Its content is given by (see Eqs. A.21 – A.23)

$$\hat{G}_{11}\left(\omega_j\right)=\left[\hat{S}_{xx}\left(\omega_j,\Delta s_{11}=0\right)\right]^{1/2}$$

$$\hat{G}_{21}\left(\omega_j\right)=\hat{S}_{xx}\left(\omega_j,\Delta s_{21}=10\right)/\hat{G}_{11}\left(\omega_j\right),$$

$$\hat{G}_{22}\left(\omega_j\right)=\left[\hat{S}_{xx}\left(\omega_j,\Delta s_{22}=0\right)-\hat{G}_{21}^2\left(\omega_j\right)\right]^{1/2}$$

$$\hat{G}_{31}\left(\omega_j\right)=\hat{S}_{xx}\left(\omega_j,\Delta s_{31}=20\right)/\hat{G}_{11}\left(\omega_j\right)$$

$$\hat{G}_{32}\left(\omega_j\right)=\left[\hat{S}_{xx}\left(\omega_j,\Delta s_{32}=10\right)-\hat{G}_{31}\left(\omega_j\right)\cdot\hat{G}_{21}\left(\omega_j\right)\right]/\hat{G}_{22}\left(\omega_j\right)$$

$$\hat{G}_{33}\left(\omega_j\right)=\left[\hat{S}_{xx}\left(\omega_j,\Delta s_{33}=0\right)-\hat{G}_{31}^2\left(\omega_j\right)-\hat{G}_{32}^2\left(\omega_j\right)\right]^{1/2}$$

Thus,

$$\omega_1=0.3\Rightarrow\left\{\begin{array}{ll}\hat{G}_{11}=1 & \hat{G}_{22}=\sqrt{1-0.6005^2}=0.7996\\ \hat{G}_{21}=0.6005 & \hat{G}_{32}=\left(0.6005-0.3606\cdot0.6005\right)/0.7996=0.4802\\ \hat{G}_{31}=0.3606 & \hat{G}_{33}=\sqrt{1-0.3606^2-0.4802^2}=0.7996\end{array}\right\}$$

$$\Rightarrow\hat{\mathbf{G}}_{xx}\left(\omega_1=0.3\right)=\begin{bmatrix}1 & 0 & 0\\ 0.6005 & 0.7996 & 0\\ 0.3606 & 0.4802 & 0.7996\end{bmatrix}$$

$$\omega_2 = 0.7 \Rightarrow \begin{Bmatrix} \hat{G}_{11} = 1 & \hat{G}_{22} = \sqrt{1-0.3042^2} = 0.9526 \\ \hat{G}_{21} = 0.3042 & \hat{G}_{32} = (0.3042 - 0.0926 \cdot 0.3042)/0.9526 = 0.2898 \\ \hat{G}_{31} = 0.0926 & \hat{G}_{33} = \sqrt{1-0.0926^2-0.2898^2} = 0.9526 \end{Bmatrix}$$

$$\Rightarrow \hat{\mathbf{G}}_{xx}(\omega_2 = 0.7) = \begin{bmatrix} 1 & 0 & 0 \\ 0.3042 & 0.9526 & 0 \\ 0.0926 & 0.2898 & 0.9526 \end{bmatrix}$$

$$\omega_3 = 1.1 \Rightarrow \begin{Bmatrix} \hat{G}_{11} = 1 & \hat{G}_{22} = \sqrt{1-0.1541^2} = 0.9881 \\ \hat{G}_{21} = 0.1541 & \hat{G}_{32} = (0.1541 - 0.0238 \cdot 0.1541)/0.9881 = 0.1522 \\ \hat{G}_{31} = 0.0238 & \hat{G}_{33} = \sqrt{1-0.0238^2-0.1522^2} = 0.9881 \end{Bmatrix}$$

$$\Rightarrow \hat{\mathbf{G}}_{xx}(\omega_1 = 1.1) = \begin{bmatrix} 1 & 0 & 0 \\ 0.1541 & 0.9881 & 0 \\ 0.0238 & 0.1522 & 0.9881 \end{bmatrix}$$

Denoting

$$\begin{bmatrix} a_1 \\ a_2 \\ a_3 \end{bmatrix} = \begin{bmatrix} \sqrt{2S_x(\omega_1 = 0.5)\cdot\Delta\omega} \\ \sqrt{2S_x(\omega_2 = 0.7)\cdot\Delta\omega} \\ \sqrt{2S_x(\omega_3 = 1.1)\cdot\Delta\omega} \end{bmatrix} = \begin{bmatrix} \sqrt{2\cdot 4\cdot 0.4} \\ \sqrt{2\cdot 7.6\cdot 0.4} \\ \sqrt{2\cdot 3\cdot 0.4} \end{bmatrix} \approx \begin{bmatrix} 1.79 \\ 2.46 \\ 1.55 \end{bmatrix}$$

then the three time series are given by (see Eq. A.19)

$$\begin{aligned} x_1(t) &= \sum_{n=1}^{1}\sum_{j=1}^{3} \left|\hat{G}_{1n}(\omega_j)\right| \cdot \sqrt{2S_x(\omega_j)\Delta\omega} \cdot \cos(\omega_j t + \psi_{nj}) \\ &= \left|\hat{G}_{11}(\omega_1)\right| a_1 \cos(\omega_1 t + \psi_{11}) + \left|\hat{G}_{11}(\omega_2)\right| a_2 \cos(\omega_2 t + \psi_{12}) + \left|\hat{G}_{11}(\omega_3)\right| \cdot a_3 \cdot \cos(\omega_3 t + \psi_{13}) \\ &= 1.79 \cdot \cos(0.3t + \psi_{11}) + 2.46 \cdot \cos(0.7t + \psi_{12}) + 1.55 \cdot \cos(1.1t + \psi_{13}) \end{aligned}$$

$$\begin{aligned} x_2(t) &= \sum_{n=1}^{2}\sum_{j=1}^{3} \left|\hat{G}_{2n}(\omega_j)\right| \cdot \sqrt{2S_x(\omega_j)\Delta\omega} \cdot \cos(\omega_j t + \psi_{nj}) \\ &= \left|\hat{G}_{21}(\omega_1)\right| a_1 \cos(\omega_1 t + \psi_{11}) + \left|\hat{G}_{21}(\omega_2)\right| a_2 \cos(\omega_2 t + \psi_{12}) + \left|\hat{G}_{21}(\omega_3)\right| \cdot a_3 \cdot \cos(\omega_3 t + \psi_{13}) \\ &+ \left|\hat{G}_{22}(\omega_1)\right| a_1 \cos(\omega_1 t + \psi_{21}) + \left|\hat{G}_{22}(\omega_2)\right| a_2 \cos(\omega_2 t + \psi_{22}) + \left|\hat{G}_{22}(\omega_3)\right| \cdot a_3 \cdot \cos(\omega_3 t + \psi_{23}) \\ &= 1.075 \cdot \cos(0.3t + \psi_{11}) + 0.748 \cdot \cos(0.7t + \psi_{12}) + 0.239 \cdot \cos(1.1t + \psi_{13}) \\ &+ 1.431 \cdot \cos(0.3t + \psi_{21}) + 2.343 \cdot \cos(0.7t + \psi_{22}) + 1.532 \cdot \cos(1.1t + \psi_{23}) \end{aligned}$$

$$x_3(t) = \sum_{n=1}^{3}\sum_{j=1}^{3}\left|\hat{G}_{3n}(\omega_j)\right| \cdot \sqrt{2S_x(\omega_j)\Delta\omega} \cdot \cos(\omega_j t + \psi_{nj})$$
$$= \left|\hat{G}_{31}(\omega_1)\right| a_1 \cos(\omega_1 t + \psi_{11}) + \left|\hat{G}_{31}(\omega_2)\right| a_2 \cos(\omega_2 t + \psi_{12}) + \left|\hat{G}_{31}(\omega_3)\right| \cdot a_3 \cdot \cos(\omega_3 t + \psi_{13})$$
$$+ \left|\hat{G}_{32}(\omega_1)\right| a_1 \cos(\omega_1 t + \psi_{21}) + \left|\hat{G}_{32}(\omega_2)\right| a_2 \cos(\omega_2 t + \psi_{22}) + \left|\hat{G}_{32}(\omega_3)\right| \cdot a_3 \cdot \cos(\omega_3 t + \psi_{23})$$
$$+ \left|\hat{G}_{33}(\omega_1)\right| a_1 \cos(\omega_1 t + \psi_{31}) + \left|\hat{G}_{33}(\omega_2)\right| a_2 \cos(\omega_2 t + \psi_{32}) + \left|\hat{G}_{33}(\omega_3)\right| \cdot a_3 \cdot \cos(\omega_3 t + \psi_{33})$$
$$= 0.646 \cdot \cos(0.3t + \psi_{11}) + 0.228 \cdot \cos(0.7t + \psi_{12}) + 0.039 \cdot \cos(1.1t + \psi_{13})$$
$$+0.86 \cdot \cos(0.3t + \psi_{21}) + 0.713 \cdot \cos(0.7t + \psi_{22}) + 0.236 \cdot \cos(1.1t + \psi_{23})$$
$$+1.431 \cdot \cos(0.3t + \psi_{31}) + 2.343 \cdot \cos(0.7t + \psi_{32}) + 1.532 \cdot \cos(1.1t + \psi_{33})$$

What then remains is to ascribe arbitrary values (between 0 and 2π) to the phase angles, ψ_{nj}. The following is chosen:

$$\boldsymbol{\psi} = 2\pi \cdot \begin{bmatrix} 0.7 & 0.6 & 0.3 \\ 0.1 & 0.4 & 0.2 \\ 0.1 & 0.7 & 0.8 \end{bmatrix}$$

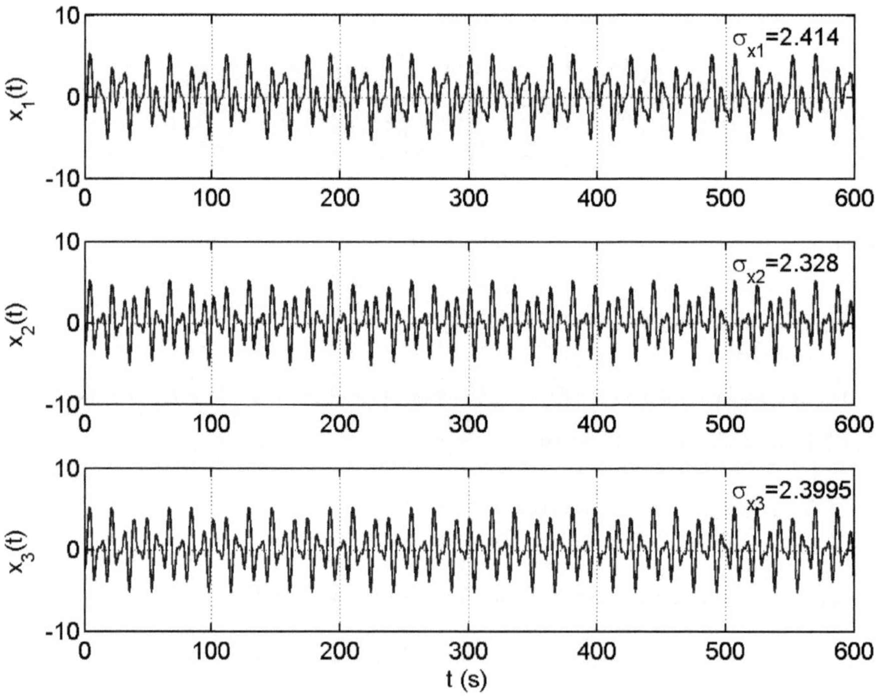

Fig. A.5 *Simulated time series*

The simulated time series are shown in Fig. A.5 ($T = 600\ s$ and $\Delta t = 0.06$ s). The standard deviation of the process as calculated from the parent spectrum is $\sigma_x = 2.3365$. The standard deviations of the three simulated time series are 2.414, 2.328 and 2.3995. The discrepancy (less than about 3 %) is caused by the unduly coarse frequency segmentation.

A.4 The Cholesky decomposition

Given a positive definite and symmetric matrix $\mathbf{X}$, the Cholesky decomposition of $\mathbf{X}$ is defined by a lower triangular matrix $\mathbf{Y}$ of the same size that satisfies the following:

$$\mathbf{X} = \mathbf{Y}\mathbf{Y}^T \tag{A.24}$$

Expanding this equation

$$\begin{bmatrix} x_{11} & \cdots & x_{1i} & \cdots & x_{1N} \\ \vdots & & \vdots & & \vdots \\ x_{i1} & \cdots & x_{ii} & \cdots & x_{iN} \\ \vdots & & \vdots & & \vdots \\ x_{N1} & \cdots & x_{Ni} & \cdots & x_{NN} \end{bmatrix} = \begin{bmatrix} y_{11} & 0 & 0 \\ \\ y_{i1} & y_{ii} & 0 \\ \\ y_{N1} & y_{Ni} & y_{NN} \end{bmatrix} \cdot \begin{bmatrix} y_{11} & y_{1i} & y_{1N} \\ \\ 0 & y_{ii} & y_{iN} \\ \\ 0 & 0 & y_{NN} \end{bmatrix} \tag{A.25}$$

and developing the matrix multiplication column by column, it is seen that the first column renders

$$\left\{ \begin{matrix} x_{11} = y_{11} \cdot y_{11} \\ x_{21} = y_{21} \cdot y_{11} \\ \vdots \\ x_{N1} = y_{N1} \cdot y_{11} \end{matrix} \right\} \Rightarrow \left\{ \begin{matrix} y_{11} = \sqrt{x_{11}} \\ y_{21} = x_{21} / y_{11} \\ \vdots \\ y_{N1} = x_{N1} / y_{11} \end{matrix} \right. \tag{A.26}$$

while the second column renders

$$\left\{ \begin{matrix} x_{22} = y_{21} \cdot y_{21} + y_{22} \cdot y_{22} \\ x_{32} = y_{31} \cdot y_{21} + y_{32} \cdot y_{22} \\ \vdots \\ x_{N2} = y_{N1} \cdot y_{21} + y_{N2} \cdot y_{22} \end{matrix} \right\} \Rightarrow \left\{ \begin{matrix} y_{22} = \sqrt{x_{22} - y_{21} y_{21}} \\ y_{32} = \left(x_{32} - y_{31} y_{21}\right) / y_{22} \\ \vdots \\ y_{N2} = \left(x_{N2} - y_{N1} y_{21}\right) / y_{22} \end{matrix} \right. \tag{A.27}$$

and so on. This can be summarized as follows:

$$
\begin{aligned}
y_{11} &= \left(x_{11}\right)^{1/2} \\
y_{ii} &= \left(x_{ii} - \sum_{k=1}^{i-1} y_{ik}^2\right)^{1/2} \quad \text{for } i = 2,.....,N-1 \\
y_{ij} &= \left(x_{ij} - \sum_{k=1}^{j-1} y_{ik} y_{kj}\right) / y_{jj} \quad \text{for all } i > j
\end{aligned}
\tag{A.28}
$$

Appendix B

DETERMINATION OF THE JOINT ACCEPTANCE FUNCTION

B.1 Closed form solutions

The calculation of wind load effects, static or dynamic, will inevitably involve the establishment of the joint acceptance function, normalised or non-normalised. As shown in chapter 2.10, it represents the statistical averaging in space, and it contains the integral

$$I(\beta) = \int_0^1 \int_0^1 f(\hat{x}_1) \cdot f(\hat{x}_2) \cdot \exp(-\beta \cdot \Delta\hat{x}) d\hat{x}_1 d\hat{x}_2 \tag{B.1}$$

where, $f(\hat{x})$ is some influence function or mode shape, $\hat{x}$ is a non-dimensional coordinate between 0 and 1, $\Delta\hat{x} = |\hat{x}_1 - \hat{x}_2|$ and

$$\beta = \begin{cases} C_{mn} \omega L_{\exp} / V & \text{if dynamic respons} \\ L_{\exp} / {}^{x}L_m & \text{if static respons} \end{cases} \quad \text{where} \quad \begin{cases} m = u \text{ or } w \\ n = y_f \text{ or } z_f \end{cases} \tag{B.2}$$

Some closed form solutions (presented by Davenport [14], see also examples 6.1 and 6.2) are given below (and plotted in Figs. B.1 – B.3):

Influence function or Mode shape, $f(\hat{x})$	Reduced integral, $I(\beta)$
1	$(2/\beta^2)[\beta - 1 + \exp(-\beta)]$
$\hat{x}$	$(2/\beta^4) \cdot [\beta^3/3 - \beta^2/2 + 1 - \exp(-\beta) \cdot (\beta + 1)]$
$2\hat{x} - 1$	$(8/\beta^4) \cdot [\beta^3/12 - \beta^2/4 + 1 - \exp(-\beta) \cdot (\beta^2/4 + \beta + 1)]$
$\sin(n\pi\hat{x})$	$\dfrac{1}{\beta^2 + (n\pi)^2} \cdot \left\{ \beta + \dfrac{2(n\pi)^2}{\beta^2 + (n\pi)^2} \cdot [1 - \exp(-\beta) \cdot \cos(n\pi)] \right\}$

B.2 Numerical solutions

In most cases a numerical integration is the most effective solution, in which case Eq. B.1 is to be replaced by:

$$I(\beta) = \frac{1}{N^2} \sum_{p=1}^{N} \sum_{k=1}^{N} f(\hat{x}_p) \cdot f(\hat{x}_k) \cdot \exp(-\beta \cdot \Delta\hat{x}) \tag{B.3}$$

where $\Delta\hat{x} = |\hat{x}_p - \hat{x}_k|$ and N is the number of integration points. It should be noted that in general a finely meshed integration scheme is required, i.e. a large N. The reason for this is of course that the exponential function is rapidly dropping at increasing values of its argument. The solution to a good number of cases has been plotted in Figs. B.1 – B.3:

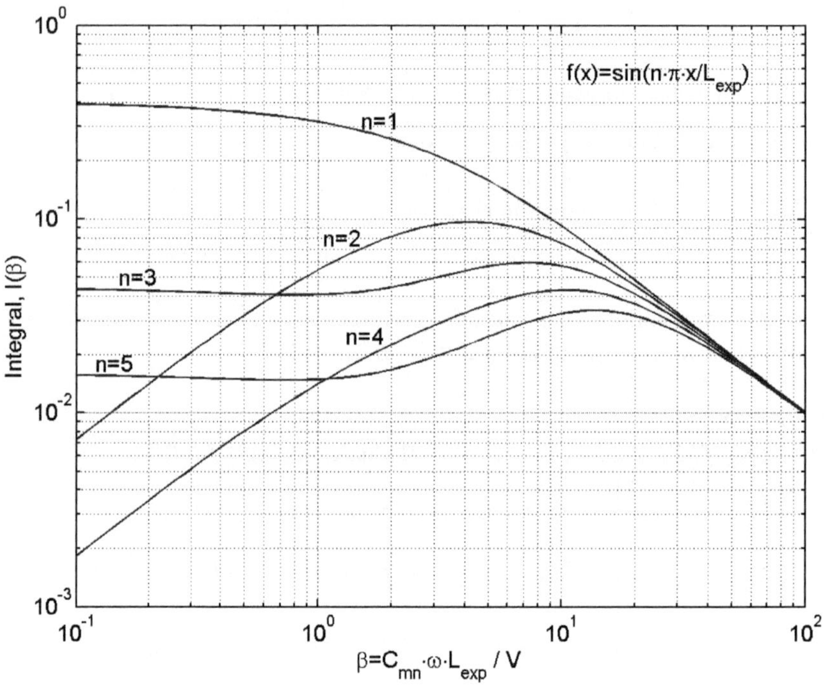

Fig. B.1 *Sinus type of typical mode shape functions*

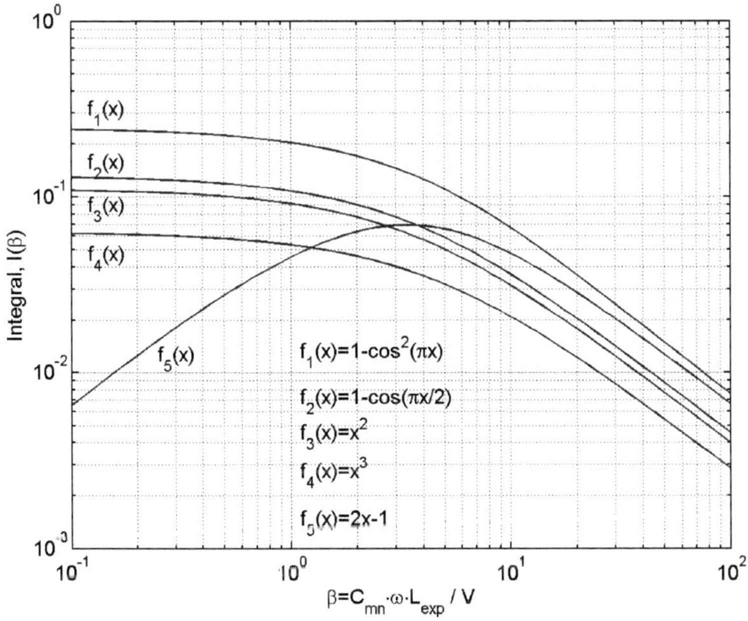

Fig. B.2 *Cosine or polynomial type of typical mode shape functions*

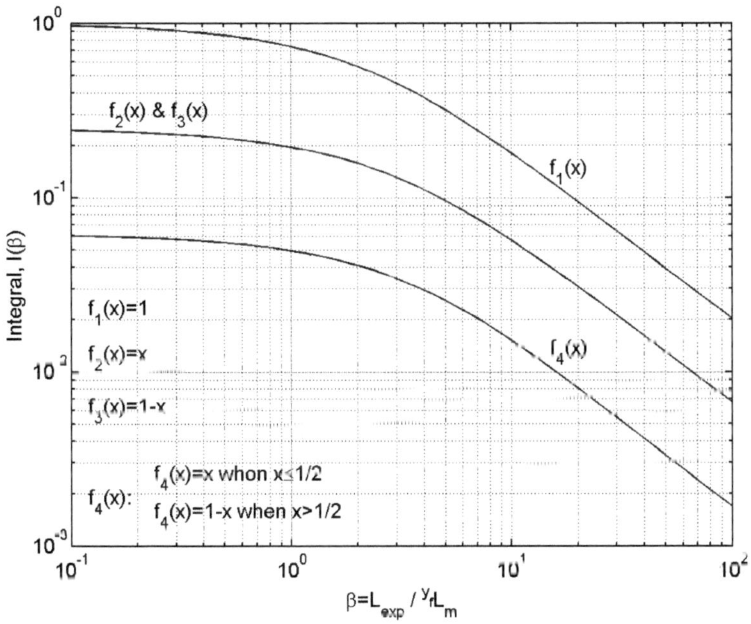

Fig. B.3 *Linear type of typical static influence functions*

Appendix C

AERODYNAMIC DERIVATIVES FROM SECTION MODEL DECAYS

From wind tunnel section model tests the aerodynamic derivatives were first quantified by the interpretation of in-wind simple decay recordings as described by Scanlan & Tomko [17]. From such testing six aerodynamic derivatives may be extracted, as shown in the following.

The section model contains two intentional modes, one in the across wind vertical direction and one with respect to torsion, i.e.:

$$\boldsymbol{\Phi}(x) = \begin{bmatrix} \boldsymbol{\varphi}_1 & \boldsymbol{\varphi}_2 \end{bmatrix} = \begin{bmatrix} \phi_z & 0 \\ 0 & \phi_\theta \end{bmatrix} \tag{C.1}$$

Internal unintentional flexibilities beyond those associated with these modes are most often insignificant, in which case $\phi_z \approx \phi_\theta \approx 1$. It is in the following taken for granted that their still-air properties

$$\left.\begin{array}{lcl} \omega_1(V=0) = \omega_z & \text{and} & \zeta_1(V=0) = \zeta_z \\ \omega_2(V=0) = \omega_\theta & & \zeta_2(V=0) = \zeta_\theta \end{array}\right\} \tag{C.2}$$

are known, and that any additional response contributions from other modes are insignificant or have effectively been filtered off. The testing strategy is to set the section model into decaying free motion at a suitable choice of mean wind velocity settings. Idealised recordings from such a test are illustrated in Fig. C.1. The velocity dependent response curves may mathematically be fitted to

$$\mathbf{r}(V,x,t) = \begin{bmatrix} r_z \\ r_\theta \end{bmatrix} = \boldsymbol{\Phi}(x) \cdot \boldsymbol{\eta}(V,t) \tag{C.3}$$

where:

$$\boldsymbol{\eta}(V,t) = \begin{bmatrix} \eta_1 \\ \eta_2 \end{bmatrix} = \exp(\lambda_r \cdot t) \cdot \begin{bmatrix} c_z \\ c_\theta \cdot \exp(-i \cdot \psi_r) \end{bmatrix} \tag{C.4}$$

and $\lambda_r(V) = -\zeta_r \cdot \omega_r + i \cdot \omega_r$, from which the in-wind damping ratio $\zeta_r(V)$, resonance frequency $\omega_r(V)$ and phase angle $\psi_r(V)$ may be quantified. The difference between observed in-wind values of ζ_r, ω_r ψ_r and their corresponding still-air counterparts will then contain all the effects of motion induced interaction between the section model and the flow. Since $\boldsymbol{\eta}(V,t)$ has been idealised into a single harmonic component it is necessary to assume that the motion induced part of the loading is dominant and narrow–banded, and that the buffeting contribution is insignificant or it has been filtered off. The general equation of motion that contains all the relevant motion induced effects as expressed by the aerodynamic derivatives is then given by

$$\tilde{\mathbf{M}} \cdot \ddot{\boldsymbol{\eta}} + \tilde{\mathbf{C}} \cdot \dot{\boldsymbol{\eta}} + \tilde{\mathbf{K}} \cdot \boldsymbol{\eta} \approx \tilde{\mathbf{C}}_{ae} \cdot \dot{\boldsymbol{\eta}} + \tilde{\mathbf{K}}_{ae} \cdot \boldsymbol{\eta} \tag{C.5}$$

where

$$\begin{bmatrix} \tilde{\mathbf{C}}_{ae} \\ \tilde{\mathbf{K}}_{ae} \end{bmatrix} = \int_{L_{\exp}} \boldsymbol{\Phi}^T \cdot \begin{bmatrix} \mathbf{C}_{ae} \\ \mathbf{K}_{ae} \end{bmatrix} \cdot \boldsymbol{\Phi} dx \tag{C.6}$$

Since the testing strategy only allows for the determination of six of the altogether eight motion induced load coefficients in the present set-up it is necessary to make a simplification. The following is adopted:

$$\mathbf{C}_{ae} = \begin{bmatrix} H_1 & H_2 \\ A_1 & A_2 \end{bmatrix} \quad \text{and} \quad \mathbf{K}_{ae} = \begin{bmatrix} 0 & H_3 \\ 0 & A_3 \end{bmatrix} \tag{C.7}$$

I.e., H_4 and A_4 are discarded. Thus,

$$\tilde{\mathbf{C}}_{ae} = \begin{bmatrix} \tilde{C}_{ae_{zz}} & \tilde{C}_{ae_{z\theta}} \\ \tilde{C}_{ae_{\theta z}} & \tilde{C}_{ae_{\theta\theta}} \end{bmatrix} = \int_{L_{\exp}} \begin{bmatrix} \phi_z^2 H_1 & \phi_z \phi_\theta H_2 \\ \phi_\theta \phi_z A_1 & \phi_\theta^2 A_2 \end{bmatrix} dx \tag{C.8}$$

and

$$\tilde{\mathbf{K}}_{ae} = \begin{bmatrix} 0 & \tilde{K}_{ae_{z\theta}} \\ 0 & \tilde{K}_{ae_{\theta\theta}} \end{bmatrix} = \int_{L_{\exp}} \begin{bmatrix} 0 & \phi_z \phi_\theta H_3 \\ 0 & \phi_\theta^2 A_3 \end{bmatrix} dx \tag{C.9}$$

The equation of motion is then given by:

$$\begin{bmatrix} \tilde{M}_z & 0 \\ 0 & \tilde{M}_\theta \end{bmatrix} \cdot \ddot{\boldsymbol{\eta}} + \begin{bmatrix} \tilde{C}_z - \tilde{C}_{ae_{zz}} & -\tilde{C}_{ae_{z\theta}} \\ -\tilde{C}_{ae_{\theta z}} & \tilde{C}_\theta - \tilde{C}_{ae_{\theta\theta}} \end{bmatrix} \cdot \dot{\boldsymbol{\eta}} + \begin{bmatrix} \tilde{K}_z & -\tilde{K}_{z\theta} \\ 0 & \tilde{K}_\theta - \tilde{K}_{ae_{\theta\theta}} \end{bmatrix} \cdot \boldsymbol{\eta} = \begin{bmatrix} 0 \\ 0 \end{bmatrix} \tag{C.10}$$

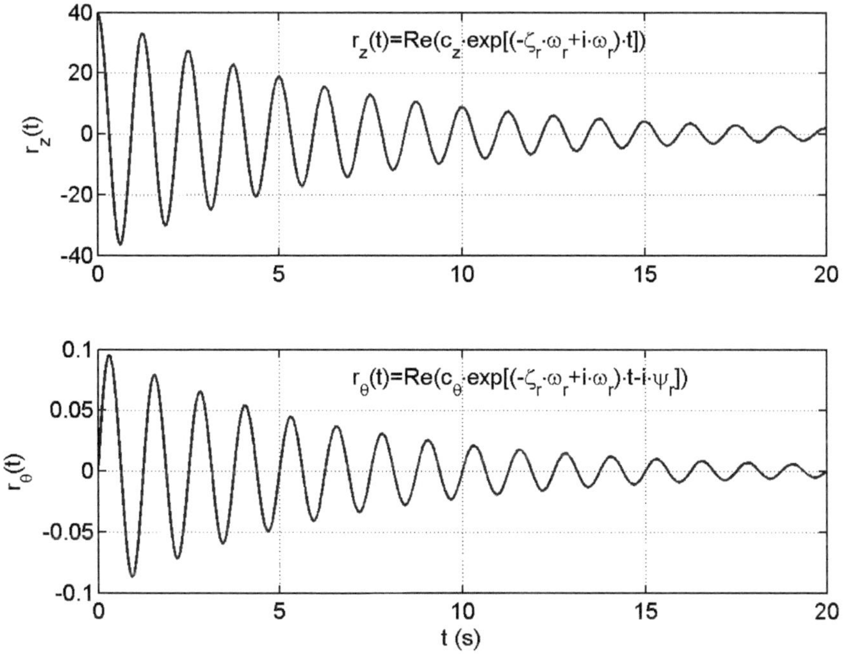

Fig. C.1 *Typical decay recordings as obtained from section model tests; top diagram: vertical displacements; lower diagram: torsion*

Introducing $\tilde{K}_z = \omega_z^2 \tilde{M}_z$, $\tilde{K}_\theta = \omega_\theta^2 \tilde{M}_\theta$, $\tilde{C}_z = 2\tilde{M}_z \omega_z \zeta_z$, $\tilde{C}_\theta = 2\tilde{M}_\theta \omega_\theta \zeta_\theta$ and that $\dot{\boldsymbol{\eta}} = \lambda_r \cdot \boldsymbol{\eta}$ and $\ddot{\boldsymbol{\eta}} = \lambda_r^2 \cdot \boldsymbol{\eta}$, then the equation of motion is reduced into

$$\left(\begin{bmatrix} 1 & 0 \\ 0 & 1 \end{bmatrix} \cdot \lambda_r^2 + \begin{bmatrix} 2\omega_z \zeta_z - \dfrac{\tilde{C}_{ae_{zz}}}{\tilde{M}_z} & -\dfrac{\tilde{C}_{ae_{z\theta}}}{\tilde{M}_z} \\ -\dfrac{\tilde{C}_{ae_{\theta z}}}{\tilde{M}_\theta} & 2\omega_\theta \zeta_\theta - \dfrac{\tilde{C}_{ae_{\theta\theta}}}{\tilde{M}_\theta} \end{bmatrix} \cdot \lambda_r + \begin{bmatrix} \omega_z^2 & -\dfrac{\tilde{K}_{ae_{z\theta}}}{\tilde{M}_z} \\ 0 & \omega_\theta^2 - \dfrac{\tilde{K}_{ae_{\theta\theta}}}{\tilde{M}_\theta} \end{bmatrix} \right) \cdot \boldsymbol{\eta} = \mathbf{0}$$

(C.11)

It is convenient to replace the aerodynamic load coefficients H_j and A_j, $j = 1,2,3$, in Eq. C.7 with the non-dimensional quantities H_j^* and A_j^* called aerodynamic derivatives and defined by:

$$\mathbf{C}_{ae} = \begin{bmatrix} H_1 & H_2 \\ A_1 & A_2 \end{bmatrix} = \frac{\rho B^2}{2} \cdot \omega_r(V) \cdot \begin{bmatrix} H_1^* & BH_2^* \\ BA_1^* & B^2 A_2^* \end{bmatrix} \tag{C.12}$$

$$\mathbf{K}_{ae} = \begin{bmatrix} 0 & H_3 \\ 0 & A_3 \end{bmatrix} = \frac{\rho B^2}{2} \cdot \omega_r^2(V) \cdot \begin{bmatrix} 0 & BH_3^* \\ 0 & B^2 A_3^* \end{bmatrix} \tag{C.13}$$

Thus,

$$\tilde{\mathbf{C}}_{ae} = \frac{\rho B^2}{2} \cdot \omega_r(V) \int_{L_{\exp}} \begin{bmatrix} \phi_z^2 H_1^* & \phi_z \phi_\theta B H_2^* \\ \phi_\theta \phi_z B A_1^* & \phi_\theta^2 B^2 A_2^* \end{bmatrix} dx \tag{C.14}$$

and

$$\tilde{\mathbf{K}}_{ae} = \frac{\rho B^2}{2} \cdot \omega_r^2(V) \int_{L_{\exp}} \begin{bmatrix} 0 & \phi_z \phi_\theta B H_3^* \\ 0 & \phi_\theta^2 B^2 A_3^* \end{bmatrix} dx \tag{C.15}$$

Defining

$$\tilde{m}_j = \tilde{M}_j \Big/ \int_L \phi_j^2 dx \qquad j = z \text{ or } \theta \tag{C.16}$$

and the abbreviations

$$\left. \begin{aligned} h_1 &= \beta_{zz} \cdot \frac{\omega_r}{2} \cdot H_1^* \\ h_2 &= \beta_{z\theta} \cdot \frac{\omega_r}{2} \cdot H_2^* \\ h_3 &= \beta_{z\theta} \cdot \frac{\omega_r^2}{2} \cdot H_3^* \end{aligned} \right\} \quad \text{where} \quad \left\{ \begin{aligned} \beta_{zz} &= \frac{\rho B^2}{\tilde{m}_z} \cdot \frac{\int_{L_{\exp}} \phi_z^2 dx}{\int_L \phi_z^2 dx} \\ \beta_{z\theta} &= \frac{\rho B^3}{\tilde{m}_z} \cdot \frac{\int_{L_{\exp}} \phi_z \phi_\theta dx}{\int_L \phi_z^2 dx} \end{aligned} \right. \tag{C.17}$$

$$\left. \begin{aligned} a_1 &= \beta_{\theta z} \cdot \frac{\omega_r}{2} \cdot A_1^* \\ a_2 &= \beta_{\theta\theta} \cdot \frac{\omega_r}{2} \cdot A_2^* \\ a_3 &= \beta_{\theta\theta} \cdot \frac{\omega_r^2}{2} \cdot A_3^* \end{aligned} \right\} \quad \text{where} \quad \left\{ \begin{aligned} \beta_{\theta z} &= \frac{\rho B^3}{\tilde{m}_\theta} \cdot \frac{\int_{L_{\exp}} \phi_\theta \phi_z dx}{\int_L \phi_\theta^2 dx} \\ \beta_{\theta\theta} &= \frac{\rho B^4}{\tilde{m}_\theta} \cdot \frac{\int_{L_{\exp}} \phi_\theta^2 dx}{\int_L \phi_\theta^2 dx} \end{aligned} \right. \tag{C.18}$$

then the equation of motion is given by

$$\left(\begin{bmatrix}1 & 0\\0 & 1\end{bmatrix}\lambda_r^2+\begin{bmatrix}2\omega_z\zeta_z-h_1 & -h_2\\-a_1 & 2\omega_\theta\zeta_\theta-a_2\end{bmatrix}\lambda_r+\begin{bmatrix}\omega_z^2 & -h_3\\0 & \omega_\theta^2-a_3\end{bmatrix}\right)\begin{bmatrix}c_z\\c_\theta\cdot\exp(-i\psi_r)\end{bmatrix}=\begin{bmatrix}0\\0\end{bmatrix} \tag{C.19}$$

Introducing $\exp(-i\psi_r)=\cos\psi_r-i\cdot\sin\psi_r$

$$\begin{bmatrix}\left\{\lambda_r^2+(2\omega_z\zeta_z-h_1)\lambda_r+\omega_z^2\right\}c_z-(h_2\lambda_r+h_3)(\cos\psi_r-i\sin\psi_r)c_\theta\\-a_1\lambda_r c_z+\left\{\lambda_r^2+(2\omega_\theta\zeta_\theta-a_2)\lambda_r+\omega_\theta^2-a_3\right\}(\cos\psi_r-i\sin\psi_r)c_\theta\end{bmatrix}=\begin{bmatrix}0\\0\end{bmatrix} \tag{C.20}$$

and that $\lambda_r=(-\zeta_r+i)\cdot\omega_r$ and $\lambda_r^2=\left(\zeta_r^2-1-i\cdot 2\zeta_r\right)\cdot\omega_r^2$, then the following is obtained:

$$\begin{bmatrix}c_z\left(\zeta_r^2-1+\frac{\omega_z^2}{\omega_r^2}-2\frac{\omega_z}{\omega_r}\zeta_z\zeta_r\right)+\frac{h_1}{\omega_r}c_z\zeta_r+\frac{h_2}{\omega_r}c_\theta(\zeta_r\cos\psi_r-\sin\psi_r)-\frac{h_3}{\omega_r^2}c_\theta\cos\psi_r\\ \\c_\theta\left(\zeta_r^2-1+\frac{\omega_\theta^2}{\omega_r^2}-2\frac{\omega_\theta}{\omega_r}\zeta_\theta\zeta_r\right)\cos\psi_r+2c_\theta\left(\frac{\omega_\theta}{\omega_r}\zeta_\theta-\zeta_r\right)\sin\psi_r+\\\frac{a_1}{\omega_r}c_z\zeta_r+\frac{a_2}{\omega_r}c_\theta(\zeta_r\cos\psi_r-\sin\psi_r)-\frac{a_3}{\omega_r^2}c_\theta\cos\psi_r\end{bmatrix}$$

$$-i\cdot\begin{bmatrix}-2c_z\left(\frac{\omega_z}{\omega_r}\zeta_z-\zeta_r\right)+\frac{h_1}{\omega_r}c_z+\frac{h_2}{\omega_r}c_\theta(\zeta_r\sin\psi_r+\cos\psi_r)-\frac{h_3}{\omega_r^2}c_\theta\sin\psi_r\\ \\c_\theta\left(\zeta_r^2-1+\frac{\omega_\theta^2}{\omega_r^2}-2\frac{\omega_\theta}{\omega_r}\zeta_\theta\zeta_r\right)\sin\psi_r-2c_\theta\left(\frac{\omega_\theta}{\omega_r}\zeta_\theta-\zeta_r\right)\cos\psi_r+\\\frac{a_1}{\omega_r}c_z+\frac{a_2}{\omega_r}c_\theta(\zeta_r\sin\psi_r+\cos\psi_r)-\frac{a_3}{\omega_r^2}c_\theta\sin\psi_r\end{bmatrix}=\begin{bmatrix}0\\ \\0\end{bmatrix} \tag{C.21}$$

The tests comprise three different conditions of motion control. First the decay tests are carried out with the physical constraint that $c_\theta=0$. Under this testing condition the imaginary part of Eq. C.21 is reduced to

$$-2c_z\left(\zeta_z\,\omega_z/\omega_r-\zeta_r\right)+h_1 c_z/\omega_r=0 \tag{C.22}$$

from which:

$$h_1=2\left(\omega_z\zeta_z-\omega_r\zeta_r\right) \tag{C.23}$$

and thus,

$$H_1^*=\frac{4}{\beta_{zz}}\left(\frac{\omega_z}{\omega_r}\zeta_z-\zeta_r\right) \tag{C.24}$$

The second series of decay tests are carried out with the physical constraint that $c_z=0$, in which case Eq. C.21 is reduced to

$$\begin{bmatrix}\left(\zeta_r^2-1+\dfrac{\omega_\theta^2}{\omega_r^2}-2\dfrac{\omega_\theta}{\omega_r}\zeta_\theta\zeta_r\right)\cos\psi_r+2\left(\dfrac{\omega_\theta}{\omega_r}\zeta_\theta-\zeta_r\right)\sin\psi_r+\\ \dfrac{a_2}{\omega_r}\left(\zeta_r\cos\psi_r-\sin\psi_r\right)-\dfrac{a_3}{\omega_r^2}\cos\psi_r\\ \\ \left(\zeta_r^2-1+\dfrac{\omega_\theta^2}{\omega_r^2}-2\dfrac{\omega_\theta}{\omega_r}\zeta_\theta\zeta_r\right)\sin\psi_r-2\left(\dfrac{\omega_\theta}{\omega_r}\zeta_\theta-\zeta_r\right)\cos\psi_r+\\ \dfrac{a_2}{\omega_r}\left(\zeta_r\sin\psi_r+\cos\psi_r\right)-\dfrac{a_3}{\omega_r^2}\sin\psi_r\end{bmatrix}=\begin{bmatrix}0\\ \\ 0\end{bmatrix} \tag{C.25}$$

Thus,
$$\begin{bmatrix}a_2\\ a_3\end{bmatrix}=\begin{bmatrix}2\left(\omega_\theta\zeta_\theta-\omega_r\zeta_r\right)\\ \omega_\theta^2-\omega_r^r-\omega_r^2\zeta_r^2\end{bmatrix} \tag{C.26}$$

from which

$$A_2^*=\frac{4}{\beta_{\theta\theta}}\left(\frac{\omega_\theta}{\omega_r}\zeta_\theta-\zeta_r\right) \tag{C.27}$$

$$A_3^*=\frac{2}{\beta_{\theta\theta}}\left(\frac{\omega_\theta^2}{\omega_r^2}-1-\zeta_r^2\right) \tag{C.28}$$

After h_1, a_2 and a_3 have been determined then the third series of decay tests are carried out with no physical constraints, such that $c_z \neq 0$ and $c_\theta \neq 0$, in which case the full version of Eq. C.21 applies. Eliminating h_3 from the first real and imaginary parts and a_2 from the second real and imaginary parts then the following equations are obtained:

$$\begin{bmatrix} c_z\left(\zeta_r^2-1+\frac{\omega_z^2}{\omega_r^2}-2\frac{\omega_z}{\omega_r}\zeta_z\zeta_r+\frac{h_1}{\omega_r}\zeta_r\right)\sin\psi_r \\ +c_z\left\{2\left(\frac{\omega_z}{\omega_r}\zeta_z-\zeta_r\right)-\frac{h_1}{\omega_r}\right\}\cos\psi_r-\frac{h_2}{\omega_r}c_\theta \\ \\ c_\theta\left(\zeta_r^2-1+\frac{\omega_\theta^2}{\omega_r^2}-2\frac{\omega_\theta}{\omega_r}\zeta_\theta\zeta_r\right)+2c_\theta\zeta_r\left(\frac{\omega_\theta}{\omega_r}\zeta_\theta-\zeta_r\right)+\frac{a_1}{\omega_r}c_z\left(\zeta_r^2+1\right)-\frac{a_3}{\omega_r^2}c_\theta \end{bmatrix}=\begin{bmatrix}0\\ \\ 0\end{bmatrix} \tag{C.29}$$

from which

$$a_1=\frac{c_\theta}{c_z}\cdot\omega_r\cdot\frac{\zeta_r^2+1-\omega_\theta^2/\omega_r^2+a_3/\omega_r^2}{\left(\zeta_r^2+1\right)\sin\psi_r} \tag{C.30}$$

$$h_2=\frac{c_z\omega_r}{c_\theta}\left\{\left(\zeta_r^2-1+\frac{\omega_z^2}{\omega_r^2}-2\frac{\omega_z}{\omega_r}\zeta_z\zeta_r+\frac{h_1}{\omega_r}\zeta_r\right)\sin\psi_r+\left[2\left(\frac{\omega_z}{\omega_r}\zeta_z-\zeta_r\right)-\frac{h_1}{\omega_r}\right]\cos\psi_r\right\} \tag{C.31}$$

Finally, h_3 may be determined from the first real part of Eq. C.21, rendering

$$h_3=\frac{\omega_r^2}{\cos\psi_r}\cdot\left[\frac{c_z}{c_\theta}\cdot\left(\zeta_r^2-1+\frac{\omega_z^2}{\omega_r^2}-2\frac{\omega_z}{\omega_r}\zeta_z\zeta_r+\frac{h_1}{\omega_r}\zeta_r\right)+\frac{h_2}{\omega_r}\cdot\left(\zeta_r\cos\psi_r-\sin\psi_r\right)\right] \tag{C.32}$$

From Eqs. C.17 and C.18

$$A_1^*=\frac{2}{\beta_{\theta z}}\cdot\frac{a_1}{\omega_r},\quad H_2^*=\frac{2}{\beta_{z\theta}}\cdot\frac{h_2}{\omega_r}\quad\text{and}\quad H_3^*=\frac{2}{\beta_{z\theta}}\cdot\frac{h_3}{\omega_r^2} \tag{C.33}$$

Appendix D

DETERMINATION OF INDICIAL FUNCTIONS FROM AERODYNAMIC DERIVATIVES

Aerodynamic load contributions in the along-wind, the across-wind and the pitching moment directions $\mathbf{q}_{ae}(t) = \begin{bmatrix} q_y & q_z & q_\theta \end{bmatrix}_{ae}^T$ generated by the interaction between the wind field and the motion of the structure $\mathbf{r}(t) = \begin{bmatrix} r_y & r_z & r_\theta \end{bmatrix}^T$ have first been developed in a quasi-steady format in chapter 5.1, where the definition of positive r_θ and q_θ comply with the usual aerodynamic conventions shown in Fig. 1.3.a. Since the determination of aerodynamic derivatives and the corresponding indicial functions are taken from wind tunnel aero-elastic section model tests, this is also the sign convention chosen for the presentation below. However, the use of indicial functions will generally only be relevant in a time domain solution in a finite element format, and in such a format all displacement and load quantities are vectors. This format has been presented in chapter 9. It should however be noted that the choice of sign convention for displacements and forces has no consequences to the determination of indicial functions as long as any displacement component and its corresponding load component has identical directions. Thus, from a quasi-steady theory (see Eq. 5.8)

$$\mathbf{q}_{ae}(t) = \mathbf{C}_{ae}\dot{\mathbf{r}}(t) + \mathbf{K}_{ae}\mathbf{r}(t) \tag{D.1}$$

where for simplicity the horizontal spanwise (axial) r_x degree of freedom and corresponding q_x load component have been omitted as they have no relevance for a line-like type of structure, and

$$\left.\begin{aligned} \mathbf{C}_{ae} &= \frac{\rho V}{2}\begin{bmatrix} -2D\bar{C}_D & -\left(DC_D' - B\bar{C}_L\right) & 0 \\ -2B\bar{C}_L & -\left(BC_L' + D\bar{C}_D\right) & 0 \\ -2B^2\bar{C}_M & -B^2C_M' & 0 \end{bmatrix} \\ \mathbf{K}_{ae} &= \frac{\rho V^2}{2}\begin{bmatrix} 0 & 0 & DC_D' \\ 0 & 0 & BC_L' \\ 0 & 0 & B^2C_M' \end{bmatrix} \end{aligned}\right\} \tag{D.2}$$

In a frequency domain formulation Eq. D.2 may be replaced by (see chapter 5.2)

$$\left.\begin{aligned}
\mathbf{C}_{ae} &= \frac{\rho B^2}{2}\omega_i \begin{bmatrix} P_1^* & P_5^* & BP_2^* \\ H_5^* & H_1^* & BH_2^* \\ BA_5^* & BA_1^* & B^2A_2^* \end{bmatrix} \\
\mathbf{K}_{ae} &= \frac{\rho B^2}{2}\omega_i^2 \begin{bmatrix} P_4^* & P_6^* & BP_3^* \\ H_6^* & H_4^* & BH_3^* \\ BA_6^* & BA_4^* & B^2A_3^* \end{bmatrix}
\end{aligned}\right\} \tag{D.3}$$

where P_k^*, H_k^*, A_k^* ($k = 1,2,...,6$) are the aerodynamic derivatives and ω_i is the frequency of motion. Theoretically, Eq. D.2 is applicable in time domain as well as in frequency domain. However, the basic hypothesis behind the quasi-steady theory was that fluctuations in the oncoming flow or in the motion of the structure will instantaneously give rise to corresponding fluctuations in the cross sectional loads. Such a hypothesis will not render reliable results in a time domain solution, and therefore, Eq. D.1 needs to be formulated at an incremental level

$$\frac{d\mathbf{q}_{ae}}{d\tau} = \mathbf{C}_{ae}(s)\frac{d\dot{\mathbf{r}}(\tau)}{d\tau} + \mathbf{K}_{ae}(s)\frac{d\mathbf{r}(\tau)}{d\tau} \tag{D.4}$$

where (see Fig. 9.8) τ is a dummy time history variable, $s = t - \tau$ and

$$\left.\begin{aligned}
\mathbf{C}_{ae}(s) &= \frac{\rho V}{2}\begin{bmatrix} -2D\bar{C}_D\Phi_{D\dot{y}}(s) & -\left(DC_D' - B\bar{C}_L\right)\Phi_{D\dot{z}}(s) & 0 \\ -2B\bar{C}_L\Phi_{L\dot{y}}(s) & -\left(BC_L' + D\bar{C}_D\right)\Phi_{L\dot{z}}(s) & 0 \\ -2B^2\bar{C}_M\Phi_{M\dot{y}}(s) & -B^2C_M'\Phi_{M\dot{z}}(s) & 0 \end{bmatrix} \\
\mathbf{K}_{ae}(s) &= \frac{\rho V^2}{2}\begin{bmatrix} 0 & 0 & DC_D'\Phi_{D\theta}(s) \\ 0 & 0 & BC_L'\Phi_{L\theta}(s) \\ 0 & 0 & B^2C_M'\Phi_{M\theta}(s) \end{bmatrix}
\end{aligned}\right\} \tag{D.5}$$

where $$\Phi_{mn}(s) \quad \begin{cases} m = D, L \text{ or } M \\ n = \dot{y}, \dot{z} \text{ or } \theta \end{cases} \tag{D.6}$$

are the indicial memory functions associated with interaction between drag, lift or moment forces and the velocity of motion in y, z or θ directions. A displacement increment and its corresponding load increment in the remote part of the response history must comply to the quasi-steady solution, and thus

$$\lim_{s\to\infty}\Phi_{mn}(s)=1 \quad \begin{cases} m = D,L \text{ or } M \\ n = \dot{y},\dot{z} \text{ or } \theta \end{cases} \tag{D.7}$$

Therefore, an exponential type of function with a limiting value of unity is usually chosen for the representation of indicial functions interpreted from experimental results (see Salvatori & Borri [31]). It is convenient to express these functions by the non-dimensional relative time parameter $\hat{s} = sV/B$. Thus,

$$\Phi_{mn}(\hat{s})=1-\sum_{j=1}^{N_j} b_j e^{-c_j\hat{s}} \quad \begin{cases} m = D,L \text{ or } M \\ n = \dot{y},\dot{z} \text{ or } \theta \end{cases} \tag{D.8}$$

where the constants b_j and c_j may be determined from experiments or, as shown below, from the aerodynamic derivatives (see Chapter 5.2).

The current motion induced load $\mathbf{q}_{ae}(t)$ may then be obtained by history integration (theoretically from $-\infty$ to the present time t)

$$\mathbf{q}_{ae}(t) = \int_{-\infty}^{t}\left[\mathbf{C}_{ae}(s)\frac{d\dot{\mathbf{r}}(\tau)}{d\tau}+\mathbf{K}_{ae}(s)\frac{d\mathbf{r}(\tau)}{d\tau}\right]d\tau \tag{D.9}$$

Integration by parts

$$\begin{aligned}\mathbf{q}_{ae}(t) &= \left[\mathbf{C}_{ae}(s)\cdot\dot{\mathbf{r}}(\tau)\right]_{\tau=-\infty}^{\tau=t} - \int_{-\infty}^{t}\frac{d\mathbf{C}_{ae}(s)}{ds}\cdot\frac{ds}{d\tau}\cdot\dot{\mathbf{r}}(\tau)d\tau \\ &+\left[\mathbf{K}_{ae}(s)\cdot\mathbf{r}(\tau)\right]_{\tau=-\infty}^{\tau=t} - \int_{-\infty}^{t}\frac{d\mathbf{K}_{ae}(s)}{ds}\cdot\frac{ds}{d\tau}\cdot\mathbf{r}(\tau)d\tau\end{aligned} \tag{D.10}$$

and assuming negligible initial displacement and velocity conditions, then

$$\mathbf{q}_{ae}(t) = \mathbf{C}_{ae}(0)\dot{\mathbf{r}}(t)+\int_{-\infty}^{t}\mathbf{C}'_{ae}(s)\dot{\mathbf{r}}(\tau)d\tau + \mathbf{K}_{ae}(0)\mathbf{r}(t)+\int_{-\infty}^{t}\mathbf{K}'_{ae}(s)\mathbf{r}(\tau)d\tau \tag{D.11}$$

where $\mathbf{C}'_{ae}(s)=\dfrac{d\mathbf{C}_{ae}(s)}{ds}$ and $\mathbf{K}'_{ae}(s)=\dfrac{d\mathbf{K}_{ae}(s)}{ds}$. Introducing a harmonic motion

$$\mathbf{r}(t) = \begin{bmatrix} a_y \\ a_z \\ a_\theta \end{bmatrix}\cdot e^{i\omega t} = \mathbf{a}\cdot e^{i\omega t} \tag{D.12}$$

then

$$\mathbf{q}_{ae}(t) = \mathbf{C}_{ae}(0)\cdot i\cdot\omega_i\cdot\mathbf{a}\cdot e^{i\omega_i t} + \int_{-\infty}^{t}\mathbf{C}'_{ae}(s)\cdot i\cdot\omega_i\cdot\mathbf{a}\cdot e^{i\omega_i\tau}d\tau$$
$$+\mathbf{K}_{ae}(0)\cdot\mathbf{a}\cdot e^{i\omega_i t} + \int_{-\infty}^{t}\mathbf{K}'_{ae}(s)\cdot\mathbf{a}\cdot e^{i\omega_i\tau}d\tau \tag{D.13}$$

Recalling that $s = t-\tau$ and using that $\dfrac{ds}{d\tau} = -1$, $\int_a^b = -\int_b^a$ and $\mathbf{r}(t) = \mathbf{a}\cdot e^{i\omega_i t}$,then

$$\mathbf{q}_{ae}(t) = \left\{\mathbf{K}_{ae}(0) + \int_0^{\infty}\mathbf{K}'_{ae}(s)\cdot e^{-i\omega_i s}ds + i\omega_i\left[\mathbf{C}_{ae}(0) + \int_0^{\infty}\mathbf{C}'_{ae}(s)\cdot e^{-i\omega_i s}ds\right]\right\}\mathbf{r}(t) \tag{D.14}$$

Introducing $\hat{s} = \dfrac{V}{B}s$, $\hat{V}_i = \dfrac{V}{B\omega_i}$, $\mathbf{K}'_{ae}(\hat{s}) = \dfrac{d\mathbf{K}_{ae}(\hat{s})}{d\hat{s}}$ and $\mathbf{C}'_{ae}(\hat{s}) = \dfrac{d\mathbf{C}_{ae}(\hat{s})}{d\hat{s}}$, then

$$\mathbf{q}_{ae}(t) = \left\{\mathbf{K}_{ae}(0) + \int_0^{\infty}\mathbf{K}'_{ae}(\hat{s})\cdot e^{-i\hat{s}/\hat{V}_i}d\hat{s} + i\frac{V}{B}\frac{1}{\hat{V}_i}\left[\mathbf{C}_{ae}(0) + \int_0^{\infty}\mathbf{C}'_{ae}(\hat{s})\cdot e^{-i\hat{s}/\hat{V}_i}d\hat{s}\right]\right\}\mathbf{r}(t) \tag{D.15}$$

Introducing Eq. D.5 renders

$$\frac{\mathbf{q}_{ae}(t)}{\frac{\rho V^2}{2}} = \sum_{j=1}^{N_j}\left\{\begin{bmatrix} 0 & 0 & DC'_D\Phi_{D\theta}(0) \\ 0 & 0 & BC'_L\Phi_{L\theta}(0) \\ 0 & 0 & B^2C'_M\Phi_{M\theta}(0)\end{bmatrix} + \int_0^{\infty}\begin{bmatrix} 0 & 0 & DC'_D\Phi'_{D\theta}(\hat{s}) \\ 0 & 0 & BC'_L\Phi'_{L\theta}(\hat{s}) \\ 0 & 0 & B^2C'_M\Phi'_{M\theta}(\hat{s})\end{bmatrix}\cdot e^{-i\hat{s}/\hat{V}_i}d\hat{s}\right.$$
$$+\frac{i}{\hat{V}_i}\left(\begin{bmatrix} -2\frac{D}{B}\bar{C}_D\Phi_{D\dot{y}}(0) & -\left(\frac{D}{B}C'_D - \bar{C}_L\right)\Phi_{D\dot{z}}(0) & 0 \\ -2\bar{C}_L\Phi_{L\dot{y}}(0) & -\left(C'_L + \frac{D}{B}\bar{C}_D\right)\Phi_{D\dot{z}}(0) & 0 \\ -2B\bar{C}_M\Phi_{M\dot{y}}(0) & BC'_M\Phi_{M\dot{z}}(0) & 0\end{bmatrix}\right.$$
$$\left.\left.+\int_0^{\infty}\begin{bmatrix} -2\frac{D}{B}\bar{C}_D\Phi_{D\dot{y}}(\hat{s}) & -\left(\frac{D}{B}C'_D - \bar{C}_L\right)\Phi'_{D\dot{z}}(\hat{s}) & 0 \\ -2\bar{C}_L\Phi_{L\dot{y}}(0) & -\left(C'_L + \frac{D}{B}\bar{C}_D\right)\Phi'_{D\dot{z}}(\hat{s}) & 0 \\ -2B\bar{C}_M\Phi_{M\dot{y}}(0) & -BC'_M\Phi'_{M\dot{z}}(\hat{s}) & 0\end{bmatrix}\cdot e^{-i\hat{s}/\hat{V}_i}d\hat{s}\right)\right\}\mathbf{a}e^{i\omega_i t} \tag{D.16}$$

From Eq. D.8 it is seen that

$$\int_0^{\infty} \Phi'_{mn}(\hat{s}) \cdot e^{-i\hat{s}/\hat{V}_i} d\hat{s} = \sum_{j=1}^{N_j} b_j c_j \int_0^{\infty} e^{-\left(c_j + i/\hat{V}_i\right)\hat{s}} d\hat{s} = \sum_{j=1}^{N_j} \frac{b_j c_j}{c_j + i/\hat{V}_i} \quad \text{(D.17)}$$

and thus

$$\begin{aligned}
\frac{\mathbf{q}_{ae}(t)}{\frac{\rho V^2}{2}} = \Bigg\{ & \begin{bmatrix} 0 & 0 & DC'_D\left(1-\sum_{j=1}^{N_j} b_j\right)_{D\theta} \\ 0 & 0 & BC'_L\left(1-\sum_{j=1}^{N_j} b_j\right)_{L\theta} \\ 0 & 0 & B^2 C'_M\left(1-\sum_{j=1}^{N_j} b_j\right)_{M\theta} \end{bmatrix} + \begin{bmatrix} 0 & 0 & DC'_D \sum_{j=1}^{N_j}\left(\frac{b_j c_j}{c_j + i/\hat{V}_i}\right)_{D\theta} \\ 0 & 0 & BC'_L \sum_{j=1}^{N_j}\left(\frac{b_j c_j}{c_j + i/\hat{V}_i}\right)_{L\theta} \\ 0 & 0 & B^2 \sum_{j=1}^{N_j}\left(\frac{b_j c_j}{c_j + i/\hat{V}_i}\right)_{M\theta} \end{bmatrix} \\
& + \frac{i}{\hat{V}_i}\left(\begin{bmatrix} -2\frac{D}{B}\bar{C}_D\left(1-\sum_{j=1}^{N_j} b_j\right)_{D\dot{y}} & -\left(\frac{D}{B}C'_D - \bar{C}_L\right)\left(1-\sum_{j=1}^{N_j} b_j\right)_{D\dot{z}} & 0 \\ -2\bar{C}_L\left(1-\sum_{j=1}^{N_j} b_j\right)_{L\dot{y}} & -\left(C'_L + \frac{D}{B}\bar{C}_D\right)\left(1-\sum_{j=1}^{N_j} b_j\right)_{L\dot{z}} & 0 \\ -2B\bar{C}_M\left(1-\sum_{j=1}^{N_j} b_j\right)_{M\dot{y}} & BC'_M\left(1-\sum_{j=1}^{N_j} b_j\right)_{M\dot{z}} & 0 \end{bmatrix} \right. \\
& \left. + \begin{bmatrix} -2\frac{D}{B}\bar{C}_D \sum_{j=1}^{N_j}\left(\frac{b_j c_j}{c_j + i/\hat{V}_i}\right)_{D\dot{y}} & -\left(\frac{D}{B}C'_D - \bar{C}_L\right)\sum_{j=1}^{N_j}\left(\frac{b_j c_j}{c_j + i/\hat{V}_i}\right)_{Dz} & 0 \\ -2\bar{C}_L \sum_{j=1}^{N_j}\left(\frac{b_j c_j}{c_j + i/\hat{V}_i}\right)_{L\dot{y}} & -\left(C'_L + \frac{D}{B}\bar{C}_D\right)\sum_{j=1}^{N_j}\left(\frac{b_j c_j}{c_j + i/\hat{V}_i}\right)_{L\dot{z}} & 0 \\ -2B\bar{C}_M \sum_{j=1}^{N_j}\left(\frac{b_j c_j}{c_j + i/\hat{V}_i}\right)_{M\dot{y}} & -BC'_M \sum_{j=1}^{N_j}\left(\frac{b_j c_j}{c_j + i/\hat{V}_i}\right)_{M\dot{z}} & 0 \end{bmatrix} \right) \Bigg\} \mathbf{a} e^{i\omega_i t}
\end{aligned}$$

(D.18)

Gathering all real terms in one matrix and all imaginary terms in another the following is obtained:

$$\frac{\mathbf{q}_{ae}(t)}{\frac{\rho V^2}{2}} = \left\{ -\begin{bmatrix} 2\frac{D}{B}\bar{C}_D \sum_{j=1}^{N_j}\left(\frac{b_j c_j}{\hat{V}_i^2 c_j^2+1}\right)_{D\dot{y}} & \left(\frac{D}{B}C_D' - \bar{C}_L\right)\sum_{j=1}^{N_j}\left(\frac{b_j c_j}{\hat{V}_i^2 c_j^2+1}\right)_{D\dot{z}} & -DC_D'\left(1-\sum_{j=1}^{N_j}\frac{b_j}{\hat{V}_i^2 c_j^2+1}\right)_{D\theta} \\ 2\bar{C}_L \sum_{j=1}^{N_j}\left(\frac{b_j c_j}{\hat{V}_i^2 c_j^2+1}\right)_{L\dot{y}} & \left(C_L' + \frac{D}{B}\bar{C}_D\right)\sum_{j=1}^{N_j}\left(\frac{b_j c_j}{\hat{V}_i^2 c_j^2+1}\right)_{L\dot{z}} & -BC_L'\left(1-\sum_{j=1}^{N_j}\frac{b_j}{\hat{V}_i^2 c_j^2+1}\right)_{L\theta} \\ 2B\bar{C}_M \sum_{j=1}^{N_j}\left(\frac{b_j c_j}{\hat{V}_i^2 c_j^2+1}\right)_{M\dot{y}} & BC_M'\sum_{j=1}^{N_j}\left(\frac{b_j c_j}{\hat{V}_i^2 c_j^2+1}\right)_{M\dot{z}} & -B^2 C_M'\left(1-\sum_{j=1}^{N_j}\frac{b_j}{\hat{V}_i^2 c_j^2+1}\right)_{M\theta} \end{bmatrix} \right.$$

$$\left. -\frac{i}{\hat{V}_i}\left(\begin{bmatrix} 2\frac{D}{B}\bar{C}_D\left(1-\sum_{j=1}^{N_j}\frac{b_j}{\hat{V}_i^2 c_j^2+1}\right)_{D\dot{y}} & \left(\frac{D}{B}C_D' - \bar{C}_L\right)\left(1-\sum_{j=1}^{N_j}\frac{b_j}{\hat{V}_i^2 c_j^2+1}\right)_{D\dot{z}} & DC_D'\sum_{j=1}^{N_j}\left(\frac{b_j c_j \hat{V}_i^2}{\hat{V}_i^2 c_j^2+1}\right)_{D\theta} \\ 2\bar{C}_L\left(1-\sum_{j=1}^{N_j}\frac{b_j}{\hat{V}_i^2 c_j^2+1}\right)_{L\dot{y}} & \left(C_L' + \frac{D}{B}\bar{C}_D\right)\left(1-\sum_{j=1}^{N_j}\frac{b_j}{\hat{V}_i^2 c_j^2+1}\right)_{L\dot{z}} & BC_L'\sum_{j=1}^{N_j}\left(\frac{b_j c_j \hat{V}_i^2}{\hat{V}_i^2 c_j^2+1}\right)_{L\theta} \\ 2B\bar{C}_M\left(1-\sum_{j=1}^{N_j}\frac{b_j}{\hat{V}_i^2 c_j^2+1}\right)_{M\dot{y}} & BC_M'\left(1-\sum_{j=1}^{N_j}\frac{b_j}{\hat{V}_i^2 c_j^2+1}\right)_{M\dot{z}} & B^2 C_M'\sum_{j=1}^{N_j}\left(\frac{b_j c_j \hat{V}_i^2}{\hat{V}_i^2 c_j^2+1}\right)_{M\theta} \end{bmatrix}\right) \right\} \mathbf{a} e^{i\omega_t t} \tag{D.19}$$

This is a frequency domain application of the indicial functions, and therefore, it must render an identical solution to that which is obtained by the use of aerodynamic derivatives. Introducing a harmonic input $\mathbf{r}(t) = \begin{bmatrix} a_y & a_z & a_\theta \end{bmatrix}^T \cdot e^{i\omega_t t} = \mathbf{a}\cdot e^{i\omega_t t}$ into Eq. D.1, and using $\mathbf{C}_{ae}$ and $\mathbf{K}_{ae}$ from Eq. D.3, then a frequency domain solution containing the aerodynamic derivatives is given by

$$\frac{\mathbf{q}_{ae}(t)}{\frac{1}{2}\rho V^2} = \left(\frac{1}{\hat{V}_i^2}\begin{bmatrix} P_4^* & P_6^* & BP_3^* \\ H_6^* & H_4^* & BH_3^* \\ BA_6^* & BA_4^* & B^2 A_3^* \end{bmatrix} + \frac{i}{\hat{V}_i^2}\begin{bmatrix} P_1^* & P_5^* & BP_2^* \\ H_5^* & H_1^* & BH_2^* \\ BA_5^* & BA_1^* & B^2 A_2^* \end{bmatrix} \right) \mathbf{a} e^{i\omega_t t} \tag{D.20}$$

Eqs. D.19 and D.20 must be identical, and thus, the following is obtained

$$\left.\begin{aligned} \left(\sum_{j=1}^{N_j}\frac{b_j c_j}{\hat{V}_i^2 c_j^2+1}\right)_{D\dot{y}} &= -\frac{1}{\hat{V}_i^2}\frac{P_4^*(\hat{V}_i)}{2\frac{D}{B}\bar{C}_D} \\ \left(\sum_{j=1}^{N_j}\frac{b_j}{\hat{V}_i^2 c_j^2+1}\right)_{D\dot{y}} &= 1+\frac{1}{\hat{V}_i}\frac{P_1^*(\hat{V}_i)}{2\frac{D}{B}\bar{C}_D} \end{aligned}\right\} \tag{D.21}$$

$$\left.\begin{aligned}\left(\sum_{j=1}^{N_j}\frac{b_j c_j}{\hat{V}_i^2 c_j^2+1}\right)_{L\dot{y}} &= -\frac{1}{\hat{V}_i^2}\frac{H_6^*\left(\hat{V}_i\right)}{2\bar{C}_L}\\ \left(\sum_{j=1}^{N_j}\frac{b_j}{\hat{V}_i^2 c_j^2+1}\right)_{L\dot{y}} &= 1+\frac{1}{\hat{V}_i}\frac{H_5^*\left(\hat{V}_i\right)}{2\bar{C}_L}\end{aligned}\right\} \tag{D.22}$$

$$\left.\begin{aligned}\left(\sum_{j=1}^{N_j}\frac{b_j c_j}{\hat{V}_i^2 c_j^2+1}\right)_{M\dot{y}} &= -\frac{1}{\hat{V}_i^2}\frac{A_6^*\left(\hat{V}_i\right)}{2\bar{C}_M}\\ \left(\sum_{j=1}^{N_j}\frac{b_j}{\hat{V}_i^2 c_j^2+1}\right)_{M\dot{y}} &= 1+\frac{1}{\hat{V}_i}\frac{A_5^*\left(\hat{V}_i\right)}{2\bar{C}_M}\end{aligned}\right\} \tag{D.23}$$

$$\left.\begin{aligned}\left(\sum_{j=1}^{N_j}\frac{b_j c_j}{\hat{V}_i^2 c_j^2+1}\right)_{D\dot{z}} &= -\frac{1}{\hat{V}_i^2}\frac{P_6^*\left(\hat{V}_i\right)}{\frac{D}{B}C_D' - \bar{C}_L}\\ \left(\sum_{j=1}^{N_j}\frac{b_j}{\hat{V}_i^2 c_j^2+1}\right)_{D\dot{z}} &= 1+\frac{1}{\hat{V}_i}\frac{P_5^*\left(\hat{V}_i\right)}{\frac{D}{B}C_D' - \bar{C}_L}\end{aligned}\right\} \tag{D.24}$$

$$\left.\begin{aligned}\left(\sum_{j=1}^{N_j}\frac{b_j c_j}{\hat{V}_i^2 c_j^2+1}\right)_{L\dot{z}} &= -\frac{1}{\hat{V}_i^2}\frac{H_4^*\left(\hat{V}_i\right)}{C_L' + \frac{D}{B}\bar{C}_D}\\ \left(\sum_{j=1}^{N_j}\frac{b_j}{\hat{V}_i^2 c_j^2+1}\right)_{L\dot{z}} &= 1+\frac{1}{\hat{V}_i}\frac{H_1^*\left(\hat{V}_i\right)}{C_L' + \frac{D}{B}\bar{C}_D}\end{aligned}\right\} \tag{D.25}$$

$$\left.\begin{aligned}\left(\sum_{j=1}^{N_j}\frac{b_j c_j}{\hat{V}_i^2 c_j^2+1}\right)_{M\dot{z}} &= -\frac{1}{\hat{V}_i^2}\frac{A_4^*\left(\hat{V}_i\right)}{C_M'}\\ \left(\sum_{j=1}^{N_j}\frac{b_j}{\hat{V}_i^2 c_j^2+1}\right)_{M\dot{z}} &= 1+\frac{1}{\hat{V}_i}\frac{A_1^*\left(\hat{V}_i\right)}{C_M'}\end{aligned}\right\} \tag{D.26}$$

$$\left.\begin{aligned}\left(\sum_{j=1}^{N_j}\frac{b_jc_j}{\hat{V}_i^2c_j^2+1}\right)_{D\theta}&=-\frac{1}{\hat{V}_i^3}\frac{P_2^*\left(\hat{V}_i\right)}{\frac{D}{B}C_D'}\\\left(\sum_{j=1}^{N_j}\frac{b_j}{\hat{V}_i^2c_j^2+1}\right)_{D\theta}&=1-\frac{1}{\hat{V}_i^2}\frac{P_3^*\left(\hat{V}_i\right)}{\frac{D}{B}C_D'}\end{aligned}\right\}\tag{D.27}$$

$$\left.\begin{aligned}\left(\sum_{j=1}^{N_j}\frac{b_jc_j}{\hat{V}_i^2c_j^2+1}\right)_{L\theta}&=-\frac{1}{\hat{V}_i^3}\frac{H_2^*\left(\hat{V}_i\right)}{C_L'}\\\left(\sum_{j=1}^{N_j}\frac{b_j}{\hat{V}_i^2c_j^2+1}\right)_{L\theta}&=1-\frac{1}{\hat{V}_i^2}\frac{H_3^*\left(\hat{V}_i\right)}{C_L'}\end{aligned}\right\}\tag{D.28}$$

$$\left.\begin{aligned}\left(\sum_{j=1}^{N_j}\frac{b_jc_j}{\hat{V}_i^2c_j^2+1}\right)_{M\theta}&=-\frac{1}{\hat{V}_i^3}\frac{A_2^*\left(\hat{V}_i\right)}{C_M'}\\\left(\sum_{j=1}^{N_j}\frac{b_j}{\hat{V}_i^2c_j^2+1}\right)_{M\theta}&=1-\frac{1}{\hat{V}_i^2}\frac{A_3^*\left(\hat{V}_i\right)}{C_M'}\end{aligned}\right\}\tag{D.29}$$

It is seen that the indicial functions are determined by the deviation between aerodynamic derivatives and their quasi-static counterparts, as could be expected. It should be noted that the aerodynamic derivatives are functions of the reduced velocity $\hat{V}_i$, i.e. experimentally they have been determined at various set values of $\hat{V}_i$. The determination of the constants b_j and c_j for each of the indicial functions $\Phi_{mn}\left(\hat{s}\right)=1-\sum_{j=1}^{N_j}b_je^{-c_j\hat{s}}$, $\begin{cases}m=D,L \text{ or } M\\ n=\dot{y},\dot{z} \text{ or } \theta\end{cases}$, will therefore require a multiple task data fitting type of approach. Often $N_j=2$ will suffice.

REFERENCES

[1] **Timoshenko, S., Young, D.H. & Weaver Jr., W.**, *Vibration problems in engineering*, 4th ed., John Wiley & Sons Inc., 1974.

[2] **Clough, R.W. & Penzien, J.**, *Dynamics of structures*, 2nd ed., McGraw–Hill, 1993.

[3] **Meirovitch, L.**, *Elements of vibration analysis*, 2nd ed., McGraw–Hill, 1993.

[4] **Simiu, E. & Scanlan, R.H.**, *Wind effects on structures*, 3rd ed., John Wiley & Sons, 1996.

[5] **Dyrbye, C. & Hansen, S.O.**, *Wind loads on structures*, John Wiley & Sons Inc., 1999.

[6] **Solari, G. & Piccardo, G.**, *Probabilistic 3 – D turbulence modelling for gust buffeting of structures*, Journal of Probabilistic Engineering Mechanics, Vol. 16, 2001, pp. 73 – 86.

[7] **ESDU Intenational**, 27 Corsham St., London N1 6UA, UK.

[8] **Batchelor, G.K.**, *The theory of homogeneous turbulence*, Cambridge University Press, London, 1953.

[9] **Tennekes, H. & Lumley, J.L.**, *A first course in turbulence*, 7th ed, The MIT Press, 1981.

[10] **Kaimal, J.C., Wyngaard, J.C., Izumi, Y. & Coté, O.R.**, *Spectral characteristics of surface–layer turbulence*, Journal of the Royal Meteorological Society, Vol. 98, 1972, pp. 563 – 589.

[11] **von Kármán, T.**, *Progress in the statistical theory of turbulence*, Journal of Maritime Research, Vol. 7, 1948.

[12] **Krenk, S.**, *Wind field coherence and dynamic wind forces*, Proceedings of Symposium on the Advances in Nonlinear Stochastic Mechanics, Næss & Krenk (eds.), Kluwer, Dordrecht, 1995.

[13] **Davenport, A.G.**, *The response of slender line – like structures to a gusty wind*, Proceedings of the Institution of Civil Engineers, Vol. 23, 1962, pp. 389 – 408.

[14] **Davenport, A.G.**, *The prediction of the response of structures to gusty wind*, Proceedings of the International Research Seminar on Safety of Structures under Dynamic Loading; Norwegian University of Science and Technology, Tapir 1978, pp. 257 – 284.

[15] **Sears, W.R.**, *Some aspects of non–stationary airfoil theory and its practical applications*, Journal of Aeronautical Science, Vol. 8, 1941, pp. 104 – 108.

[16] **Liepmann, H.W.**, *On the application of statistical concepts to the buffeting problem*, Journal of Aeronautical Science, Vol. 19, 1952, pp. 793 – 800.

[17] **Scanlan, R.H. & Tomko, A.**, *Airfoil and bridge deck flutter deriva-tives*, Journal of the Engineering Mechanics Division, ASCE, Vol. 97, No. EM6, Dec. 1971, Proc. Paper 8609, pp. 1717 – 1737.

[18] **Vickery, B.J. & Basu, R.I.**, *Across–wind vibrations of structures of circular cross section. Part 1, Development of a mathematical model for two–dimensional conditions*, Journal of Wind Engineering and Industrial Aerodynamics, Vol. 12 (1), 1983, pp. 49 – 73.

[19] **Vickery, B.J. & Basu, R.I.**, *Across–wind vibrations of structures of circular cross section. Part 2, Development of a mathematical model for full–scale application*, Journal of Wind Engineering and Industrial Aerodynamics, Vol. 12 (1), 1983, pp. 79 – 97.

[20] **Ruscheweyh, H.**, *Dynamische windwirkung an bauwerken*, Bauverlag GmbH, 1982, Wiesbaden und Berlin.

[21] **Dyrbye, C. & Hansen, S.O.**, *Calculation of joint acceptance function for line – like structures*, Journal of Wind Engineering and Industrial Aerodynamics, Vol. 31, 1988, pp. 351 – 353.

[22] **Selberg, A.**, *Oscillation and aerodynamic stability of suspension bridges*, Acta Polytechnica Scandinavica, Civil Engineering and Building Construction Series No. 13, Oslo, 1961.

[23] **Shinozuka, M.**, *Monte Carlo solution of structural dynamics*, Computers and Structures, Vol. 2, 1972, pp. 855 – 874.

[24] **Deodatis, G.**, *Simulation of ergodic multivariate stochastic processes*, Journal of Engineering Mechanics, ASCE, Vol. 122 No. 8, 1996, pp. 778 – 787.

[25] **Hughes, T.J.R.**, *The finite element method*, Prentice-Hall, Inc., 1987.

[26] **Scanlan, R.H.**, *Roll of indicial functions in buffeting analysis of bridges*, Journal of Structural Engineering, Vol.110 No. 7, 1984, pp. 1433 – 1446.

[27] **Chen, W.F. & Atsuta, T.**, *Theory of beam–columns, Volume 2, Space behaviour and design*, McGraw–Hill Inc., 1977.

[28] **Theodorsen, T.**, *General theory of aerodynamic instability and the mechanism of flutter*, NACA Report No. 496, Washington DC, 1934.

[29] **Den Hartog, J.P.**, *Mechanical vibrations*, 4th ed., McGraw–Hill, New York, 1956.

[29] **Cook, R.D., Malkus, D.S., Plesha, M.E. & Witt, R.J.**, *Concepts and applications of finite element analysis*, 4th ed., John Wiley & Sons Inc., 2002.

[30] **Millikan, C.B.**, *A critical discussion of turbulent flows in channels and circular tubes*, Proceedings of the 5th International Congress of Applied Mechanics, Cambridge, MA, 1938, pp. 386 – 392.

[31] **Salvatori, L. & Borri, C.**, *Frequency- and time domain methods for the numerical modelling of full-bridge aeroelasticity*, Computers and structures 85 (2007), pp. 675 – 687.

[32] **Newmark, N.M.**, *A method of computation for structural dynamics*, Journal of Engineering Mechanics Division, ASCE Vol. 85, EM3 (1959).

[33] **Hilber, H.M., Hughes, T.J.R. & Taylor, R.L.**, *Improved numerical dissipation for time integration algorithms in structural dynamics*, Earthquake engineering and structural dynamics Vol. 5 (1977)

INDEX

MIX
Papier aus verantwortungsvollen Quellen
Paper from responsible sources
FSC® C105338

If you have any concerns about our products,
you can contact us on
ProductSafety@springernature.com

In case Publisher is established outside the EU,
the EU authorized representative is:
Springer Nature Customer Service Center GmbH
Europaplatz 3, 69115 Heidelberg, Germany

Printed by Libri Plureos GmbH
in Hamburg, Germany